Fluid
Mechanics

RAYMOND C. BINDER

Professor of Mechanical Engineering
University of Southern California

Fluid
Mechanics

FIFTH EDITION

PRENTICE-HALL, Inc. Englwood Cliffs, N.J.

Library of Congress Cataloging in Publication Data

BINDER, RAYMOND CHARLES.
    Fluid mechanics.

    Includes bibliographical references.
    1.–Fluid mechanics. 2.–Hydraulics.
QA901. B5–1972   532'.5   72–7242
ISBN   0–13–322594–1

10 9 8 7 6 5 4 3 2 1

Printed in the United States of America

PRENTICE-HALL INTERNATIONAL, INC., London
PRENTICE-HALL OF AUSTRALIA, PTY. LTD., Sydney
PRENTICE-HALL OF JAPAN, INC., Tokyo
PRENTICE-HALL OF CANADA, LTD., Toronto
PRENTICE-HALL OF INDIA PRIVATE LTD., New Delhi

Contents

PART II SELECTED TOPICS IN FLUID MECHANICS

Preface

Fluid mechanics is that study of fluid motion involving a rational method of approach based on general physical laws and consistent with the results of modern experimental study. The word *fluid* implies a treatment of both liquids and gases. Fluid mechanics uses the same principles employed in the mechanics of solids. The modern trend is to avoid a collection of specialized empirical data of limited applicability; the trend is to develop general relations, and to organize experimental observations in a form suitable for use over a wide range of conditions.

There is real economy and value in studying at one time the same principles underlying the flow of different fluids. Such a study tends to develop a sound background and to make one versatile in approaching problems new to him. Studying fluid mechanics is analogous to studying applied mechanics and thermodynamics as a broad, unified preparation for subsequent specialized courses.

The aim of this book is to present an introduction to the fundamentals of fluid mechanics. The wisdom of concentrating on all the minute details of all existing applications is open to some question, in view of the time usually available in a course and the possible development of new and unexpected applications.

A serious attempt has been made to provide a balanced treatment in a logical fashion, and to keep physical concepts and basic quantitative relations in the foreground. Physical concepts are stressed with the hope that once the student has a good physical picture he can proceed of his own accord, with interest, in understanding and analyzing flow phenomena.

Experience has shown that problem work on the part of the student is very helpful, if not necessary, to give him a working knowledge of the subject. Problem work offers the student a definite test and challenge to supplement his reading.

This book is divided into two parts:

Part I—*Basic Relations*
Part II—*Selected Topics in Fluid Mechanics*

Part I presents an introduction to various fundamental physical relations and also provides tools which can be used to answer directly many practical problems. Part II gives information about different cases of flow and shows how the various basic relations can be applied. With Part I as a background, each chapter in Part II is essentially self-contained. This arrangement provides flexibility in arranging a course. Various chapters in Part II can be taken in a different sequence without difficulty.

It is not possible to give adequate, explicit credit to all those individuals and organizations who have directly and indirectly given aid in the preparation of this book. The writer, however, is deeply grateful to all of them for help, and wishes to acknowledge it as best he can. Many fellow workers, in industrial and research work, have made suggestions particularly as to material content. Various instructors and students in different schools have offered suggestions as to method of presentation. Dr. Theodore von Kármán and Dean Andrey A. Potter, each in his own particular way, has provided wise counsel and an outstanding stimulus. Dr. George V. Chillingar has made specific suggestions as to subject matter and arrangement of topics.

RAYMOND C. BINDER

Los Angeles, California

Fluid
Mechanics

PART I

Basic
Relations

In approaching a problem we must establish what factors are known or given, and what factor we wish to determine. Then there is this question: What fundamental relations, equations, or tools are available for solving our problem?

There are certain relations that must hold in any case of flow. For example, we should be able to account for all material, all energy and work. In any flow case there is a definite relation between force, mass and acceleration.

This book is divided into two main parts. Part I discusses the tools or basic relations. Chapter One defines some terms. Chapter Two covers fluids at rest. Chapter Three deals with the geometry of motion, a material accounting, a relation between force, mass and acceleration, and an energy accounting. Chapter Four outlines techniques for organizing experimental data.

Part II, starting with Chapter Five, serves several purposes. It gives information about different cases of flow. It also shows how basic relations are applied. With Part I as a background, each chapter in Part II is essentially self-contained. Various chapters in Part II can be studied in a different sequence without any difficulty.

CHAPTER ONE

Introduction—Some Fluid Properties

It may be permitted to doubt whether the public at large has any adequately realizing sense of the part which fluid mechanics plays, not only in our daily lives, but throughout the entire domain of Nature. Matter as we know it is either solid or fluid. . . .
Thus the flow of rivers and streams in their boundaries, . . . the circulation of the blood in our arteries and veins . . . the flight of the insect, the bird and the airplane; the movement of a ship in the water or of a fish in the depths, . . . these are all, in major degree, varied expressions of the laws of fluid mechanics . . . everywhere we find fluids and solids in reactive contact, usually in relative motion, and everywhere in this domain, the laws of fluid mechanics must control.

—W. F. DURAND.[1]

This chapter defines some terms and discusses units; these terms and units will be used in subsequent parts of the book.

Mechanics is a science concerned with the motion of bodies and the conditions governing such motion. The subject of mechanics is frequently divided into two general sections, one *kinematics* and the other *dynamics*. Kinematics deals with the geometry of motion without consideration of the forces causing that motion; kinematics is concerned with a description of *how* bodies move. Dynamics, on the other hand, is concerned with the action of forces on bodies.

1–1. Fluids

The term "fluid" can be characterized in different ways, as by molecular spacing and activity. In a fluid the spacing between molecules is greater than that in a solid. In a fluid the range of motion of the molecules is greater than that in a solid. For example, if a solid such as iron is heated sufficiently, the

[1] "The Outlook in Fluid Mechanics," *Journal of the Franklin Institute*, vol. 228, No. 2. August, 1939, page 183.

molecules become more violently agitated and less closely bound together, and the molten or fluid state finally results.

A more precise and useful definition of the word "fluid" can be formed on the basis of action under stress. A *fluid* is a substance which when in static equilibrium cannot sustain tangential or shear forces. A fluid yields continuously to tangential forces, no matter how small they are. This property of action under stress distinguishes between the two states of matter, the solid and the fluid. Imagine two plates of metal joined by a solid rivet, as shown in Fig. 1-1. The two parallel pulls tend to slide one plate with respect to the other; a shearing tendency is developed. For small pulls, the solid rivet sustains shear forces in static equilibrium. If a fluid, such as oil, water, or air, were subjected to a shear action, there would be continuous relative motion, even for small forces.

Fig. 1-1. Plates joined by a rivet.

Fluids are commonly divided into two subclasses: liquids and gases. A liquid occupies a definite volume, independent of the dimensions of the vessel in which it is contained. A liquid can have a free surface, like the surface of a lake. A gas, on the other hand, tends to expand to fill any container in which it is placed. In a gas the spacing between molecules is greater than that in a liquid. Sometimes a distinction is drawn between gases and vapors. A vapor, such as steam or ammonia, differs from a gas by being readily condensable to a liquid.

Gases are frequently regarded as compressible, liquids as incompressible. Strictly speaking, all fluids are compressible to some extent. Although air is usually treated as a compressible fluid, there are some cases of flow in which the pressure and density changes are so small that the air may be assumed to be incompressible. Illustrations are the flow of air in ventilating systems and the flow of air around aircraft and other craft at low speeds. Liquids, like oil and water, may be considered as incompressible in many cases; in other cases, the compressibility of such liquids is important. For instance, common experience shows that sound or pressure waves travel through water and other liquids; such pressure waves depend upon the compressibility or elasticity of the liquid.

Fluids will be discussed as continuous media. Actually, liquids and gases consist of molecues and atoms; fluid properties and phenomena are intimately related to molecular behavior. In many engineering problems, however, the mean free path of the molecule is small in comparison with the distances involved, and the flow phenomena can be studied by reference to bulk properties, without a detailed consideration of the behavior of the molecules.

1–2. Pressure

Shear force, tensile force, and compressive force are the three kinds of force which may act on any body. Fluids move continuously under the action of shear or tangential forces. It is well established that fluids are capable of withstanding a compressive stress, which is usually called pressure.

The term "pressure" will be used to denote a *force per unit area*. Probably a more descriptive label would be "pressure-intensity," but the briefer term "pressure" will be employed. Sometimes the term "pressure" is used in the sense of a total force, but this practice in fluid mechanics is apt to be confusing, and will not be followed in this book.

Atmospheric pressure is the force exerted on a unit area due to the weight of the atmosphere. Many pressure-measuring instruments indicate relative or *gage* pressure. Gage pressure is the difference between the pressure of the fluid measured and atmospheric pressure. *Absolute* pressure is the sum of gage pressure plus atmospheric pressure. The word "vacuum" is frequently used in referring to pressures below atmospheric.

Example. Assume that the atmospheric pressure is 14.7 pounds per square inch. A pressure of 5 pounds per square inch gage would mean an absolute pressure of 14.7 + 5 or 19.7 pounds per square inch. A vacuum of 4 pounds per square inch would mean an absolute pressure of 14.7 − 4 or 10.7 pounds per square inch.

Note in the foregoing example that the *pound* is used as a unit of force. When using the American system, the *pound* will be taken *only* as a unit of *force*.

1–3. Force and mass

A quantity which has both magnitude and direction is called a *vector* quantity. A quantity which has magnitude only is called a *scalar* quantity. The linear *displacement* of a moving point, or a very small particle, is its change of position. Displacement is a directed distance or a vector quantity; displacement has both magnitude and direction. The linear *velocity* of a moving particle is defined as the time rate at which the particle is changing position, or the rate of displacement with respect to time. Velocity is a vector quantity with both magnitude and direction; the magnitude of velocity is frequently called *speed*. Linear *acceleration* is defined as the rate of change of linear velocity with respect to time; acceleration also is a vector quantity.

If the linear velocity of a moving particle changes, some force is causing that change. For example, imagine a particle or body moving in a certain direction along a straight line. If the magnitude of the velocity is increasing in the direction of motion, the body is accelerating, and a force is acting in the direction of acceleration to cause the velocity change. A particle may be moving in a curved path, with constant speed or constant magnitude of velocity. There is an acceleration, however, because the velocity direction is changing, and a force must be acting on the particle to change the velocity direction. A basic relation in both solid-body and fluid mechanics is the dynamic equation

Force equals mass times acceleration

Note that mass is a scalar quantity; it has magnitude only. The foregoing is a vector equation, with the vector force in the same direction as the vector acceleration.

A consistent set of units, following modern practice, will be employed in order to avoid confusion. When using the American system, the *pound* will be taken as a unit of force and the *slug* as a unit of mass. As an example, one pound is the force acting on a mass of one slug which will accelerate the body by one foot per second each second. Inertia or sluggishness is a mass tendency to resist any force causing acceleration. The word "slug" was probably prompted by the word "sluggishness."

In the CGS (centimeter-gram-second) system the dyne is a force acting on a mass of one gram which will accelerate it one centimeter per second per second. In the MKS (meter-kilogram-second) system the force of one newton acting on a mass of one kilogram accelerates it one meter per second per second.

1–4. Density and specific weight

Each body in the universe exerts a force of gravitational attraction on every other body. The earth attracts a body on its surface, the earth attracts the moon, and the sun attracts the earth and other planets of the solar system. The force of gravitational attraction which the earth exerts on body is called the *weight* of the body. Weight is *not* the same as mass. A body of mass m is attracted by the earth with a force of magnitude mg, where g is the gravitational acceleration. The weight of a given mass changes as the gravitational acceleration changes.

Density ρ (Greek letter rho) is defined as *mass* per unit volume. For example, water, at a certain temperature and pressure, has a density ρ of 1.94 slugs per cubic foot. For standard sea-level air, at 59 degrees Fahrenheit and 14.7 pounds per square inch absolute, the density ρ is 0.002378 slug per cubic foot.

Specific weight γ (Greek letter gamma) is defined as weight per unit volume. Specific volume v is defined as volume per unit weight and is the reciprocal of specific weight. Since force equals mass times acceleration,

$$\gamma = \rho g \qquad \rho = \frac{\gamma}{g} \qquad\qquad (1\text{-}1)$$

For example, if g at a particular locality is 32.174 feet per second^2, the water with a density of 1.94 slugs per cubic foot would have a specific weight of (32.174) (1.94) or about 62.42 pounds per cubic foot. Unless otherwise specified, the subsequent discussions and problem work in this text will take g as 32.2 feet per second^2 (about 980 centimeters per second^2) and the specific weight of fresh water at the earth's surface as 62.4 pounds per cubic foot. Standard sea-level air has a specific weight of 0.0765 pound per cubic foot.

1–5. Density of liquids

In many flow problems it is necessary to know the density of the fluid. The density can be calculated if the specific gravity is known. The dimensionless specific gravity of a substance is defined as the ratio of its density (or specific weight) to the density (or specific weight) of some standard substance. For liquids the standard usually employed is either water at 4 degrees centigrade (39.2 degrees Fahrenheit), or water at 60 degrees Fahrenheit. The specific gravities of some liquids at atmospheric pressure are listed in Table 1–1.

Example. If the specific gravity of a certain oil is 0.80, then the specific weight of this oil is 0.80(62.4) or 49.92 pounds per cubic foot.

TABLE 1–1

SPECIFIC GRAVITIES OF SOME COMMON LIQUIDS*
(Referred to water at 39.2 degrees Fahrenheit)

	Specific gravity	*Temperature, degrees Fahrenheit*
Alcohol, ethyl............	0.807	32
Benzene	0.899	32
Gasoline	0.66–0.69	
Glycerine................	1.260	32
Mercury	13.546	68
Oil, castor	0.969	59
Oil, linseed (boiled)	0.942	59
Turpentine	0.873	60.8

* *Smithsonian Physical Tables*, 9th rev. ed., vol. 120, Smithsonian Institution, Washington, 1954.

1–6. Equation of state for gases

If t is the thermometer reading on the Fahrenheit scale, the absolute temperature $T = t + 459.3 = t + 460$ (approximately). The absolute temperature is often expressed in degrees Rankine. For example, a temperature reading of 70 degrees Fahrenheit would corresponds to an absolute temperature of $460 + 70 = 530$ degrees Rankine. If t is the thermometer reading on the Centigrade scale, the absolute temperature T is given by $t + 273$ degrees. This absolute temperature using the Centigrade scale is expressed in degrees Kelvin. For example, a temperature reading of 30 degrees Centigrade would correspond to an absolute temperature of $273 + 30 = 303$ degrees Kelvin.

The density of a gas can be calculated from the equation of state, or pressure–density–temperature relation. The "ideal" or "perfect" gas law is commonly employed.[3] This simple equation of state is

$$p = \rho RT \qquad (1\text{--}2)$$

Some average values of the gas factor R are given in Table 1–2 for the case in which p is expressed in pounds per square foot, T in degrees Rankine and ρ in slugs per cubic foot. For example, for air at 59 degrees Fahrenheit, and 14.7 pounds per square inch absolute, R is 1716 (feet)^2/(second)^2(degree Rankine), and the density ρ is 0.002378 slug per cubic foot.

TABLE 1–2

SOME AVERAGE R FACTORS FOR GASES

Gas	R
Air	1716
Ammonia	2918
Argon	1246
Carbon dioxide	1130
Carbon monoxide	1775
Ethane	1654
Helium	12428
Hydrogen	24659
Methane	3101
Nitrogen	1774
Oxygen	1553
Sulfur dioxide	776

For real gases, Equation (1–2) is accurate at ordinary temperatures and

[3] In thermodynamics the term "ideal" or "perfect" gas is sometimes defined as one that obeys the relation $p = \rho RT$. In hydrodynamics an "ideal" or "perfect" fluid is sometimes defined as one which is frictionless. Although this book will avoid this double use, the distinction should be noted in reading current literature.

for relatively small changes in pressure or volume. No gas obeys this simple relation in the region of liquefaction. Suitable vapor tables and charts should be consulted for such cases.

1–7. Definition of viscosity

The flow of any real fluid gives rise to tangential frictional forces which are called *viscous* forces. The action of such internal shearing forces results in a degradation of mechanical energy into heat or unavailable thermal energy. The following discusses the definition and units of viscosity and gives numerical values of this fluid property. An understanding of viscosity is of direct help in studying subsequent sections dealing with fluid resistance.

When a shear stress is applied to an elastic solid material, there is a definite angular deformation which is proportional to the stress. The modulus of elasticity in shear, G, is defined as the ratio of the shear stress to the shear strain (angular deformation in radians). G is a characteristic property of the material; for steel G is about 11.5×10^6 pounds per square inch. The shear modulus G for a fluid is zero; a fluid in static equilibrium cannot sustain a shear stress. *Viscosity* is a similar modulus, or factor of proportionality, which is a characteristic property of a fluid. Viscosity, however, differs from G in one important respect in that viscosity involves a *rate* of shearing strain.

Fluid fills the space between the two parallel plates is Fig. 1–2. The upper

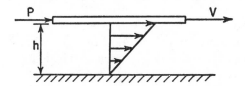

Fig. 1-2. Flow between parallel plates. The lower plate is stationary.

plate moves with the velocity V, while the lower plate is stationary. A very thin layer of fluid adheres to the lower plate; the velocity of this layer is zero. A thin layer of fluid adheres to the upper plate; this layer of fluid has the velocity V. It will be assumed that the fluid flows in parallel layers or laminas and that there are no secondary irregular fluid motions and fluctuations superimposed on the main flow. Such flow is called *laminar*. P is the force required to maintain the flow, to slide the fluid layers relative to each other by overcoming the internal fluid resistance. If A is the area of the plate in contact with the fluid, then the shear stress is $P/A = \tau$ (Greek letter tau).

The linear velocity distribution in the fluid is shown in Fig. 1–2. The *rate of shearing strain* of the fluid is V/h. During each unit of time there is an angular change equal to V/h. The coefficient of viscosity of the fluid is defined as follows:

$$\text{viscosity} = \frac{\text{shearing stress}}{\textit{rate} \text{ of shearing strain}} \tag{1-3}$$

Sometimes the foregoing term is called *absolute viscosity*. Probably a better term would be *dynamic viscosity*. The symbol μ (Greek latter mu) will be used for dynamic viscosity. Thus

$$\text{dynamic viscosity} = \mu = \frac{P/A}{V/h} = \frac{\tau}{V/h}$$

$$\tau = \mu \frac{V}{h} \tag{1-4}$$

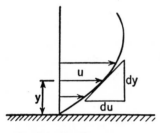

If the velocity distribution is nonlinear, as indicated in Fig. 1-3, the shear stress varies from point to point in the fluid. The shear stress at a point is

$$\tau = \mu \frac{du}{dy} \tag{1-5}$$

Fig. 1-3. Nonlinear velocity distribution.

The ratio du/dy is the velocity gradient or rate of shearing strain. *Fluidity* is defined as the reciprocal of dynamic viscosity.

1-8. Units of dynamic viscosity

The units of dynamic viscosity can always be determined by reference to the fundamental definition given in Equation (1-3). Let the symbols F, M, L, and T represent the primary or fundamental dimensions force, mass, length, and time, respectively. Area is then represented by L^2, and velocity by LT^{-1}. Since force equals mass times acceleration, the symbol for force F can be replaced by its equivalent MLT^{-2}. The rate of shearing strain has the dimensions of T^{-1}, because the ratio of velocity to length has the dimensions of T^{-1}. Then the dimensions of dynamic viscosity are FT/L^2, or the dimensions of dynamic viscosity are $MLT^{-2}T/L^2$ or M/LT. Dynamic viscosity may be expressed in pounds-second per square foot or slugs per foot-second:

$$1 \frac{\text{pound-second}}{\text{square foot}} = 1 \frac{\text{slug}}{\text{foot-second}}$$

In the metric system, dynamic viscosity may be expressed in dynes-second per square centimeter or grams per centimeter-second.

$$1 \frac{\text{dyne-second}}{\text{square centimeter}} = 1 \frac{\text{gram}}{\text{centimeter second}} = 1 \text{ poise}$$

The term "poise" is in honor of Poiseuille, a French scientist. The centipoise, or 0.01 poise, is a common unit. The dynamic viscosity of water at 20 degrees centigrade is approximately 1 centipoise.

Some conversions between the American and metric systems are as follows:

$$1 \text{ inch} = 2.54 \text{ centimeters}; \qquad 1 \text{ dyne} = 2.248 \times 10^{-6} \text{ pound}$$

Then

$$1 \text{ poise} = 1 \frac{\text{dyne-second}}{\text{centimeter squared}}$$

$$= 2.248 \times 10^{-6}(2.54 \times 12)^2 \frac{\text{pounds-second}}{\text{foot squared}}$$

$$1 \text{ poise} = 2.089 \times 10^{-3} \frac{\text{pound-second}}{\text{foot squared}}$$

$$= 2.089 \times 10^{-3} \frac{\text{slug}}{\text{foot-second}}$$

1-9. Kinematic viscosity

Kinematic viscosity is defined as the ratio of dynamic viscosity to density.

$$\text{kinematic viscosity} = v \text{ (Greek letter nu)} = \frac{\mu}{\rho}$$

Dimensions of v are

$$\frac{ML^3}{LTM} = \frac{L^2}{T}$$

Kinematic viscosity can be expressed in terms of feet squared per second or centimeters squared per second. In honor of Sir George Stokes, an English scientist, 1 centimeter squared per second is sometimes called a *stoke*.

1-10. Numerical values of viscosity

Figure 1-4 shows the dynamic viscosity of some liquids as a function of temperature. Figure 1-5 shows the dynamic viscosity of some gases, at atmospheric pressure, and of steam as a function of temperature. As an example of reading the vertical scale of each chart, if the vertical ordinate is 50, that means that $\mu \times 10^5 = 50$ and the μ is 50×10^{-5} slugs per foot-second. Figure 1-4 shows a general trend for liquids, that the dynamic viscosity decreases as the temperature increases. Figure 1-5 shows a general feature for gases, that the dynamic viscosity increases as the temperature increases. Table 1-3 gives some values at 59 degrees Fahrenheit and 29.92 inches of mercury. The effect of small pressure changes on the dynamic viscosity of fluids is usually considered negligible for ordinary conditions.

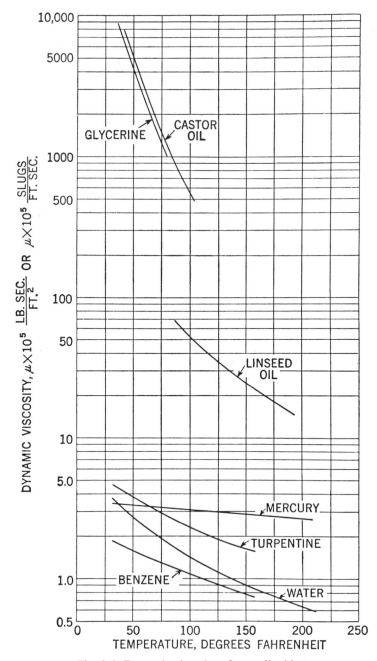

Fig. 1-4. Dynamic viscosity of some liquids.[4]

[4] Data for water from *Properties of Ordinary Water-Substance* by N. E. Dorsey. Reinhold, New York, 1940. Data for other liquids from *Smithsonian Physical Tables*, 9th rev. ed., Smithsonian Institution, Washington, D. C., 1954.

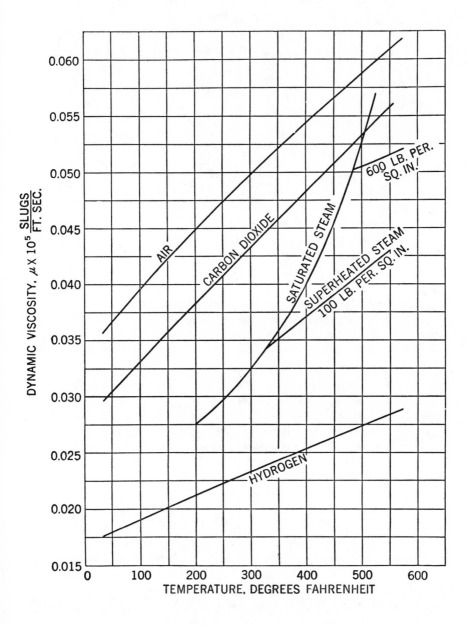

Fig. 1-5. Dynamic viscosity of some gases, at atmospheric pressure, and of steam.

13

TABLE 1–3

SOME VALUES OF VISCOSITY AT STANDARD
CONDITIONS

Fluid	Dynamic viscosity, $\frac{\text{slugs}}{\text{ft-sec}}$	Kinematic viscosity, $\frac{\text{ft}^2}{\text{sec}}$
Air	0.0373×10^{-5}	1.57×10^{-4}
Water	2.391×10^{-5}	1.233×10^{-5}
Castor oil	3160×10^{-5}	1.692×10^{-2}

1–11. The viscosity of an ideal gas

The exact nature or mechanism of the internal friction between adjacent layers of moving fluids has not been completely established. The explanation of the viscosity of gases is probably more fully developed than that for liquids.

The kinetic-molecular theory of gases postulates a number of features regarding an ideal gas. A chemically homogeneous gas is assumed to consist of identical molecules which are moving in random fashion; on the average, the number moving in any one direction is the same as that moving in any other direction. The space occupied by the molecules themselves is regarded as negligible in comparison with the space between the molecules. It is assumed that the molecules exert forces on each other or a wall *only* when they actually collide with each other or with a wall. It is assumed that, on the average, the impacts between molecules are perfectly elastic or without permanent deformation of the molecules and therefore without loss of energy.

According to this theory, the internal energy of the ideal gas is primarily kinetic. The internal energy of an ideal gas is directly proportional to the absolute temperature. We might infer, then, that the average kinetic energy of the molecules is directly proportional to the absolute temperature.

The term linear "momentum" of a body is defined as the product of the mass of that body times the velocity of the body. The action of momentum transfer of diffusion can be illustrated by imagining two trains, with open freight cars, moving parallel to each other in the same direction. Say train *A* has a speed of 10 miles per hour, and train *B* has a speed of 20 miles per hour. Masses, as coal, rock, or sand, are thrown from train *B* across to train *A*. As the masses enter train *A* they have a speed of 20 miles per hour, whereas the train *A* itself has a speed of 10 miles per hour. There is a tendency for the masses to increase the speed of the slower train. Also, masses thrown from train *A* to train *B* tend to reduce the speed of the faster train. We can say that there is a momentum exchange between trains *A* and *B*. This momentum exchange tends to reduce the relative motion between the trains.

According to the kinetic-molecular theory, the viscous action in a gas may be regarded as due to a process of momentum exchange or diffusion between adjacent gas layers which have different velocities. This momentum exchange is a molecular action and results in a tendency to reduce the relative motion between neighboring layers. This theory agrees with the trend shown by experimental measurements. The dynamic viscosity of a gas increases with an increase in temperature. As the temperature rises, the molecular activity or kinetic energy increases, the momentum exchange is greater, and the resistance to relative motion, and consequently the dynamic viscosity, increases.

1–12. Fluids and other substances

The behavior of materials under shear stress seems to offer the best means for distinguishing between a fluid and a solid. A substance is a fluid if it is continuously and permanently deformed by a shear stress, no matter how small the stress. There are substances—for example tar, sealing wax, and some glues—which appear to behave like solids under certain circumstances; these substances, however, may be classified as very viscous liquids with very low rates of deformation. At times some plastic substances are confused with fluids. A soft or plastic substance like lead, soap, sewage sludge, clay slurry, or a tallow candle may flow, but it flows only after a certain minimum stress has been exceeded. Such a plastic substance is therefore not a true fluid as normally defined.

Figure 1–6, a plot of shearing stress against rate of shearing strain, brings out these distinctions. Figure 1–6 shows a straight line passing through

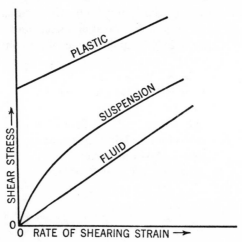

Fig. 1-6. Types of diagrams of shear stress against rate of shearing strain for various classes of substances.

the origin for fluids. The ratio between shearing stress and rate of shearing strain is the same for all rates of shearing strain. The dynamic viscosity of a particular fluid is uniquely determined by the temperature and pressure, regardless of the rate of shearing strain. For the plastic, as shown in Fig. 1–6, the minimum stress necessary before flow starts is sometimes called the *yield* value.

In many industrial processes the engineer is concerned with the movement of a combination consisting of a fluid and suspended particles. Sometimes the prediction of friction losses for such suspensions becomes an important and difficult problem. There is a scarcity of experimental data on the flow properties of suspensions. There is some question as to the meaning and measurement of the "viscosity" of a suspension. Taking the meaning as defined by Equation (1–3), experiments show that a suspension may have a "viscosity" considerably higher than that of the fluid carrier. Further, the "viscosity" of a suspension may be different for different rates of shearing strain. This introduces a complication not found with fluids. This book will be concerned only with fluids, namely with substances which have the same ratio between shear stress and rate of shearing strain for all rates of shearing strain.

Before discussing the methods for measuring viscosity, it will be necessary to discuss pipe flow. Chapter Five covers pipe flow, and the measurement of viscosity is treated in Chapter Seven, Flow Measurement.

1–13. Some conversion factors

When the term "ton" is used in this book without qualification, it will mean the short ton of 2000 pounds. Naval architects and marine engineers use the long ton, which is 2240 pounds. Some other common conversion factors are

$$1 \text{ mile} = 5280 \text{ feet}$$
$$1 \text{ knot} = 1 \text{ nautical mile per hour}$$
$$1 \text{ nautical mile} = 6076.1 \text{ feet}$$
$$1 \text{ U.S. gallon} = 231 \text{ cubic inches}$$
$$1 \text{ cubic foot} = 7.48 \text{ U.S. gallons}$$
$$1 \text{ inch} = 2.54 \text{ centimeters}$$
$$1 \text{ dyne} = 2.248 \times 10^{-6} \text{ pound}$$

PROBLEMS

1-1. What is the density of benzene? of mercury?

1-2. A liquid has a specific gravity of 1.8. What is the specific volume?

1-3. What is the specific weight of carbon dioxide at 90 degrees Fahrenheit and an absolute pressure of 110 pounds per square inch?

1-4. Calculate the density of air at 45 degrees Fahrenheit and 60 pounds per square inch absolute.

1-5. Five pounds of carbon dioxide gas at 40 degrees Fahrenheit fills a closed container having a volume of 10 cubic feet. What is the gas pressure?

1-6. What is the gas constant R_0 for air with pressure in pounds per square foot, specific volume in cubic feet per pound, and temperature in degrees centigrade?

1-7. A submarine is closed at the surface, with the air content at 80 degrees Fahrenheit and 14.7 pounds per square inch absolute. After submerging, the air temperature drops to 58 degrees Fahrenheit. What is the air pressure for the submerged conditions if the hull has not changed in size?

1-8. Hydrogen, initially at 40 pounds per square inch absolute and 60 degrees Fahrenheit, expands isothermally, or at constant temperature, to 15 pounds per square inch absolute. What is the final specific volume?

1-9. The air in an automobile tire is at 42.7 pounds per square inch absolute and 80 degrees Fahrenheit. Assuming no change in volume, what would be the pressure if the temperature rose to 150 degrees Fahrenheit?

1-10. A tank with a volume of 1.2 cubic feet contains air at 114.7 pounds per square inch absolute and 84 degrees Fahrenheit. What is the weight of the air in this tank?

1-11. A pressure is expressed as 500 dynes per square centimeter. What would be this pressure in pounds per square foot?

1-12. A tank with a fixed volume initially contains nitrogen at 15 pounds per square inch absolute and 70 degrees Fahrenheit. Three pounds of nitrogen are added to the tank. The final conditions are 25 pounds per square inch absolute and 75 degrees Fahrenheit. What is the volume of the tank?

1-13. Castor oil at 59 degrees Fahrenheit fills the space between two parallel horizontal plates which are $\frac{3}{8}$ inch apart. If the upper plate moves with a velocity 5 feet per second, and the lower one is stationary, what is the shear stress in the oil?

1-14. A liquid has a dynamic viscosity of 1.85 centipoises and a specific gravity of 1.046. What are the dynamic and kinematic viscosities in American units, and the kinematic viscosity in stokes?

1-15. What is the kinematic viscosity of mercury at 68 degrees Fahrenheit in American units? in metric units?

1-16. The kinematic viscosity of an oil is 0.20 foot squared per second, and the specific gravity is 0.87. Determine the dynamic viscosity.

1-17. What is the kinematic viscosity of carbon dioxide at atmospheric pressure and 200 degrees Fahrenheit in American units? in metric units? What is the dynamic viscosity of this gas in pound-hour-inch units?

1-18. A long vertical cylinder 3.000 inches in diameter rotates concentrically inside a fixed tube having a diameter of 3.002 inches. The uniform annular space between tube and cylinder is filled with water at 59 degrees Fahrenheit. Assuming a linear velocity distribution, what is the resistance to motion at a relative velocity of 0.3 foot per second, for a 6-inch length of cylinder? What would be the resistance if the fluid were castor oil at 59 degrees Fahrenheit?

1-19. With the notation shown in Fig. 1–3, the velocity profile is given by the relation $u = 5y + 10y^2$, where u is in feet per second and y is in feet. What is the velocity gradient at the boundary and at 8 inches away from the boundary? What is the shear stress at the wall if the fluid is standard air?

1-20. What is the kinematic viscosity of air at atmospheric pressure and 350 degrees Fahrenheit in American units? in metric units?

1-21. There is two-dimensional flow of water between two parallel fixed walls 12 inches apart. The velocity distribution is parabolic; the maximum velocity is 20 feet per second. What is the pressure drop in the direction of flow for a length of 100 feet?

CHAPTER TWO

Fluid
Statics

*A solid heavier than a fluid will, if placed in it, descend to the bottom
of the fluid, and the solid will, when weighed in the fluid, be lighter
than its true weight by the weight of the fluid displaced.*

—ARCHIMEDES.[1]

Statics refers to a study of the conditions under which a particle or a
body remains at rest. If a particle or body is at rest under a given set of
forces, the forces are said to be in equilibrium or the body is said to be in
static equilibrium.

2–1. Pressure in a fluid

A body of fluid in static equilibrium is free in every part from tangential
or shear forces. An example is a quantity of water or oil at rest in a container.
One fluid layer does not slide relative to an adjacent layer; there is no distor-
tion of the fluid elements. Absence of shear means that friction need not be
considered. Frictional effects arise only when there is relative sliding or shearing, as with a fluid in motion.

Pascal's important law, that the pressure in a static fluid is the same
in all directions, can be shown by studying the forces acting on any infinitesimal element in a body of fluid. In Fig. 2–1 the distance between the two
triangular faces is unity, and the sides dx and dy are infinitesimal. Each of
the pressures p_1, p_2, and p_3 is normal to the face upon which it acts. Imagine
that the volume of the prism converges to zero in such a manner that the
leg ds always moves parallel to itself. In the limit both weight and surface
force go to zero, but $dxdy$ is an infinitesimal of higher order than dx, dy, or
ds. Since the forces balance in equilibrium, taking vertical and horizontal
components of the forces gives

[1] As translated by T. L. Heath in *The Works of Archimedes*, Cambridge University
Press, London, 1897.

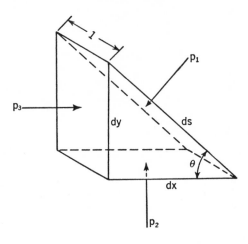

Fig. 2-1. Equilibrium of a small fluid element.

$$p_2 dx - p_1 ds \cos \theta = 0, \qquad p_3 dy - p_1 ds \sin \theta = 0$$

Since $dx = ds \cos \theta$ and $dy = ds \sin \theta$,

$$p_2 - p_1 = 0, \qquad p_3 - p_1 = 0, \qquad p_1 = p_2 = p_3$$

In words, the pressure at a point in a static fluid is the same in all directions. This result is different from that obtained for a stressed solid in static equilibrium; in such a solid the stress on a plane depends on the orientation of that plane. The preceding analysis can be extended to show that the pressure at a point within a fluid is the same for any state of motion, provided there is no shear stress (as with a frictionless fluid).

It is common to refer to a liquid surface in contact with the atmosphere as a "free" surface. A static liquid has a horizontal free surface if gravity is the only force acting. Imagine a surface other than horizontal. Shear forces would be necessary in order to hold one point on the surface above another. A sloping surface may exist as a transient flowing condition, but when static equilibrium is finally reached, the free surface of the liquid is horizontal.

2–2. Static fluid in a gravitational field

Consider a mass of static fluid m in a gravity field. The weight of this mass is mg. This mass at the surface of the earth has a certain weight corresponding to the local value of g; this same mass m in some craft in outer space will have a much different weight if the g in outer space is much different from that at the surface of the earth.

Figure 2–2(a) shows an infinitesimal element in the body of any static fluid. Figure 2–2(b) shows the element in detail. z is a vertical distance measured

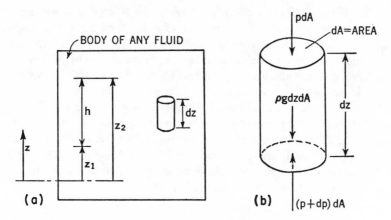

Fig. 2-2. Vertical forces on an infinitesimal element.

positively in a direction of decreasing pressure, and dA is an infinitesimal area. p is the pressure on the top surface, and $(p + dp)$ is the pressure on the bottom surface, the pressure increase being due solely to the fluid weight. The weight of the element is $\rho g\,dz\,dA$. Balancing forces on the free body element in the vertical direction gives

$$dp\,dA = -\rho g\,dz\,dA \qquad \text{or} \qquad dp = -\rho g\,dz \qquad (2\text{–}1)$$

Since z is measured positively upward, the minus sign indicates that the pressure *decreases* with an *increase* in height. *Equation (2–1) is a fundamental equation of fluid statics.* It states that the pressure decreases in the upward direction, the decrement per unit length being equal to the weight per unit volume; dp is zero when dz is zero; the pressure is constant over any horizontal plane in a fluid. In integral form Equation (2–1) becomes

$$\int_1^2 \frac{dp}{\rho g} = -\int_1^2 dz = -(z_2 - z_1) \qquad (2\text{–}2)$$

The functional relation between pressure and specific weight (or ρg) must be established before Equation (2–2) can be integrated further. The following articles consider separately the case for incompressible fluids and the case for compressible fluids.

2–3. Pressure-height relation for incompressible fluids

Liquids can be treated as incompressible in many practical cases. In some cases, as for small differences in height, gases can be regarded as incompressible. For an incompressible fluid the density ρ is constant. Thus, for a constant g, Equation (2–2) becomes

$$p_2 - p_1 = -\rho g(z_2 - z_1) \qquad \text{or} \qquad \Delta p = \gamma h \qquad (2\text{–}3)$$

where Δp is the pressure difference and h is the difference in levels. If γ is expressed in pounds per cubic foot, and h in feet, then Δp is in pounds per square foot. h is sometimes called a *pressure head*, and may be expressed in feet or inches of water, inches of mercury, or some height of any liquid.

Several pressure and head equivalents for common liquids are convenient for problem solutions. The use of special conversion factors whose physical significance may be obscure or easily forgotten, is not necessary, if it is kept in mind that standard atmospheric or barometric pressure is 14.7 pounds per square inch, 29.92 inches of mercury, or about 33.9 feet of water. One unit of pressure derived from the barometer is the *atmosphere;* 1 atmosphere equals 14.7 pounds per square inch. A pressure of 4 atmospheres, for example, is $4 \times 14.7 = 58.8$ pounds per square inch.

Example. Figure 2–3 shows a container with oil (specific gravity $= 0.90$) in static equilibrium. Standard atmospheric pressure exists at point A. At A the *gage* pressure is zero, whereas the *absolute* pressure is 14.7 pounds per square inch. The pressure at C equals that at A, because C is at the same level as A. The pressure at a particular point will be designated by a subscript for that point. For example, the pressure at point B will be designated by p_B. Since

$$p_B - p_C = \gamma h = \frac{0.9(62.4)30}{144} = 11.7 \text{ pounds per square inch}$$

the pressure at B is 11.7 pounds per square inch gage, or $11.7 + 14.7 = 26.4$ pounds per square inch absolute. Since

$$p_C - p_D = \frac{0.9(62.4)20}{144} = 7.8 \text{ pounds per square inch}$$

the absolute pressure at D is $14.7 - 7.8 = 6.9$ pounds per square inch.

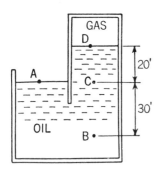

Fig. 2-3. Container with oil.

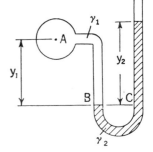

Fig. 2-4. Open manometer.

2–4. Manometers

The term "manometer" is applied to a device that measures pressure by balancing the pressure against a column of liquid in static equilibrium.

A wide variety of manometers is in use: vertical, inclined, open, differential, and compound. Fundamentally, the use of a manometer is good technique; the instrument is simple, and may be employed for precise pressure measurements. Both the specific weight of a liquid and the height of a liquid column can be measured accurately.

The basic equation for calculating the pressures indicated by liquid manometers is simply $\Delta p = \gamma h$. It is suggested that each manometer installation be worked out as a separate problem, using this basic relation. A few examples illustrating the method of approach follow.

In the *open* manometer, Fig. 2–4, the right leg is open to the atmosphere and contains a liquid of specific weight γ_2. Consider the problem of determining the pressure at point A in a liquid of specific weight γ_1. This liquid partially fills the left leg. Starting with the right leg, the gage pressure at point C is

$$p_C = \gamma_2 y_2$$

Because the manometer contains a fluid in static equilibrium, the pressure at B equals the pressure at C. Then

$$p_B = p_C = \gamma_2 y_2$$

The pressure at A is less than that at B:

$$p_B - p_A = \gamma_1 y_1$$

From the last two relations, the gage pressure at A is

$$p_A = \gamma_2 y_2 - \gamma_1 y_1 \tag{2–4}$$

Example. Assume that the liquid in the right leg is mercury, that the liquid at A is water, that $y_1 = 5$ feet, and $y_2 = 10$ feet. The pressure at A is

$$p_A = 62.4(13.55)10 - 62.4(5)$$
$$= 8140 \text{ pounds per square foot gage}$$

or the pressure at A is

$$\tfrac{8140}{144} + 14.7 = 71.2 \text{ pounds per square inch absolute}$$

The name "piezometer" is sometimes applied to the type of manometer shown in Fig. 2–4, in which only the liquid at A is used in the $\cup$ tube. The *differential* manometer in Fig. 2–5 is used to measure the pressure difference $p_A - p_B$. The gage can be analyzed in steps. Starting with point B and proceeding along the tube, the pressure at D is greater than that at B:

$$p_D = p_B + \gamma_2(y_3 - y_2)$$

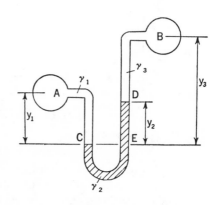

Fig. 2-5. Differential manometer.

The pressure at E is greater than that at D:

$$p_E = p_D + \gamma_2 y_2$$
$$p_E = p_B + \gamma_3(y_3 - y_2) + \gamma_2 y_2$$

The pressure at C equals the pressure at E because each point is at the same level. The pressure at A is less than that at C

$$p_A = p_C - \gamma_1 y_1 = p_E - \gamma_1 y_1 = p_B + \gamma_3(y_3 - y_2) + \gamma_2 y_2 - \gamma_1 y_1$$

From the last relation we get the result

$$p_A - p_B = \gamma_3(y_3 - y_2) + \gamma_2 y_2 - \gamma_1 y_1 \qquad (2-5)$$

Equation (2–5) provides a direct calculation of the pressure difference. In solving a manometer problem, one can start at one end of the gage, work through in steps to the other end of the gage, and use the relation $\Delta p = \gamma h$.

The widely employed mercury barometer is essentially a manometer for measuring atmospheric pressure. The glass tube of the instrument is closed at one end and open at the other. The tube is first filled completely with pure mercury and is then inverted into a vertical position with the open end submerged in a small vessel (cistern) containing mercury. The principle is illustrated in Fig. 2–6. The direct reading of the barometer gives the height of a column of mercury; the weight of this mercury column balances the weight of the air column above the barometer.

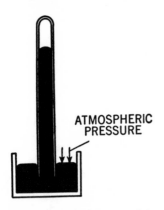

Fig. 2-6. Main features of the mercury barometer.

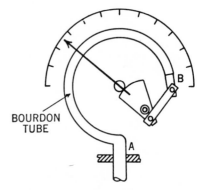

Fig. 2-7. Essential features of a Bourdon gage.

2–5. Bourdon gages

The Bourdon tube is used in various forms of instruments for measuring pressure differences. The essential feature of this type of gage is shown diagrammatically in Fig. 2–7. The Bourdon tube is a hollow metal tube, of elliptical cross section and bent in the form of a circle. One end of the Bourdon tube is fixed to the frame at A, whereas the other end B is free to move. The free end actuates a pointer through a suitable linkage. As pressure inside the tube increases, the elliptical cross section tends to become circular, and the free end of the Bourdon tube (point B) moves outward. A pressure scale or dial can be devised from a calibration of the instrument.

Note that the position of the free end of the Bourdon tube depends upon the difference in pressure between the inside and the outside of the tube. If the outside of the Bourdon tube is exposed to atmospheric pressure, the instrument responds to *gage* pressure.

2–6. Total force on plane submerged surfaces

This article will be limited to flat surfaces in liquids or incompressible fluids. The force exerted on any elementary surface by a static fluid is normal to that surface because the surface is plane, and there are no shear or tangential forces. The force on an elementary area dA is pdA. The total force $F = \int pdA$ is perpendicular to the plane.

For a nonhorizontal plane surface in a liquid, as shown in Fig. 2–8, the pressure varies directly with depth according to the relation $\Delta p = \gamma h$.

In Fig. 2–9 a submerged plane area BC makes an angle θ with the free surface of the liquid. The problem is to calculate the total force exerted by the liquid on one side of the area. The gage pressure p at depth h is $p = \gamma h$; the force dF on the elementary area dA is

$$dF = pdA = \gamma h dA$$

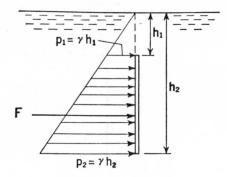

The distance y is measured from point D; D represents the end view of a line of intersection of the free liquid surface and the plane; y, then, is a perpendicular distance from this line. In terms of y, $dF = \gamma y \sin \theta dA$.

Fig. 2-8. Vertical surface in a liquid.

The resultant force F is

$$F = \int dF = \gamma \sin \theta \int ydA$$

The distance to the center of gravity (c.g.) or the centroidal distance $\bar{y}$ is defined by the relation $\bar{y}A = \int ydA$. Since $\bar{h} = \bar{y} \sin \theta$,

$$F = \gamma A \bar{y} \sin \theta = \gamma \bar{h} A \qquad (2\text{–}6)$$

Equation (2–6) shows that the *total force on a plane area is the product of the pressure at the centroid multiplied by the total area.* Equation (2–6)

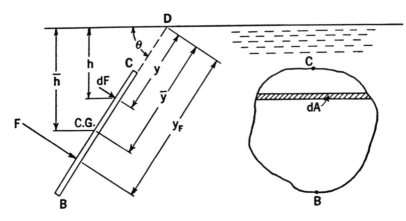

Fig. 2-9. Total force on a plane area.

provides a means for calculating only the *magnitude* of the total force; the
direction of this force is normal to the surface.

As an example, Fig. 2–10 shows a flat plate immersed in water. The area

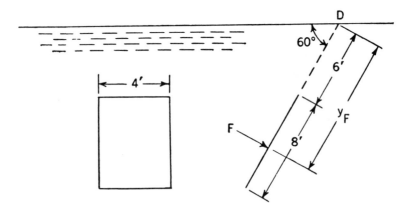

Fig. 2-10. Plate in water.

A is $8 \times 4 = 32$ square feet. The centroid of the rectangular area is at the
intersection of the diagonals of the plate; the centroidal distance $\bar{y}$ is $6 + 4 =$
10 feet. Therefore the total force F is

$$F = 62.4(32)10 \sin 60° = 17,300 \text{ pounds}$$

2–7. Location of total force on a plane submerged area

This article, like the foregoing article, will be limited to flat surfaces in
liquids or incompressible fluids. In Fig. 2–9 the plane area is acted upon

by a system of parallel forces, each normal to the plate. The force dF exerts a moment ydF with respect to D. The total of all these moments is given by the integral

$$\int ydF$$

This integral will be replaced by a moment due to the total force F acting at a certain distance y_F from the point D. Thus

$$Fy_F = \int ydF$$

Since

$$F = \int dF = \int \gamma dAy \sin\theta$$

and

$$Fy_F = \int ydF = \int \gamma dAy^2 \sin\theta$$

Then

$$y_F = \frac{\int y^2 dA}{\int ydA} = \frac{I}{A\bar{y}} \tag{2–7}$$

in which I is the moment of inertia of the area about the axis from which y is measured, and $\bar{y}$ is the centroidal distance. Equation (2–7) can be arranged in another form that is frequently more convenient for practical calculations.

$$I = I_G + A\bar{y}^2$$

where I_G is the moment of inertia of the area about a parallel axis through the centroid. Then

$$y_F = \frac{I_G}{A\bar{y}} + \frac{A\bar{y}^2}{A\bar{y}} = \bar{y} + \frac{I_G}{A\bar{y}} \tag{2–8}$$

Equation (2–8) shows that the line of action of the resultant force is always *below* the centroid by the distance $I_G/A\bar{y}$. Since the radius of gyration k of an area is defined as equal to $\sqrt{I/A}$, Equation (2–8) can be written as

$$y_F = \bar{y} + \frac{k_G^2}{\bar{y}}$$

Example. Figure 2–10 shows a flat plate immersed in water. In the previous article the calculation gave a total force F of 17,300 pounds. The centroidal distance y is 10 feet, and the area A is 32 square feet. The moment of inertia I_G for the rectangular plate is $4(8)^3/12$. Therefore

$$y_F = y + \frac{I_G}{A\bar{y}} = 10 + \frac{4(8)^3}{12(32)10} = 10.53 \text{ feet}$$

The total force of 17,300 pounds passes through a point 10.53 feet below D measured down the plane.

2–8. Forces on submerged irregular surfaces

In some cases the calculation of the total forces acting on an irregular surface becomes involved. Frequently such a calculation can be expedited by dealing with components of a total force rather than with the total force itself. As an example, refer to the curved surface ABC in Fig. 2–11. The area DC is the projection of ABC on a horizontal plane, whereas the area AD is the projection of ABC on a vertical plane.

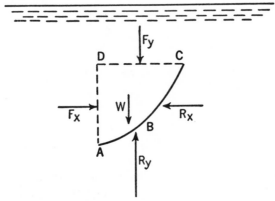

Fig. 2-11. Curved surface submerged in a liquid.

A study can be made of the forces acting *on* the body of fluid $ABCD$. The forces F_x and F_y can be computed by the methods previously outlined. W is the weight of the fluid in the volume $ABCD$, and acts through the center of gravity or centroid of this volume. The total force of the surface ABC *on* the body of fluid is represented by the combination of the vertical force R_y and the horizontal force R_x. For equilibrium the sum of all the forces in any direction must equal zero, and the sum of all the moments about any point must equal zero. In the horizontal direction, then, $R_x = F_x$. The line of action of R_x coincides with that of F_x. This condition explains the rule sometimes stated, that the total horizontal component of the forces acting on any area is equal to the horizontal force acting on a plane which is the vertical projection of the curved area. In the vertical direction, $F_y + W - R_y = 0$.

If F_y and W are known, the magnitude of R_y can be determined by the foregoing relation. The location of R_y can be calculated by taking moments about any convenient point, as A, or D, or C. As an example, the moment of F_y about A plus the moment of W about A must balance the moment of R_y about A.

The forces acting on a wall or a gravity dam might be cited as another example. In a gravity dam the weight of the structure is used to resist the

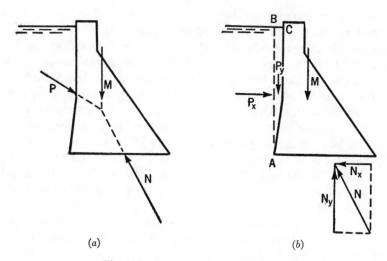

(a)   (b)

Fig. 2-12. Forces acting on a gravity dam.

fluid forces. Consider the gravity dam illustrated in Fig. 2–12(a) as acted upon by three forces; the total force P of the water on the upstream side of the dam, the vertical force or weight M of the dam, and the resultant force N of the ground on the base of the dam. If the structure is in static equilibrium, these three forces must balance. The reaction N on the base can be broken up into the horizontal component N_x and the vertical component N_y, as illustrated in Fig. 2–12(b). The total fluid force P can be broken up into the components P_x and P_y. P_x is calculated as a force acting over the plane vertical surface AB. The vertical force P_y is the weight of the fluid in the space $ABCA$ and acts through the centroid of this area.

2–9. Buoyancy and static stability

When a body is partly or completely immersed in a static fluid, as shown in Fig. 2–13, every part of the surface in contact with the fluid is pressed on by the latter, the pressure being greater on the parts more deeply immersed. The resultant of all these forces exerted by fluid pressure is an upward, buoyant, or lift force.

The pressure on each part of the surface of the body is independent of the body material. Imagine the body, or as much of it as is immersed, to be replaced by fluid like the surrounding mass. This substituted fluid will experience the pressures that acted on the immersed body, and this substituted fluid will be at

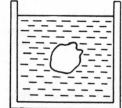

Fig. 2-13. Body in a fluid.

rest. Hence the resultant upward force on the substituted fluid will equal its weight, and will act vertically through its center of gravity.

Imagine a body partly or completely immersed in a static fluid, and assume that the body is at rest or in static equilibrium. There are two forces acting *on* this body: the upward force of buoyancy and the downward force of the body weight. The buoyant force acts vertically upward through the center of buoyancy (the center of gravity of the displaced fluid volume), whereas the body weight acts vertically downward through the center of gravity of the body. If the body is at rest, these two forces must be equal and opposite, and the center of buoyancy must lie on a vertical line through the center of gravity of the body. The term *static stability* refers to the initial tendency of the body to return to or move away from the equilibrium condition after a disturbance.

A body is said to be statically *stable* or to be in *stable equilibrium* when there is an initial tendency to return toward its equilibrium position after being slightly displaced. An example is a pendulum pivoted above the center of gravity of the mass. When the mass is at rest there is static equilibrium. Further, this equilibrium is *stable;* if the mass is displaced slightly from its equilibrium position and then released, there is a restoring force or a tendency to bring the mass back toward the equilibrium position.

A pencil just balanced on its point is in static equilibrium. The body is *unstable*, however, because if slightly displaced the pencil continues moving away from the balanced state. A sphere or ball resting on a perfectly horizontal surface is an example of *neutral* equilibrium. If the ball is displaced slightly and the displacing force is removed, the ball has no tendency to return to or move away from its former position.

In Fig. 2–14, B is the center of buoyancy and G is the center of gravity of the body. Static equilibrium is reached when the buoyant force F_B equals the weight W and when the forces are in a vertical line. The equilibrium of a completely immersed body is stable, as for the balloon and the submarine, when the center of buoyancy is *above* the center of gravity. If such a body is tilted, a couple is developed which tends to swing the body back to its original position.

Figure 2–15(a) illustrates a floating body, such as a surface vessel. The buoyant force F_B equals the weight W. The weight W acts through the center of gravity of the ship, whereas the buoyant force acts through the center of gravity of the displaced water. For equilibrium it is necessary that G and B lie on a vertical line. G is usually above B for surface ships except some fast sailing boats. If, as indicated in Fig. 2–15(b), the body is so displaced that the center of buoyancy is not in a vertical line directly over or under the center of gravity G, the body will not remain as rest in that position; the ship will move.

In Fig. 2–15(b) the line of action of the buoyant force intersects the center

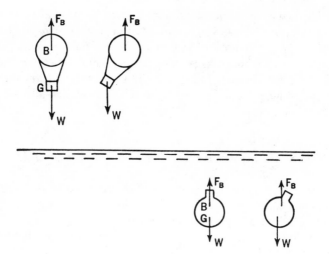

Fig. 2-14. Equilibrium and displaced positions of bodies immersed in a fluid.

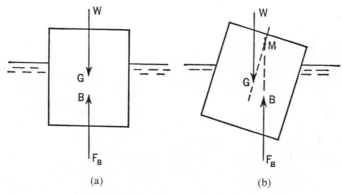

Fig. 2-15. Equilibrium of a floating body.

line of the body at point M. There is a couple tending to rotate the body back to the original position; there is a righting moment, and the ship is statically *stable*. The magnitude of the couple equals the product of the weight and the perpendicular distance between the line of action of the weight and the line of action of the buoyant force. For the same inclination but a different condition of the ship, imagine that the buoyant force F_B acts on the left side of the weight force; in this case there would be a moment tending to incline the ship more, and the ship would be *unstable*.

The curve drawn through successive positions of the center of buoyancy, for different angles of inclination (or "heel"), is termed the locus of the centers of buoyancy. The center of curvature for any infinitesimally small portion of this curve is called a *metacenter*. For an infinitesimally small angular displacement of the ship, the ship is swinging about an imaginary pivot or instantaneous center of rotation; this center is the metacenter. For very small angles of inclination, point M is taken as the metacenter. Thus sometimes the rule is stated that for a floating body to be stable, it is necessary that the metacenter lie above the center of gravity. If the metacenter lies below the center of gravity, the body is unstable.

2-10. Pressure-height relations for compressible fluids

The fundamental equation of fluid statics is derived in Section 2-2. The functional relation between pressure and specific weight or ρg must be established before this basic relation can be integrated. The case for incompressible fluids gave a simple integration. This article will illustrate two further cases, for compressible fluids.

As one example, consider an isothermal or constant-temperature layer of gas. Then the simple equation of state for gases shows that

$$RT_1 = \frac{p_1}{\rho_1} = \text{CONSTANT} = \frac{p}{\rho}$$

Substituting in Equation (2-2) gives

$$\int_1^2 \frac{dp}{\rho g} = -\int_1^2 dz \qquad \frac{RT_1}{g} \int_1^2 \frac{dp}{p} = -\int_1^2 dz$$

$$z_2 - z_1 = \frac{RT_1}{g} \log_e \frac{p_1}{p_2} \qquad (2\text{-}9)$$

For dry air R is 1716, whereas for medium damp air R is about 1726.

Equation (2-9) is sometimes called the *barometric-height relation*. For an isothermal atmosphere a measurement of the temperature and the pressure (as by a barometer) at two different levels will provide data for the calculation of the height difference.

The relation between pressure and density may be different from that for the isothermal. The pressure, density, and temperature all vary in the general case; this general case is commonly called the *polytropic process*. The general equation for the polytropic process is

$$\frac{p}{\rho^n} = \text{CONSTANT}$$

where n is some constant for the particular process under consideration. For the expansion or compression process in an engine, a vibrating air column, or a sound wave, n may be 1.4 or some other value different from unity. The isothermal process might be considered as a special case of the polytropic in which $n = 1$.

One process of many is the "adiabatic" process. An adiabatic process is one in which no heat (thermal energy in transition) is added to or removed from the fluid mass. Energy considerations (see Chapter Three) and application of the equation of state show that for ideal gases $n = k$ for an adiabatic process, where

$$k = \frac{\text{specific heat at constant pressure}}{\text{specific heat at constant volume}}$$

The specific heat is the amount of heat required to change the temperature of a unit mass of a substance one degree. For such gases as air, carbon monoxide, hydrogen, and oxygen, experiments show an average value of $k = 1.40$. A discussion of this $p/\rho^k = $ constant relation is given in Chapter Six.

Using the relations

$$\frac{p}{\rho^n} = \frac{p_1}{\rho_1^n} \qquad \frac{p_1}{\gamma_1} = \frac{RT_1}{g}$$

substituting in the fundamental Equation (2-2); and integrating gives the pressure–height relation:

$$z_2 - z_1 = \frac{n}{n-1}\frac{RT_1}{g}\left[1 - \left(\frac{p_2}{p_1}\right)^{\frac{n-1}{n}}\right] \qquad (2\text{-}10)$$

For a dry adiabatic atmosphere, $n = k = 1.4$. For a moist atmosphere n is lower than 1.4, an average approximate value being 1.2.

This section and Section 2-3 illustrate how the pressure-height relation for *any* fluid can be obtained. The functional relation between p and ρg is inserted in the fundamental equation of statics, and the integration is carried out.

2-11. Atmosphere

The atmosphere, that body of fluid in which we live, is an envelope of gas surrounding the earth. In the lower layer of the atmosphere, or *troposphere*, the temperature usually decreases with height. In the upper layer, or *stratosphere*, the temperature is practically constant. The *tropopause* is the zone between the troposphere and the stratosphere; the tropopause is the gradual transition between the two. The height of the base of the stratosphere varies with latitude and season but averages about 35,000 feet.

Besides the units already mentioned, the unit *bar* is sometimes used for pressure in sound and weather work. For example, the United States Weather Bureau and the weather bureaus of some of the other countries have adopted the bar unit for reporting pressure on daily weather maps. The following conversions apply:

$$1 \text{ millibar} = 1000 \text{ dynes per square centimeter}$$
$$1000 \text{ millibars} = 29.53 \text{ inches of mercury}$$

1013.2 millibars corresponds to the "standard" pressure of 29.92 inches of mercury.

2-12. Convective stability of masses in fliuds

Some important features of stability, other than those previously treated, can be conveniently illustrated by reference to the vertical movement of air masses (convection) in the atmosphere.

If an air mass at the surface were moved aloft by some thermal or mechanical means, as by a gust of wind or a mountain, the air mass would cool as it expanded. Such cooling is a common occurrence and may be sufficient to cause condensation in moist air with the subsequent formation of rain, fog, or cloud. If the air mass moved earthward from aloft, the compression would result in a temperature rise, in a manner similar to the familiar action in a tire pump.

It is commonly considered that such vertical movements in the atmosphere follow the adiabatic law; the mass motion is so rapid that the heat addition or removal is negligible. It is usually assumed that when no condensation occurs, the process follows the dry adiabatic law; that is, $n = 1.4$ in the relation $p/\rho^n =$ constant. The temperature change per unit height, or *temperature gradient*, can be computed with the aid of relations already established. In meteorology the decrease of temperature with height is frequently called the *lapse rate*.

Eliminating ρ from the equations $p/\rho^n =$ constant and $p/\rho = RT$ gives

$$\frac{T_2}{T_1} = \left(\frac{p_2}{p_1}\right)^{\frac{n-1}{n}}$$

Substitution of the foregoing in Equation (2–10) yields

$$z_2 - z_1 = \frac{n}{n-1} \frac{RT_1}{g}\left(1 - \frac{T_2}{T_1}\right)$$

$$\frac{T_2 - T_1}{z_2 - z_1} = -\frac{n-1}{n} \cdot \frac{g}{R_0} \tag{2–11}$$

For dry air following the adiabatic process $n = 1.4$ and $R = 1716$. With these units the temperature gradient is

$$\frac{T_2 - T_1}{z_2 - z_1} = -0.0054 \text{ (nearly)}$$

Imagine a mass of air at the surface, as represented by point A in Fig. 2–16. If this air mass were to move aloft it would experience a drop in temperature of about 5.4 degrees Fahrenheit for each 100 feet, as indicated by the dotted line marked "dry adiabatic." The line labeled "unstable" indicates an existing condition of the surrounding atmosphere, as might be determined

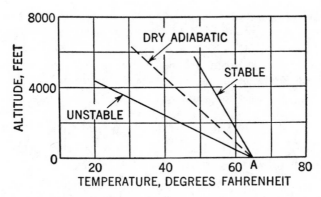

Fig. 2-16. Types of equilibrium in the atmosphere.

by a ballon or airplane sounding. At a certain level, say 3000 feet, the surrounding atmosphere is *colder* than the air mass which has just moved to this altitude. A lower temperature means denser air; the mass moved from the surface would be pushed upward further by the byoyant force; the condition is *unstable*.

The line in Fig. 2–16 labeled "stable" represents a temperature gradient less than the dry adiabatic. The stability can be reasoned out in a manner similar to that of the previous case. An "inversion" refers to a condition in which the temperature rises with an increase in height; an inversion is a very stable condition. If the observed temperature gradient in the atmosphere were equal to the adiabatic, the equilibrium would be neutral. The temperature of the moved air mass would be the same as the surrounding air; there would be no force pushing the air mass up or down.

Cumulus clouds, characterized by a horizontal flat base and a fluffy cauliflower top, may be formed by a vertical movement of moist air in an unstable atmosphere, with condensation starting at the base. If a mild shower results, it is sometimes called an *instability shower*. A violent instability shower may develop into a thunderstorm, a condition accompanied by thunder, lightning, and sometimes hail. Vertical observations in the atmosphere together with studies of stability, may be very helpful in the forecasting of thunderstorms. Questions of convective stability are also involved in applications dealing with heat-transfer equipment.

PROBLEMS

2-1. A diver is at a depth of 15 fathoms in the sea (1 fathom is 6 feet). What is the gage pressure at this depth?

2-2. What is the pressure equivalent of 6 atmospheres in tons per square yard?

2-3. How many feet of water are equivalent to 25 pounds per square inch? How many inches of mercury are equivalent to 25 pounds per square inch?

2-4. A large closed tank is partially filled with linseed oil. The pressure on the surface of the liquid is 11.0 pounds per square inch absolute. What is the absolute pressure in the oil 10 feet below the surface of the oil?

2-5. Inside of a television picture tube the absolute pressure is equivalent to 10^{-6} millimeter of mercury. A flat surface has a rectangular area of 13 inches by 17 inches. What is the net, total force of the atmosphere on this flat surface?

2-6. The gage on a condenser shows a vacuum of 24 inches of mercury. What is the absolute pressure in pounds per square foot?

2-7. The pressure 8 feet below the free surface of a liquid is 2.4 pounds per square inch. What is the specific gravity of the liquid?

2-8. A tank, having a volume of 20 cubic feet, contains air at a vacuum of 18 inches of mercury and 100 degrees Fahrenheit. What is the weight of the air in this tank?

2-9. A tube of short length is closed at one end and open at the other. This tube, originally containing air at standard conditions, was lowered vertically into the sea (specific gravity is 1.025), with the closed end up. When the tube was raised, it was found that the water had risen 0.80 of the length of the tube. Find the depth reached.

2-10. Two cylindrical diving bells of the same size are held immersed in water. Each is closed at the top and open at the bottom. The water inside bell *A* stands 4 feet below the free surface outside, while the water in bell *B* stands 6.5 feet below the free surface outside. If the interiors are connected, what is the new water level for the same weight of atmospheric air?

2-11. A cylindrical diving bell 10 feet high, orginally full of air, is lowered until the fresh water rises 4 feet in the interior. The cross-sectional area of the bell is 10 square feet. What is the depth of the bell below the surface? How many cubic feet of air at atmospheric pressure must be pumped in, in order that the water may be expelled from the interior?

2-12. Piston *B* in the fluid press shown in Fig. 2–17 has a diameter of 2 inches.

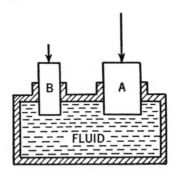

Fig. 2-17.

A force of 20 pounds at piston B exerts a force of 1000 pounds at piston A. Neglecting friction in the piston guides, what is the diameter of piston A?

2-13. A vertical glass U tube is employed to measure the pressure of air in a pipe. The water in the arm of the tube open to the atmosphere stands 7.5 inches higher than that in the arm connected to the pipe. Barometric pressure is 30.2 inches of mercury, and the temperature is 60 degrees Fahrenheit. What is the absolute pressure in the pipe?

2-14. For the arrangement shown in Fig. 2–18, determine the gage pressure at A. The kerosene has a specific gravity of 0.82.

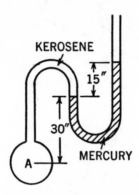

KEROSENE

15"

30"

A

MERCURY

Fig. 2-18.

2-15. For the inclined-tube draft gage shown in Fig. 2–19, compute the gage pressure at B if the right leg is open to the atmosphere. The oil has a specific gravity of 0.87.

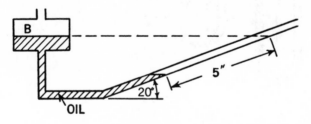

B

5"

20°

OIL

Fig. 2-19.

2-16. For the arrangement in Fig. 2–20, calculate the pressure difference between points A and B. Specific gravity of the oil is 0.85.

2-17. For the arrangement shown in Fig. 2–21, calculate the pressure difference between A and B.

2-18. The type of manometer shown in Fig. 2–22 is very useful for measuring small differences. The inside diameter of tube C is 7 millimeters, and the inside diameter of tube B is the same as that of tube A, namely 44 millimeters. The oil

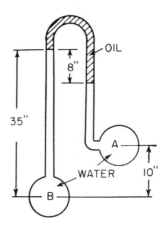

Fig. 2-20.

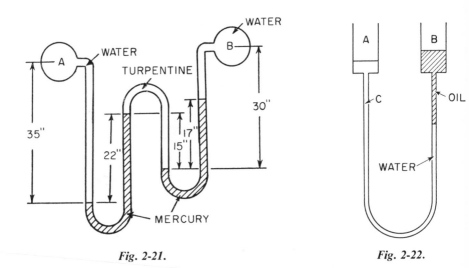

Fig. 2-21.   Fig. 2-22.

has a specific gravity of 0.83. Let p_A represent the pressure above the liquid in tube A, and p_B the pressure above the liquid in tube B. If the surface of separation between oil and water moves one inch, what is the corresponding change in the pressure difference $p_A - p_B$, in terms of inches of water?

2-19. If a mercury barometer reads 29.0 inches at 68 degrees Fahrenheit, what height would a benzene barometer read at 32 degrees Fahrenheit?

2-20. A rectangular tank 4 feet long, 3 feet wide, and 5 feet deep is filled with glycerine. What is the total force on the bottom, on one side, and one end? What is the location of the total force on one end?

2-21. A rectangular plate of height h and width b is submerged vertically in any incompressible fluid. The top edge of the plate is at the free surface of the liquid. What is the distance from the free surface to the total force?

2-22. The triangular tank in Fig. 2–23 is filled with turpentine. What is the total force on one end and one side? What is the location of the force on one end?

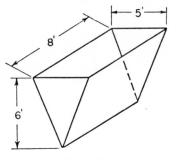

Fig. 2-23.

2-23. A flat plate, 2 feet by 10 feet is immersed in a liquid with the 10-foot side vertical. The total force on one side of the plate acts 10 inches below the centroid of the plate. What is the distance from the free surface to the upper edge of the plate?

2-24. A cylindrical tank 3 feet in diameter has its axis horizontal. At the middle of the tank, on top, is a pipe 2 inches in diameter, which extends vertically upward. The tank and pipe are filled with castor oil, with the free surface in the 2-inch pipe at a level 10 feet above the tank top. What is the total force on one end of the tank?

2-25. A cylindrical container 10 inches in diameter, with axis horizontal, is filled with mercury up to the cylinder center. Find the total force acting on one end.

2-26. A circular disk 8 feet in diameter is in a plane sloping 30 degrees from the vertical. A lye solution (specific gravity = 1.10) stands above the disk center to a depth of 10 feet. Calculate the magnitude, direction, and location of the total force of the solution on the disk.

2-27. A rectangular gate 7 feet high and 5 feet wide is placed vertically on the side of an open rectangular container of oil (specific gravity = 0.84). The free oil surface is 4 feet above the upper edge of the gate. What force must be applied at the upper edge of the gate to keep it closed if the gate is hinged at the lower edge?

2-28. A tank, separated in the center by a vertical partition, contains water on one side to a depth of 10 feet. The other side contains nitric acid (specific gravity = 1.50) to a depth of 12 feet. A rectangular opening in the center partition, 1.5 feet wide by 2 feet high, is closed by a flat plate. The plate is hinged at its upper edge, which is 2 feet above the bottom of the tank. What force applied at the lower edge of the plate is necessary to keep it closed?

2-29. Refer to Fig. 2–11. The surface ABC is a circular arc shell (one-fourth of a complete circle) with a radius of 10 feet. The distance perpendicular to the plane of the diagram is one foot. The fluid is water and the area DC is 12 feet below the liquid free surface. What are the magnitude and location of the components R_x and R_y?

2-30. A semicircular horizontal trough 8 feet in diameter and 20 feet long is closed at each end. It is filled with water. What is the total force of the water on the curved surface?

2-31. Refer to Fig. 2–11. The surface ABC is a parabolic curve with the vertex at A. The distance AD is 12 feet and the distance DC is 10 feet. The distance perpendicular to the plane of the diagram is one foot. The fluid is water, and the area DC is 15 feet below the free surface. What are the magnitude and location of the components R_x and R_y?

2-32. A sphere 4 inches in diameter just closes a hole 4 inches in diameter in the horizontal bottom of a container. If water in the container is at a height of 8 inches, what is the total force of the water on the hemispherical surface?

2-33. A cylindrical tank 5 feet in diameter is mounted with its axis horizontally. The tank is filled completely with oil having a specific gravity of 0.85. Atmospheric pressure acts on the topmost point of the oil. For a foot length along the axis, what is the resultant force of the fluid on the curved surface of a vertical half of the tank?

2-34. Refer to Fig. 2–12. Consider a slice of the dam one foot thick perpendicular to the plane of the diagram. The water height is 700 feet and the area $ABCA$ is 55,000 square feet. What is the total force of the water on the dam?

2-35. Water acts on each side of a vertical wall 5 feet wide. The surface at the bottom of the body of water is horizontal. On one side of the wall the water is 12 feet deep. On the other side of the wall the water is 6 feet deep. What is the net moment at the bottom end of the wall due to the water forces?

2-36. A vertical rectangular gate is 4 feet wide and 6 feet high. There is turpentine on one side only, and the upper edge of the gate lies in the liquid surface. The sole support of the gate is a shaft attached along the bottom edge. What torque must be applied to hold the gate against the liquid?

2-37. A rectangular tank is 10 feet wide, 20 feet long, and 10 feet deep. The tank is half full of water and half full of oil (specific gravity is 0.85). What are the magnitude and location of the total force on the 10- by 10-foot end?

2-38. A butterfly valve is mounted in the vertical end wall of a large water basin. This valve consists of a circular disk 6 feet in diameter pivoted about a horizontal axis passing through its center. Water stands on one side of the valve to a level 4 feet above the top of the disk. Atmospheric air only acts on the other side of the disk. What is the necessary torque about the pivot axis to hold the valve in position?

2-39. A circular, cylindrical log, 24 inches in diameter and 20 feet long, rests on the bottom of a creek bed. The water is stationary. On one side of the log the depth is 2 feet; on the other side of the log the depth is 1 foot. What are the magnitude and location of the net horizontal force of the water on the log? What is the location of the net total force of the water on the log?

2-40. A circular log, 36 inches in diameter and 30 feet long, rests on the bottom of a channel. On one side of the log is water 36 inches deep; on the other side of the log there is no water. What is the magnitude and location of the resultant water force on the log?

2-41. A cylindrical tank 8 feet in diameter, with axis horizontal, has hemispherical ends. The tank is filled with water, and a pressure gage at the top reads 5 pounds per square inch gage. What is the magnitude of the force tending to pull one of the hemispherical ends from the cylindrical part?

2-42. A rectangular bin is 10 feet long, 5 feet wide, and 6 feet deep. This bin is filled with wheat grain which has a specific weight of 48.0 pounds per cubic foot. Assuming this grain acts as a fluid, what is the magnitude and location of the force on the 5- by 6-foot wall?

2-43. A bin has a partition wall in the middle which is 5 feet wide. On one side is barley grain (specific weight 38.4 pounds per cubic foot) at a depth of 7 feet. On the other side is rye grain (specific weight 44.8 pounds per cubic foot) at a depth of 6 feet. Assume each grain acts as a fluid. What is the net moment of all the grain on the partition wall with respect to the bottom edge of the partition?

2-44. A rock weighs 90 pounds in air and 62 pounds in water. Find its volume and specific gravity.

2-45. The volume of the displacement of a ship in sea water is 4500 cubic feet. What is the weight of the ship?

2-46. A rectangular barge 60 feet by 30 feet sank $1\frac{1}{4}$ inches in fresh water when an elephant was taken aboard and located centrally. What was the elephant's weight?

2-47. A closed rectangular steel box, 40 feet by 20 feet by 25 feet high, and weighing 80 tons, is to be sunk in water. How deep will it sink when launched? If the water is 18 feet deep, what weight must be added to cause it to sink to the bottom?

2-48. A closed rectangular box 20 feet long, 9 feet wide, 10 feet high, and weighing 22 tons is lowered into water. What is the magnitude and location of the water force acting on the 10-by-9 vertical side?

2-49. An iceberg has a specific weight of 57.2 pounds per cubic foot. What portion of its total volume will extend above the surface if it is in fresh water?

2-50. An airship contains 5 tons of hydrogen at the surface of the earth. What will be the net lift in air, and the volume of the hydgrogen at 56 degrees Fahrenheit and 30.2 inches of mercury? Assume that the air and hydrogen are at the same temperature.

2-51. An airship is to have a net lift of 40 tons when the gas temperature equals that of the air. What weight of helium must it have in its cells?

2-52. The hull of a certain ship has sides sloping outward and upward at the water line. Explain why this design is more stable than a hull with parallel sides.

2-53. It is common practice to place small metal balls or lead shot at the bottom of a hydrometer. What is the advantage of this particular location?

2-54. For a sphere and a cylinder the center of gravity is on the geometrical axis? Why would each floating be in a condition of neutral equilibrium?

2-55. A closed cylindrical drum 2 feet in diameter and 4 feet high weighs 70 pounds. Would this drum float stably in water in an upright position?

2-56. What is the apparent weight of a 6-inch steel cube suspended in water? Assume steel weighs 490 pounds per cubic foot.

2-57. A spherical balloon is 10 feet in diameter. The balloon is filled with helium. Barometric pressure is 29.92 inches of mercury. The air and helium are at the same temperature of 70 degrees Fahrenheit. What load would the balloon lift?

2-58. Would oxygen be suitable in a balloon to be used for lifting purposes? Explain why.

2-59. For an assumed isothermal atmosphere, show that the pressure becomes zero at an infinite height.

2-60. The barometric pressure at sea level is 30.10 inches of mercury when that on a mountain top is 29.00 inches. If the air temperature is constant, at 58 degrees Fahrenheit, what is the elevation at the mountain top?

2-61. Atmospheric pressure at the ground is 14.70 pounds per square inch at a temperature of 60 degrees Fahrenheit. Determine the absolute pressure 12,000 feet above the ground: (a) if the air is incompressible, (b) if the air follows the isothermal relation, and (c) if the air follows the dry adiabatic relation.

2-62. At a certain altitude in a dry adiabatic atmosphere the pressure is one-half of that at sea level. Calculate the absolute pressure at this same altitude for an isothermal atmosphere, assuming standard conditions at sea level.

2-63. An air mass at 70 degrees Fahrenheit is quickly forced up vertically a distance of 8000 feet. At this altitude the surrounding atmosphere has a temperature of 20 degrees Fahrenheit. At this level, will the air mass be pushed up or down?

2-64. What is the pressure, in pounds per square inch, at 10,000 feet in the standard atmosphere? Assume a lapse rate $-0.00356°F/ft$.

2-65. In meteorology the so-called "potential temperature" is defined as the temperature an air particle would have if its pressure were changed to 1000 millibars in a dry adiabatic process. Consider a particle of air at 890 millibars and an absolute temperature of 293 degrees Kelvin (centrigrade scale). What is the potential temperature for this air particle?

2-66. If standard conditions exist at sea level, calculate the altitude in an isothermal atmosphere at which the specific weight is 0.80 the specific weight at sea level.

2-67. In the atmosphere, at an altitude of 4000 feet, the temperature is -5.9 degrees Fahrenheit and the pressure is 860 millibars. What is the temperature at sea level for a dry adiabatic process?

2-68. Readings at a weather station give a barometric pressure of 1000 millibars and a temperature of 68 degrees Fahrenheit. An air mass at the ground rises aloft quickly following a dry adiabatic process. At an elevated position the air mass has a pressure of 868 millibars. What is the temperature and elevation above ground of the elevated air mass?

2-69. At a location which is 2000 feet above sea level, the barometer reads 29.80 inches of mercury at 70 degrees Fahrenheit. What would be the barometric pressure at sea level assuming a dry adiabatic process?

CHAPTER THREE

Kinematic, Dynamic, and Energy Equations

*All that we see distinctly in the motion of a body is that the body
traverses a certain distance and that it takes a certain time to traverse
that distance. It is from this one idea that all the principles of
mechanics should be drawn, if we wish to demonstrate them in a clear
and accurate way.*

—D'ALEMBERT.[1]

In approaching a problem we must establish what factors are known or
given and what factor we wish to determine. Then there is this question:
What fundamental relations, equations, or tools are available for use in solving
our problem? Four useful basic relations are: (a) the equation of state, a
pressure–temperature–density relation; (b) the equation of continuity, a
kinematic relation dealing with the geometry of motion without regard to
the forces causing that motion; (c) the dynamic or momentum equation;
and (d) the energy equation.

KINEMATICS OF FLUID FLOW

3–1. Methods of describing motion

Two methods of representing fluid motion have been devised. The
Lagrangian method (after J.L. Lagrange) refers to a description of the
behavior of individual fluid particles. A description of the paths of individual
fluid particles would be an example of the Lagrangian method. Similar de-
scriptions could be given for velocities, accelerations, and other characteristics.
The Eulerian method (after L. Euler) refers to a description of the velocity,
pressure, and other characteristics at certain points or a section in the fluid.
The Eulerian method is probably the more common and familiar. For example, consider the flow in a pipe, as shown in Fig. 3–1; the velocity distribution

---

[1] *A Source Book in Physics*, by W. F. Magie. McGraw-Hill, New York, 1935.

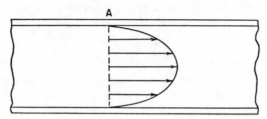

Fig. 3-1. Example of Eulerian method of description.

at section A is given. Different fluid particles pass this section. It is not known just *which particles* pass this section, but the description gives the velocity at various points at the particular station or section.

A *path line* is a line made by a single particle as it moves during a period of time. The trace made by a single smoke particle as it issues from a chimney would be a path line. A path line may be obtained with a long exposure on a fixed photographic plate. A *streamline* is a line which gives the velocity direction of the fluid at each point along the line. Consider a series of particles at an instant in a flowing fluid. Imagine that at each particle a line is drawn showing the direction of the instantaneous velocity at each point. A smooth curve tangent to each of these lines would be a streamline. Note that a path line refers to the path of a *single particle*, whereas a streamline refers to an instantaneous picture of the velocity direction of a *number* of particles.

As illustrated in Fig. 3–2, imagine a fluid particle at point A on a streamline at a particular time. Let s represent distance along the streamline, let t represent time, and let R represent the radius of curvature. The distance s is sometimes called a natural coordinate. Acceleration is the time rate of change of velocity. The acceleration at point A has a *radial* or *normal component* and a *tangential component*. The normal or radial component of acceleration a_n is

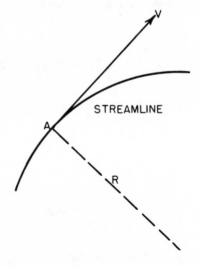

$$a_n = -\frac{V^2}{R} \qquad (3\text{-}1)$$

Next, consider only the tangential acceleration. The tangential velocity V is a function of both the distance s along the streamline and the time t, that is

Fig. 3-2. Point on a streamline.

$$V = f(s, t) \qquad (3\text{-}2)$$

The tangential velocity V could change from point to point in space at one

instant of time and also from moment to moment of time at any one point in space. Thus

$$dV = \frac{\partial V}{\partial s} ds + \frac{\partial V}{\partial t} dt \tag{3-3}$$

The partial derivative is taken with respect to one variable only, keeping all other variables constant. For example $\partial V/\partial s$ is the derivative of V with respect to s only, with time constant.

Acceleration is the time rate of change of velocity, Thus the tangential acceleration dV/dt is

$$\frac{dV}{dt} = \frac{\partial V}{\partial s}\frac{ds}{dt} + \frac{\partial V}{\partial t} = V\frac{\partial V}{\partial s} + \frac{\partial V}{\partial t} \tag{3-4}$$

The term $\partial V/\partial t$ is frequently called the "local acceleration," and the term $V(\partial V/\partial s)$ the "convective acceleration." The local acceleration represents the change in local velocity with time at a fixed point. If the flow is steady, then $\partial V/\partial t$ equals zero.

Sometimes it is possible and convenient to change a nonsteady flow into a steady flow by shifting the reference system or the position of an observer. Picture the flow around a bridge pier. An observer on the pier will note steady flow if the streamlines do not change with time. On the other hand, an observer in a rowboat moving with the same stream will find the flow unsteady. If an observer on some aircraft witnesses steady flow, another observer on the ground would find an unsteady flow of the fluid as the aircraft approaches and passes.

3-2. One-dimensional method of flow analysis

Figure 3-3 shows a rectangular coordinate system in space; let x, y, and z be the coordinates. Let V represent the resultant velocity at any point in a body of fluid, u the x-component of this resultant velocity, v the y-component, and w the z-component. In the most general case there are u, v, and w velocity components, there is flow in the three directions, and the motion is classed as *three-dimensional*.

In numerous engineering applications it is sufficiently accurate to regard the motion as *one-dimensional*. Such a motion is an approximation or simplification of two- and three-dimensional distributions. Figure 3-4 illustrates one-

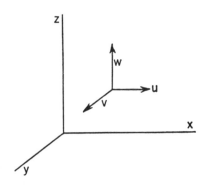

Fig. 3-3. Coordinate system.

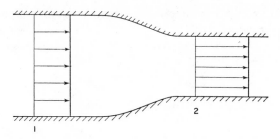

Fig. 3-4. Illustration of one-dimensional flow.

dimensional flow; the fluid passes through a tube having circular cross sections. The vectors at sections 1 and 2 are velocity vectors. At section 1 the velocity is assumed as constant over the cross section normal to the flow; let this velocity be represented by V_1. Actually the velocity may vary over the section; the assigned value of V_1 is a certain "average" or "mean" value. This mean value represents a real simplification in making calculations. At section 2 the velocity is also constant over the cross section; the mean velocity may vary, however, from section 1 to section 2.

An average fluid property at a cross section may be determined by a graphical or an analytical integration.

As an example, for the case shown in Fig. 3–5, at section 1 consider an

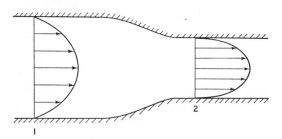

Fig. 3-5. Illustration of flow.

infinitesimal area dA normal to the flow. The fluid velocity at this point is V. The volume of fluid crossing this infinitesimal area per unit time, or the *flux*, is VdA. The total volume of fluid per unit time, or Q, passing through the entire channel cross section at station 1 is the sum of all the individual amounts, or

$$Q = \int VdA$$

Let A represent the total area of the channel normal to the flow. Then an average velocity V_1 can be formed by writing $Q = AV_1$ or $V_1 = Q/A$.

Various parts of this book will be devoted to one-dimensional motion.

This method of analysis is sufficient in many engineering applications. Care should be taken however, to avoid extending the one-dimensional analysis beyond its limits of application, as to cases in which the two- and three-dimensional features are important.

3–3. Equation of continuity-steady flow

We will consider a matter or material accounting for flow through a *stream-tube*, or tubular space bounded by a surface consisting of streamlines. It may be an imaginary surface in a large body of fluid, which behaves like a tube with rigid walls inside which fluid flows. On the other hand, a stream-tube may be formed by the walls of a pipe, the walls of a converging nozzle, the walls of a diverging conical channel, or by the walls of some other structure. No fluid crosses the walls of the stream-tube.

The *equation of continuity* for flow in this stream-tube is a special case of the general physical law of the conservation of matter. Figure 3–6 illustrates

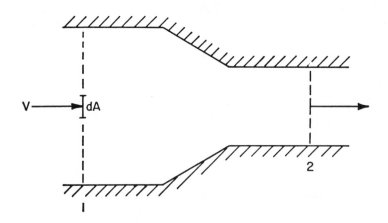

Fig. 3-6. Stream tube.

a stream-tube, in which fluid flows from section 1 to section 2. At section 1 the velocity at any point is V; this velocity is normal to the infinitesimal area dA, and the density is ρ; the mass rate of flow through the area dA is $\rho V dA$ (as slugs per second). The velocity V may vary over the section 1; the total mass rate of flow through section 1 is given by the integral

$$\int_1 \rho V dA$$

which is taken over the entire section 1. In a similar fashion, for section 2 the mass rate of flow through the entire section 2 is given by the integral

$$\int_2 \rho V dA$$

where ρ, V, and dA are taken at a point in section 2. For the case of steady flow, the mass rate of flow across section 1 equals the mass rate of flow through section 2; thus the equation of continuity becomes

$$\int_1 \rho V dA = \int_2 \rho V dA = \text{constant} \tag{3-5}$$

Consider next the case of steady, one-dimensional flow, in which V_1 is the uniform, or average, velocity across section 1, ρ_1 is the uniform density over section 1, and A_1 is the entire area of section 1; this type of flow is illustrated in Fig. 3–4. Let the subscript 2 be used similarly for the corresponding factors at section 2. Then the equation of continuity takes the form

$$\rho_1 V_1 A_1 = \rho_2 V_2 A_2 = \text{constant} \tag{3-6}$$

For liquids and gases which can be treated as incompressible, the density is constant, and Equation (3-6) takes the special form

$$Q = A_1 V_1 = A_2 V_2 = \text{constant} \tag{3-7}$$

where the product $Q = AV$ is a volume rate of flow.

3–4. Equation of continuity-unsteady flow

The case for nonsteady or unsteady flow will be set up in differential form. The *control volume* is an arbitrary volume, fixed in space, which is bounded by a closed surface called the *control surface*. The control volume may be infinitesimally small, or it may be large, of finite size. Although the control volume is fixed in space, fluid may enter or leave the control volume by crossing the control surface.

The solid lines in Fig. 3–7 illustrate a wall or surface formed by stream lines; no fluid crosses the streamline surface. The control volume indicated by the dotted lines consists of the area dA, through which fluid enters with the velocity V and density ρ, the surface of infinitesimal length ds along the streamline surface, and the exit area. The mass rate of flow entering the control volume is $\rho V dA$. As the fluid passes through, there is a change in mass rate in the distance ds by the amount

$$\frac{\partial}{\partial s}(\rho V dA)ds$$

Thus, the total exit mass rate is

$$\rho V dA + \frac{\partial}{\partial s}(\rho V dA)ds \tag{3-8}$$

The fluid in the small control volume has a mass equal to $\rho dA ds$. For unsteady

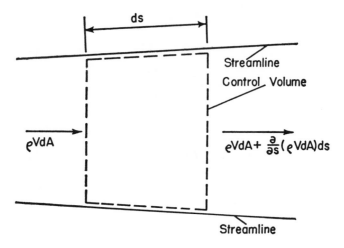

Fig. 3-7. Control volume.

flow the rate at which this mass is increasing in the control volume is

$$\frac{\partial}{\partial t}(\rho\, dA\, ds)$$

This increase in mass rate of flow must equal the difference between the mass rate of inflow and mass rate of outflow, that is,

$$\frac{\partial}{\partial t}(\rho\, dA\, ds) + \frac{\partial}{\partial s}(\rho V\, dA)\, ds = 0$$

Since ds does not depend on time, we can write

$$\frac{\partial}{\partial t}(\rho\, dA) + \frac{\partial}{\partial s}(\rho V\, dA) = 0 \qquad (3\text{-}9)$$

For the special case of steady flow, the first left-hand term in Equation (3-9) is zero, and the equation becomes

$$\frac{d}{ds}(\rho V\, dA) = 0 \qquad (3\text{-}10)$$

which is the differential form of Equation (3-4). Equation (3-10) can be written in the form

$$\frac{d\rho}{\rho} + \frac{dV}{V} + \frac{dA}{A} = 0 \qquad (3\text{-}11)$$

A more general treatment of the continuity equation is given in Chapter Twelve.

3–5. Velocity distribution

Streamlines can be made visible in various ways, by using smoke or aluminum powder. Figures 3–8 and 3–9 show some smoke-tunnel photo-

Fig. 3-8. Flow around a tapered section.

Fig. 3-9. Flow through a nozzle.

graphs. The two-dimensional air flow, between two glass plates $\frac{1}{4}$ inch apart, is from left to right. Titanium tetrachloride smoke was introduced at equidistant points before the model to indicate the streamlines. For this case, and if it is assumed that the fluid is incompressible, the continuity equation shows that a converging of the streamlines is associated with an increase in the velocity of the fluid between the streamlines. There is a crowding of streamlines above the tapered section and in the nozzle throat. At the nozzle throat, particularly, the velocity is considerably higher than that some distance upstream.

A general problem is to map out the velocity distribution in the field of a flowing fluid. If the velocity at each point in the fluid is known, an application of a dynamic equation yields the pressure at each point. If a body is in the fluid, the pressure distribution around the body can be used in determining the total force acting on the body. A common ultimate problem is to determine the force acting on a body or structure.

DYNAMIC OR MOMENTUM EQUATION

3–6. Laws of motion

Galileo and Newton made important contributions to basic mechanics. Some of these contributions are frequently formulated into three laws of motion. Stated in modern terms, and considering only linear motion or translation, these laws are as follows:

1. If a body is not acted upon by any external force, the linear momentum of the body remains constant.

2. The time rate of change of the linear momentum of a body is directly proportional to the force acting on the body; the change of linear momentum is in the direction of the force acting.

3. For every force acting *on* a body there is a corresponding force exerted *by* the body; these two forces are equal in magnitude but opposite in direction.

Let m represent the mass of a particle, F the net or resultant of the external forces acting on the particle, V the instantaneous linear velocity, and t time. The product mV is called linear momentum. Then the basic dynamic or momentum equation becomes

$$F = \frac{d}{dt}(mV) \qquad (3\text{–}12)$$

Force equals the time rate of change in linear momentum. For a constant mass m, this equation becomes

$$F = m\frac{dV}{dt} \qquad (3\text{–}13)$$

3–7. Euler equation for a frictionless fluid

We will study first, in this article, the case of a frictionless fluid. As illustrated in Fig. 3–10, imagine an infinitesimally small cylinder, a control volume, inside a stream-tube. The length of this element is ds, and the area normal to the flow at each end is dA. The fluid density is ρ, and the gravitational acceleration is g. The elementary control volume is fixed in space; we will study the fluid flowing through it, and write the equation of motion for the mass which occupies the control volume at a given instant. The mass

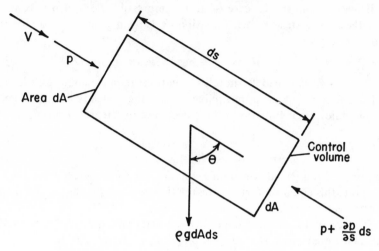

Fig. 3-10. Forces on an element of a frictionless fluid.

involved is $\rho dAds$, the acceleration of this mass is dV/dt. The pressure on the left or entering face of the control volume is p. The pressure on the right or exit face is $p + (\partial p/\partial s)ds$. The force pdA acts on the left face, and the force $(p + (\partial p/\partial s)ds)\,dA$ acts on the right face. The gravitational force $\rho gdAds$ adds vertically downward as shown. There is no friction force opposing the motion. The flow is along the streamline. Thus, the dynamic equation, with F as the net force in the direction of motion and acceleration, becomes

$$F = m\frac{dV}{dt}$$

$$F = pdA - \left(p + \frac{\partial p}{\partial s}ds\right)dA + \rho gdAds \cos\theta = \rho dAds\frac{dV}{dt} \quad (3\text{-}14)$$

Note that $\cos\theta = -\,\partial z/\partial s$, where z is vertical height. Note also that in the most general case the acceleration is

$$\frac{dV}{dt} = V\frac{\partial V}{\partial s} + \frac{\partial V}{\partial t} \quad (3\text{-}15)$$

Then Equation (3-14) becomes

$$\rho\left(V\frac{\partial V}{\partial s} + \frac{\partial V}{\partial t}\right) = -\rho g\frac{\partial z}{\partial s} - \frac{\partial p}{\partial s} \quad (3\text{-}16)$$

Equation (3–16), for frictionless flow, is called Euler's equation of motion; it applies to both steady and unsteady flow.

For steady flow, $\partial V/\partial t = 0$, the remaining derivatives become total derivatives, and Equation (3–16) can be then written in the form

$$\rho VdV + dp + g\rho dz = 0 \quad (3\text{-}17)$$

If we take the special case of an incompressible fluid, if ρ is constant, and if the flow is steady, then Equation (3–17) can be integrated along the streamline to give

$$\tfrac{1}{2}\rho V^2 + p + \rho gz = \text{constant} \qquad (3\text{-}18)$$

Equation (3-18) is called Bernoulli's equation for an incompressible fluid.

Picture next the case of negligible gravity effect, steady flow, but a compressible fluid. Then Equation (3–17) becomes, in integral form,

$$\frac{V^2}{2} + \int \frac{dp}{\rho} = \text{constant} \qquad (3\text{-}19)$$

Equation (3–19) is called Bernoulli's equation for compressible flow. In order to integrate this equation further, the relation between pressure p and density ρ must be established.

The case of a static fluid might be regarded as the special case of no flow; in this case Equation (3–16) or Equation (3–17) gives the relation

$$dp = -\rho g dz$$

which is the same as Equation (2–1).

3–8. Some illustrations

Consider the flow around the body shown in Fig. 3–11, for a frictionless incompressible fluid and negligible changes in elevation. The pressure is

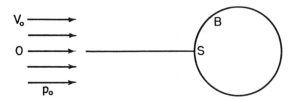

Fig. 3-11. Flow around body.

p_0 and the velocity is V_0 in the undisturbed stream approaching the body. Applying Bernoulli's Equation (3–18) between point 0 and point B on the body gives

$$p_0 + \tfrac{1}{2}\rho V_0^2 = p_B + \tfrac{1}{2}\rho V_B^2 \qquad (3\text{-}20)$$

where p_B is the pressure at point B and V_B is the velocity at B. If p_0 and V_0 are known, then p_B can be calculated if V_B is known. V_B might be determined by a kinematic study, as by a study of the streamline pattern.

At the *stagnation point* S on the body the fluid velocity is zero and the stagnation pressure is p_s. Applying Bernoulli's equation along the small stream-tube O to S gives

$$p_s - p_0 = \tfrac{1}{2}\rho V_0^2 \tag{3-21}$$

The term $\tfrac{1}{2}\rho V_0^2$ is called the *dynamic* or *stagnation pressure*. The pressure p_0 in the undisturbed stream is called the *static pressure*. The dynamic pressure term occurs frequently in studies of fluid meters and in studies dealing with the total forces acting on moving bodies in a fluid. With some fluid meters, p_s and p_0 are measured, and V_0 is calculated by Equation (3-21).

Example. A submarine moves through salt water at a depth of 55 feet with a speed of 10.42 knots (12 land miles per hour). Determine the gage pressure on the bow stagnation point of the submarine. Specific weight of the water is 64.0 pounds per cubic foot.

$$p_s = p_0 + \tfrac{1}{2}\rho V_0^2$$

$$p_s = 55(64) + \frac{1}{2}\left(\frac{64.0}{32.2}\right)\left(\frac{12 \times 5280}{3600}\right)^2 \text{ pounds per square foot}$$

$$p_s = \frac{3828}{144} = 26.6 \text{ pounds per square inch}$$

3-9. Dynamic equation with friction

The previous Section 3-7 presented Euler's equation for a frictionless fluid. In various cases friction, and other forces, can be neglected, whereas in some cases friction, along with other forces, must be included. The general method is to start with the basic dynamic equation and include all the forces involved.

As illustrated in Fig. 3-12 imagine an infinitesimally small cylinder, a fixed control volume, inside a stream-tube. The length of this element is ds and the area normal to the flow at each end is dA. As with the previous case, there is a pressure force at each end and a gravitational force. Imagine, in addition, a friction force dF_F acting on the surface of length ds of the control surface; this friction force opposes motion. Let us include also another force dF_D due to a body or surface (other than the force dF_F) which acts on the element; this force dF_D may be opposite or it may be in the direction of the flow. The dynamic equation, with F as the net force, is

$$F = m\frac{dV}{dt}$$

The mass involved is $\rho dAds$. The net force F is

$$F = pdA - \left(p + \frac{\partial p}{\partial s}ds\right)dA + \rho gds\cos\theta - dF_F - dF_D \tag{3-22}$$

Then the basic relation becomes

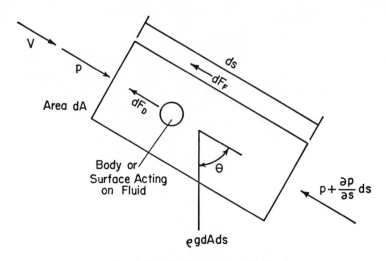

Fig. 3-12. Fluid element with friction.

$$-\frac{\partial p}{\partial s}dsdA + \rho gds \cos \theta - dF_F - dF_D = \rho dAds\left(V\frac{\partial V}{\partial s} + \frac{\partial V}{\partial t}\right) \quad (3\text{--}23)$$

In a particular case we would have to consider the given conditions and integrate the foregoing relation accordingly.

3-10. Linear momentum relation for steady flow

In some cases it is convenient to employ a momentum relation. Momentum relations are useful in attacking flow problems because these relations give information about conditions along the boundaries. Certain conclusions about the flow can be drawn without a detailed knowledge of the flow inside the boundaries. This article will be restricted to steady flow.

As indicated by Equation (3–12), force equals the time rate of change in linear momentum. Figure 3–13(a) illustrates steady one-dimensional flow through a channel with side walls. The fixed control surface consists of the side walls of the channel, the inlet area, and the exit area. The velocity V_1 is normal to the inlet area A_1, and the velocity V_2 is normal to the outlet area A_2. The force F is the resultant of all the external forces acting on the fluid; this resulant force includes the weight of the walls and fluid (if significant), all the external forces applied by channel walls on the fluid, such as friction and others, and the pressure forces acting on the inlet and exit area sections.

Figure 3–13(b) illustrates the system at time t; the system consists of the channel side walls, the fluid within it, and the fluid mass m which will enter the channel in the time dt. After the time interval dt, the system has the

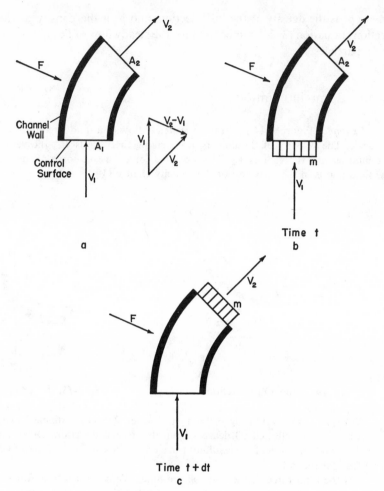

Fig. 3-13. Flow through a channel.

position shown in Fig. 3–13(c), in which the mass of amount m has just left the exit section. The momentum of the fluid within the channel between the two ends at time t is identical with that between the same two ends of the channel at time $t + dt$; thus the difference between the momentum of the system at the two times is $V_2 m - V_1 m$. Since this change in momentum takes place in the time dt, the time rate of change in momentum can be expressed in the form

Force = (mass rate of flow)(velocity change)

$$F = \text{(mass rate)}(V_2 - V_1) \tag{3–24}$$

The mass rate of flow m/dt can be expressed in the form

$$A_1 V_1 \rho_1 = A_2 V_2 \rho_2 \tag{3–25}$$

where ρ_1 is the density at the inlet section, and ρ_2 is the density at the exit section. Equation (3–24) can be used for the steady flow of liquids, gases, and solid particles.

3–11. Some illustrations

Example. Figure 3–14 illustrates the steady flow of a fluid stream or jet from a nozzle. The velocity of the fluid before striking the vane is V_1; the velocity of the fluid leaving the vane is V_2. The indicated force F is the force of the vane *on* the fluid and is in the direction of the velocity change ($V_2 - V_1$).

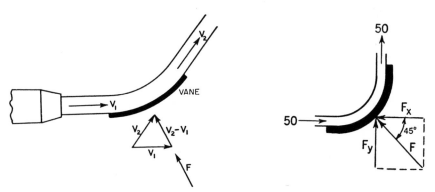

Fig. 3-14. Vane and jet action.   **Fig. 3-15.** Jet example.

Example. As shown in Fig. 3–15, a jet of water, 2 inches in diameter, strikes a fixed vane and is deflected 90 degrees from its original direction. Determine the magnitude and direction of the resultant force on the vane or blade if the jet velocity is 50 feet per second.

F is the total force of the vane *on* the fluid. We can break this force F into two components F_x and F_y. F_x is a force in the direction indicated which decelerates the fluid from 50 feet per second to zero velocity. F_y is a force in the direction indicated which accelerates the fluid from zero velocity tc 50 feet per second. Let us arbitrarily call the direction to the right as positive, and the direction upward as positive. Then

$$F_x = \frac{\pi}{4}\left(\frac{2}{12}\right)^2 50\left(\frac{62.4}{32.2}\right)[0 - 50] = -105.7 \text{ pounds}$$

$$F_y = \frac{\pi}{4}\left(\frac{2}{12}\right)^2 50\left(\frac{62.4}{32.2}\right)50 - 0 = 10.57 \text{ pounds}$$

The magnitude of the total force is

$$F = \sqrt{F_x^2 + F_y^2} = \sqrt{2(105.7)^2}$$
$$= 149.5, \text{ say } 150 \text{ pounds}$$

The resultant force of 150 pounds acts at 45 degrees with the original jet direction.

Example. As illustrated in Fig. 3–16, picture one-dimensional steady flow in an enclosed channel. Assume that each section normal to the axis of the channel is circular, and that the flow is parallel to the axis of the channel at sections 1 and 2. At section 1 the static pressure is p_1, the uniform velocity is V_1, the density is ρ_1, and the area is A_1. In a similar fashion the subscript 2 is used for the factors at section 2. The walls of the diverging channel exert a total force on the fluid. Let R represent the axial component of this total force or the component in the direction of motion. Figure 3–16(b) illustrates the forces acting on the free body or control surface consisting of the two circular ends and the conical transition area.

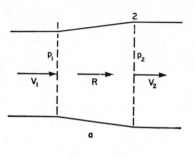

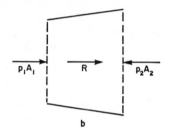

The mass rate of flow times the velocity change equals the sum of the external forces acting on the body of fluid. Thus

$$p_1A_1 - p_2A_2 + R = \frac{m}{dt}(V_2 - V_1)$$

Fig. 3-16. Flow in an enclosed channel.

We can write the continuity equation in the form

$$\frac{m}{dt} = A_2V_2\rho_2 = A_1V_1\rho_1$$

Thus the force R can be expressed as

$$R = A_2(p_2 + \rho_2V_2^2) - A_1(p_1 + \rho_1V_1^2)$$

Example. Figure 3–17 shows a jet striking a fixed plate which is not normal to

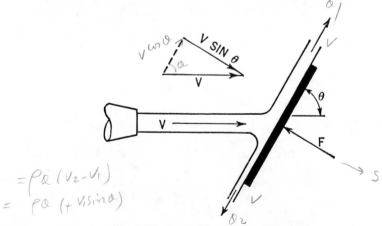

$= \rho \dot{Q}(V_2 - V_1)$
$= \rho Q(+V_1\sin\alpha)$

Fig. 3-17. Jet division by an oblique plate.

the initial stream direction. We assume two-dimensional steady incompressible flow parallel to the plane of the diagram. The dimensions of the jet are small in comparison with those of the plate. The total change in linear momentum is normal to the plate if friction along the plate is neglected. The total force F is normal to the plate and has the magnitude

$$F = \rho Q V \sin \theta \qquad (3\text{-}26)$$

where $V \sin \theta$ is the velocity change, Q is volume rate of flow approaching the plate (as cubic feet per second), ρ is density, and $Q\rho$ is mass rate of flow.

Let Q_1 represent the volume rate of flow moving upward along the plate, and Q_2 the flow rate downward along the plate. The continuity equation indicates that $Q = Q_1 + Q_2$. For the case of no shock and no loss, the linear velocity upward along the plate is V and the linear velocity downward along the plate is V. Since there is no force parallel to the plate, there is no momentum change in the direction parallel to the plate. Thus the initial momentum parallel to the plate equals the fluid momentum after the plate. Thus,

$$Q\rho V \cos \theta = Q_1 \rho V - Q_2 \rho V \qquad (3\text{-}27)$$

Combining the continuity equation with Equation (3–27) gives the relations

$$Q_1 = \frac{Q}{2}(1 + \cos \theta) \qquad (3\text{-}28)$$

$$Q_2 = \frac{Q}{2}(1 - \cos \theta) \qquad (3\text{-}29)$$

Example. A *single* moving surface or vane in the path of a jet deflects only that portion of the total discharge which overtakes the vane. If Q is the discharge from the fixed nozzle in Fig. 3–18, the discharge Q' overtaking the vane is

$$Q' = Q\left(\frac{V - v}{V}\right) \qquad (3\text{-}30)$$

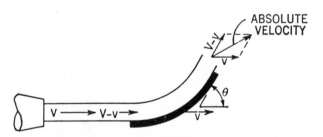

Fig. 3-18. Deflection of a jet by a moving vane.

where v is the absolute vane velocity. The fluid velocity relative to the blade at entrance is $V - v$. If there is no friction along the blade, the relative velocity of the fluid at exit has a *magnitude* equal to $V - v$. The absolute exit velocity of the fluid is the vector sum of $(V - v)$ and v. Taking the jet axis as the x-axis, the magnitudes of the force components for a *single* vane are

$$F_x = \rho Q'\{V - [v + (V - v) \cos \theta]\}$$
$$F_y = \rho Q'(V - v) \sin \theta \qquad (3\text{-}31)$$

Since power is defined as the rate of doing work, the power $E = F_x v$. Note that the force F_x is in the direction of v. No work is done in a direction perpendicular to v.

Example. A *series* of vanes or blades might be so arranged that the entire jet discharge is deflected by the vanes; if so, Q' would equal Q. An arrangement of vanes on the periphery of a wheel, as in an ideal impulse turbine, is one example. The useful torque producing force F_x for the arrangement in Fig. 3–19 is

$$F_x = Q\rho\{V_1 - [v + (V_1 - v)\cos\theta]\} \tag{3-32}$$

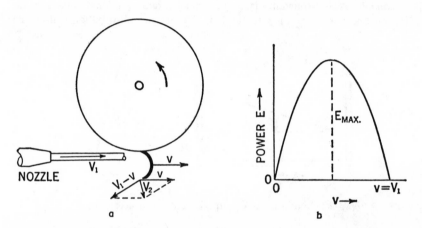

Fig. 3-19. Deflection of a jet by a series of vanes, as in an ideal impulse turbine.

The power E becomes

$$E = F_x v = Q\rho\{V_1 v - [v^2 + (V - v)v\cos\theta]\} \tag{3-33}$$

For a given Q and jet size, the ideal power developed by the wheel, as expressed by Equation (3–33), will depend only on v. The power variation with v shows a maximum value of power for a certain value of v. This maximum value $E_{\max}$ can be found by differentiating E with respect to v, and setting the result equal to zero;

$$\frac{dE}{dv} = \frac{d}{dv}[Q\rho(V_1 - v)v(1 - \cos\theta) = 0$$

Solving for v gives the result that $v = V_1/2$ for maximum power.

3–12. D'Alembert's principle

Frequently reference is made to an "inertia" force. For a constant mass, the basic dynamic equation is

$$F = m\frac{dV}{dt} \tag{3-34}$$

D'Alembert's principle states that this equation can be written as

$$F - m\frac{dV}{dt} = 0 \qquad (3\text{-}35)$$

The fictitious force $(-m(dV/dt))$ is sometimes called the reversed effective force or the *inertia* force. D'Alembert's principle thus makes it possible to reduce a problem in dynamics to an equivalent problem in statics by the introduction of an inertia force.

    Example. As illustrated in Fig. 3–20, a tank containing a liquid has a constant horizontal linear acceleration *a* to the left. The pressure distribution will be the

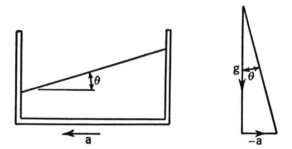

Fig. 3-20. Liquid in tank with constant linear acceleration.

same as if the liquid were at rest under gravitational acceleration and an acceleration $-a$. The effect is the same as if the direction of gravity were turned an angle θ back from the vertical, such that

$$\tan \theta = \frac{a}{g}$$

The free surface is a plane tilted through an angle θ. The pressure at any point on the bottom of the container is found by the relation $\Delta p = \gamma h$, where h is the height to the free surface at the point in question.

    Imagine a vertical cylindrical vessel with a liquid at rest. Assume first that the liquid is frictionless. For a rotation of the vessel at constant angular velocity about a central vertical axis, the frictionless liquid, theoretically, would remain stationary. If the liquid were real and viscous, however, rotation of the vessel would rotate the liquid through a viscous shear action. After an initial transient, all of the liquid would rotate steadily as a solid body or a flywheel. This motion might be classed as a solid-body rotation, a rotational vortex motion, or a forced vortex. The streamlines are concentric circles.

    Example. Consider the action of a liquid in a container rotating about a vertical axis with constant angular velocity ω (Greek letter omega). ω is expressed in radians per unit time. Relative equilibrium is reached a short time after starting; the liquid then rotates as a solid body. Such a motion is a solid-body, rotational, or *forced vortex.* The free surface of the liquid is curved, as indicated in Fig. 3–21.

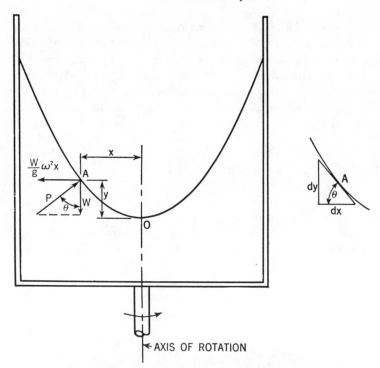

Fig. 3-21. Rotating body of liquid.

There are three forces acting on the fluid element at point A in Fig. 3–21. One force is the weight of the element W. Each element experiences an acceleration toward the axis of rotation; this centripetal acceleration is $\omega^2 x$. The inertia force $(W/g)\omega^2 x$ acts in a radial direction away from the axis of rotation. The force P is the resultant force due to the pressure of the surrounding fluid particles. Because there is no relative sliding between the particles, P is normal to the curved surface. Since these three forces are in equilibrium,

$$P \sin \theta = \frac{W}{g}\omega^2 x \qquad P \cos \theta = W$$

$$\tan \theta = \frac{\omega^2 x}{g} = \frac{dy}{dx}$$

Integration gives the equation of the curved free surface:

$$y = \frac{\omega^2 x^2}{2g} \tag{3-36}$$

which is the equation of a parabola. The pressure at any point on the bottom of the container is found from the relation $\Delta p = \gamma h$, where h is the height to the free surface above that point.

If fluids of different densities are placed in a stationary vessel, static equilibrium is reached when the density in any horizontal layer is constant and the less dense fluid is above the more dense fluid. There are important engineering applications involving the settling and separation of fluids and particles, for example oil and water or oil and foreign particles. In a stationary vessel, with only gravity force acting, the settling may take place slowly. By rotating the vessel at a high speed, however, an acceleration much greater than the gravitational acceleration can be obtained and a much quicker separation effected.

3-13. Flow in a curved path

Flow in a curved path is so common in nature and in machines and equipment that it is well to emphasize some features of this type of flow.

Consider the flow between two concentric streamlines an infinitesimal distance apart, as represented in Fig. 3-22. The radius of curvature of the

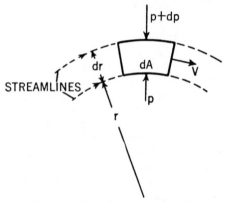

Fig. 3-22. Flow in a curved path.

path is r, and the tangential linear velocity is V. The infinitesimal element has a height dr, and an average area dA along the curved surface. The mass of this element is $\rho\, dr\, dA$. The normal or radial acceleration is V^2/r. The centrifugal force acting on the element has the magnitude $\rho\, dA\, dr (V^2/r)$. The pressure varies from p to $p + dp$ as the radius varies from r to $r + dr$. The centrifugal force on the fluid element is just balanced by the resultant force due to the pressures over the surfaces. If infinitesimals of higher order than the first are neglected, a force balance in the radial direction gives

$$dp\, dA = \rho \frac{V^2}{r}\, dr\, dA, \qquad dp = \rho \frac{V^2}{r}\, dr \qquad (3\text{-}37)$$

The pressure increases with radius in curved flow. There is a fall in pressure per unit radial distance toward the center of curvature by the amount $\rho(V^2/r)$; the pressure gradient $dp/dr = \rho(V^2/r)$.

The dynamic equation was applied to flow along a streamline. Equation (3–37), which gives the fundamental relation $dp = \rho(V^2/r)dr$, provides means for studying conditions in a direction normal to the streamlines. If the streamline is straight, the pressure change normal to the streamline is zero, because r is infinitely large. For streamlines of finite curvature, the pressure varies from p to $p + dp$ as the radius varies from r to $r + dr$. Since dp is positive if dr is positive, Equation (3–37) shows that the pressure increases for successive points from the concave to the convex side of the stream. The exact variation in pressure depends upon the variation in V with radius.

Imagine a vertical cylindrical vessel with liquid rotating about a central vertical axis. The streamlines are concentric circles. If there is no motion of the fluid relative to the container, the fluid rotates like a solid body. Such a flow is a *rotational, solid-body* or *forced vortex;* a torque is applied to the body of fluid. Let ω be the constant angular speed of the vessel, and r the radial distance to any point. As indicated in Fig 3–23, the linear velocity at any point in the fluid is given by the relation $V = r\omega$. The parabolic pressure distribution for this type of flow is given in Section 3–12. The relations presented in Section 3–12 can be derived by starting with Equation (3–37). A forced or solid-body vortex might be found in a rotating container fitted with radial vanes or partitions, in the rotating impeller of a centrifugal pump, or in the rotating runner of a turbine.

Imagine a frictionless fluid moving in a horizontal circular path with *no* torque applied; such a flow is called a *free* or irrotational *vortex*. The streamlines are concentric circles, but the velocity variation is different from that of a forced or solid-body vortex. Consider the case in which the elevation changes are negligible, the fluid is incompressible, and there is no friction. Then, for one stream-tube or streamline, Bernoulli's equation becomes

$$p + \tfrac{1}{2}\rho V^2 = \text{constant} \qquad (3\text{--}38)$$

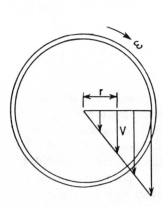

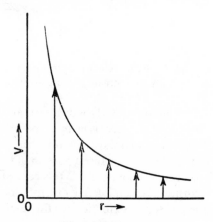

Fig. 3-23. Forced vortex.   *Fig. 3-24.* Free vortex.

Say the constant in this relation is the same for all neighboring streamlines. No energy is added to the fluid by a torque and no energy is dissipated by friction. Then we can combine Equation (3–37) and the derivative of Equation (3–38) with respect to r to give

$$\frac{dp}{dr} = \rho \frac{V^2}{r} \tag{3–39}$$

$$\frac{dp}{dr} = -\rho V \frac{dV}{dr} \tag{3–40}$$

Combining these last two relations gives a differential equation which can be separated and integrated to give

$$\frac{dV}{V} + \frac{dr}{r} = 0$$

$$Vr = \text{constant} = K \tag{3–41}$$

where K is some constant. K can be determined from the known velocity and radius at a particular point. As illustrated by Fig. 3–24, Equation (3–41) shows that the velocity increases as the radius decreases.

A free or irrotational vortex might be found in the casing (volute) outside the rotating impeller of a centrifugal pump, or in the two-dimensional flow in the main body of a stream flowing around a bend, as shown in Fig 3–25. Experimental results obtained from Pitot traverses in pump volutes and in bends show fair agreement with foregoing relations.

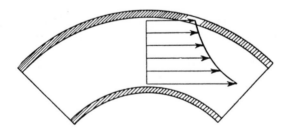

Fig. 3-25. Velocity distribution for two-dimensional flow in the main body of a stream flowing around a bend.

Equation (3–41) indicates that the velocity becomes infinitely large as the radius is decreased to zero. Infinite velocities are not reached in real fluids. In actual flows, viscous shear effects, which were neglected in deriving Equation (3–41), cause a portion of the fluid in the region near $r = 0$ to rotate like a solid body. Outside this central core is a transition region; outside the transition region is a free or irrotational vortex. The velocity distribution and the pressure distribution for this combination flow are indicated in Fig. 3–26.

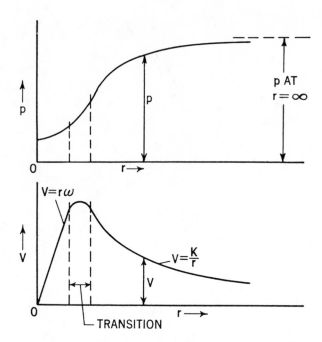

Fig. 3-26. Pressure and velocity distribution in a vortex flow.

The tornado and the waterspout (a tornado that forms at sea) are examples of such a circular whirl in the atmosphere. There are indications that the horizontal velocities in some tornadoes have exceeded 200 miles per hour. The low pressure in the central core or "eye" and the high velocities of a tornado may be very destructive.

Under some circumstances, as in the flow around a bend in a pipe, there may be a *secondary flow* superimposed upon the primary or main flow. The additional energy dissipation resulting from the secondary flow may be undesirable, or the secondary flow may affect the operation of a machine or meter. This secondary flow, on the other hand, may be useful in separating particles, like silt, carried by the fluid.

If, for example, fluid is circulating in a stationary flat-bottom cylindrical vessel, as shown in Fig. 3–27, the primary pattern can be indicated by concentric circles (shown dotted in the figure). The velocity in a layer next to the bottom of the vessel, such as AA, is retarded by the bottom; the layer near the bottom moves at a lower velocity than some layer above it, such as BB. The pressure of one layer, however, is transmitted to the layer below it; the pressure gradient dp/dr in each adjacent layer is the same, namely $\rho V^2/r$. In a layer near the bottom the radius of curvature is smaller than that in the layer above it, because the velocity in the lower layer is smaller. Besides a

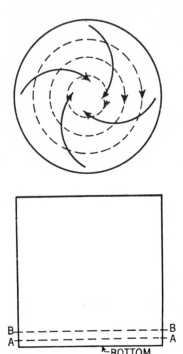

concentric primary flow, there is thus superimposed a secondary flow; the resultant flow follows an inward spiral path.

Observation shows that small particles carried by such a stream are heaped toward the center of the bottom. In natural rivers and streams the secondary flow at bends tends to pile up sand and gravel at the inner side of the bend and to deepen the bend at the outer side. Secondary flow thus tends to make the bend more pronounced or more meandering.

Figure 3–28 illustrates diagrammatically the secondary flow observed in some pipe bends. The side AD is at the inner radius, whereas the side BC is at the outer radius. Viewing one half of the channel only, for example, the velocity in the layers close to the wall AB is less than that in the central plane EE. The pressure gradient, however, is transmitted from one layer to the next, parallel to AB. Thus there is a tendency for the secondary flow to proceed from B to A. The rest of the fluid, by continuity, is forced to stream slowly, to complete the circuit. The secondary flow, consisting of two spirals, is superimposed upon the primary flow.

Fig. 3-27. Secondary flow in a flat-bottom cylindrical vessel.

The secondary flow in a bend may be reduced or eliminated by suitable guide vanes. Also, since the walls AB and CD play such a prominent role in developing the secondary flow, the secondary flow may be reduced by making

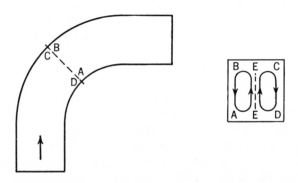

Fig. 3-28. Second flow in a pipe bend.

the sides AB and DC small in comparison with BC and AD. Two dimensional flow in a curved channel may be realized with such a deep, narrow channel.

3–14. Action between runner and fluid

There are various cases in which a fluid passes through a rotating member which may be called a runner, wheel, rotor, or impeller. Figure 3–29 shows velocity diagrams at the exit from a runner. In Fig. 3–29(a) the velocity U

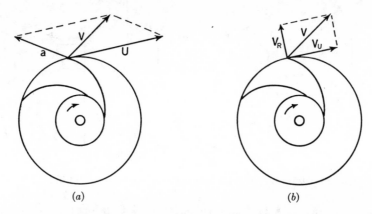

(a)   (b)

Fig. 3-29. Velocity triangle at exit from an impeller vane.

is the velocity of a point on the runner. Let R be the radius to this point and ω the angular speed of the runner in radians per unit time. Then $U = R\omega$. The velocity a is the velocity of the fluid relative to the impeller. The absolute or resultant velocity of the fluid V is the vector sum of a and U. Figure 3–29(b) illustrates components of V. V_R is the radial component of V, and V_U is the tangential component of V.

The linear momentum of a particle (or body) of mass m is defined as the product of mass and linear velocity V. The resultant force F acting on the body is equal to the time rate of change in linear momentum, that is,

$$F = \frac{d}{dt}(mV) \tag{3–12}$$

For fluid machines with a runner, it is necessary to extend the foregoing dynamic equation. Consider a body of mass m moving at a radius R about an axis of rotation. Let V_U be the tangential component of the resultant V of the body. Equation (3–12) can be extended, by the multiplication of each side by R, to give

$$FR = T = \frac{d}{dt}(mRV_v) \tag{3–42}$$

where T is the moment (tangential force times radial distance or moment arm) or torque producing the momentum change. The product $MV_U R$ is called "angular momentum" or "moment of momentum." In words, torque equals the time rate of change in angular momentum.

Refer to Fig. 3–30. The fluid flows through some channel as the channel

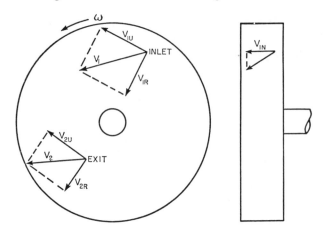

Fig. 3-30. Notation for work transfer between runner and fluid.

rotates about an axis. Steady flow will be assumed. Consider the action during a short time interval dt. The subscript 1 will be used for quantities at runner inlet. During the time interval dt a quantity of fluid, of mass m, enters the rotor at a point at radius R_1 with the resultant velocity V_1. The tangential component of this resultant velocity is V_{1U}. At inlet the radial component is V_{1R} and the axial (in the direction of the axis of rotation of the runner) component is V_{1N}. The angular momentum of this mass of fluid is $mR_1 V_{1U}$. The subscript 2 will be used for quantities at runner exit. In the time interval dt an equal amount of fluid of mass m leaves the rotor at point 2, at the radius R_2 with the resultant velocity V_2; the angular momentum of this fluid is $mR_2 V_{2U}$, where V_{2U} is the tangential component of the resultant velocity at exit. The increase in angular momentum during the time interval dt is

$$m(R_2 V_{2U} - R_1 V_{1U}) \tag{3–43}$$

The rotor acts on the fluid with the torque T to increase the angular momentum of the fluid, or

$$T = \frac{m}{dt}(R_2 V_{2U} - R_1 V_{1U}) \tag{3–44}$$

where m/dt is mass rate of flow.

If the angular speed of a runner is given, then the values of U_1 and U_2 can be calculated for each radius. If the velocities V_1 and V_2 are given in direction and magnitude, Equation (3–44) can be used to calculate the torque. For the notation taken, the torque T is positive for a pump, fan, or compressor runner; the torque T is negative for a turbine runner.

ENERGY EQUATION FOR THE STEADY FLOW
OF ANY FLUID

In some problems an energy accounting, or an energy equation, is useful. The following will consider an energy relation for steady, one-dimensional flow.

3–15. Some terms

Work is defined as the product of a force times displacement (or distance traversed), with the force acting in the direction of the displacement. *Power* is defined as the time rate of doing work. Work may be expressed in such units as foot-pounds or dyne-centimeters (ergs) or newton-meters (joules). Power may be expressed in such units as foot-pounds per second or ergs per second. One horsepower equals 550 foot-pounds per second. One horsepower equals 746 watts.

The *energy* of a body is defined as the capacity of the body to do work. Energy is expressed in the same units as work. A body may possess energy because of a variety of states or conditions of the body. Energy may be classified in such groups as mechanical, thermal, electrical, and atomic.

Mechanical energy is frequently subdivided into potential energy and kinetic energy. The potential energy of a body is that energy of the body due to its position. A body of weight W ($W = mg$, where m is mass) at a height z above some arbitrary datum has potential energy equal to Wz with respect to that datum. If the body were moved to a lower level, potential energy of that body could be converted into work.

The kinetic energy of a body is its energy due to its motion. Imagine a mass m with the velocity V with respect to some reference such as the earth. With respect to this reference, this mass has a kinetic energy equal to $\frac{1}{2}mV^2$. A moving body having kinetic energy is capable of doing work against forces which change its motion.

The *internal energy* of a substance is the energy stored within the substance and is due to the activity and spacing of the molecules.

The term *heat* or *heat transfer* is defined as thermal energy in transition as a result of a temperature difference between two points. The distinction between internal energy and heat should be noted. Internal energy is *stored* thermal energy; it refers to a *single* condition or state. Heat, on the other hand, is thermal energy in transition; heat transfer requires a temperature difference between two points, which means two different conditions or states.

A unit of quantity of heat is the British thermal unit, Btu, which is defined as the quantity of heat transferred in order to raise the temperature of unit weight of water one degree Fahrenheit, at a specified temperature. Many experiments have shown that 1 Btu is approximately equal to 778 foot-pounds; this equivalence will be used in subsequent problem work in this book.

3–16. Energy equation for steady flow

We will examine the case in which the fluid may pass through any kind of apparatus. The flow may be through a piece of pipe, a nozzle, a pump, turbine, some other machine, a rocket engine, a propulsion system, or the passage between streamlines.

The lines in Fig. 3–31 illustrate the control volume, which consists of

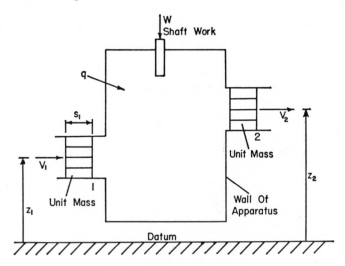

Fig. 3-31. Notation for steady flow.

the walls of the apparatus, the inlet area A_1 at section 1, with the velocity V_1 normal to A_1, and the exit area A_2 at section 2 with the velocity V_2 normal to A_2.

As illustrated, picture a unit mass of fluid about to cross section 1 at inlet. For steady flow there is a corresponding amount of unit mass of fluid about to leave at exit 2 in a certain time interval. We will make an energy accounting; we can state that

$$\begin{bmatrix} \text{heat added to} \\ \text{unit mass} \end{bmatrix} + \begin{bmatrix} \text{total work done} \\ \text{on unit mass} \end{bmatrix} = \begin{bmatrix} \text{increase in energy} \\ \text{of unit mass} \end{bmatrix} \quad (3\text{-}45)$$

Let the heat added per unit mass be represented by q (in suitable units). The total work done on unit mass consists of three parts. One part is the shaft work done on or by the fluid by means of some external machine (such as a pump, compressor, fan, or turbine). Let this shaft work per unit mass be indicated by W; this work is positive for a pump and negative for a turbine. At inlet, work is done in pushing a unit mass of fluid across the inlet; this work is sometimes called "flow-work" or "work-transfer." This flow-work is the product of a force times a distance. The force equals $p_1 A_1$, where p_1 is the pressure at section 1. The distance S_1 through which this force acts is $1/A_1\rho_1$ for unit mass. Thus the flow-work at inlet is p_1/ρ_1. In a similar fashion the magnitude of the flow-work at exit, or the work done in pushing unit mass across section 2, is p_2/ρ_2. Thus the total work done per unit mass is

$$W + \frac{p_1}{\rho_1} - \frac{p_2}{\rho_2}$$

At inlet the potential energy per unit mass is gz_1. At inlet the kinetic energy per unit mass is $V_1^2/2$. Let u_1 be the internal energy per unit mass at inlet. The subscript 2 will be used for corresponding quantities at exit. Then the energy relation, or accounting, Equation (3-45) becomes

$$q + \frac{p_1}{\rho_1} - \frac{p_2}{\rho_2} + W = u_2 - u_1 + \frac{V_2^2 - V_1^2}{2} + g(z_2 - z_1) \quad (3\text{-}46)$$

Some of the terms in Equation (3-46) are frequently regrouped by using a quantity called *enthalpy*, which is defined as follows:

$$\text{enthalpy} = h = u + \frac{p}{\rho} \quad (3\text{-}47)$$

Enthalpy is simply a quantity established by arbitrary definition for the sake of convenience. Then Equation (3-46) can be written in the alternate form

$$q + W = h_2 - h_1 + \frac{V_2^2 - V_1^2}{2} + g(z_2 - z_1) \quad (3\text{-}48)$$

For numerical calculations each term in Equation (3-46) and (3-48) should be expressed in consistent units. For example, if p is expressed in pounds per square foot, ρ in slugs per cubic foot, V in feet per second, z in feet, and g in feet per second^2, then $V^2/2$, p/ρ, and gz each are in foot-pounds per slug. If these units are used, then q, u, and W each should be expressed in similar consistent units.

In some cases the flow-work is written in the form pv, where v is specific volume. As an example, p could be expressed in pounds per square foot and v in cubic feet per pound. Then the product pv would be in foot-pounds per pound. If u and h each were expressed in Btu per pound, then the enthalpy could be written in the form

$$h = u + \frac{pv}{J} \tag{3-49}$$

where J is the conversion factor approximately equal to 778; that is, one Btu is approximately equal to 778 foot-pounds.

Any other set of consistent units could be used in making an energy accounting.

The energy relation can be put in the following alternate differential forms

$$dq + dW = d\left(u + \frac{p}{\rho}\right) + VdV + gdz \tag{3-50}$$

$$dq + dW = dh + VdV + gdz \tag{3-51}$$

3-17. Work done against friction

There is a question as to how friction is taken into account by the energy relation given by Equations (3–46) and (3–48). Work done against friction is dissipated as thermal energy at the rubbing point. There are two possibilities:

1. The thermal energy thus developed may pass through the walls of the apparatus, either wholly or in part. That part of the thermal energy which passes through the walls of the apparatus is included in the q term. If the heat transfer q were measured, frictional heat could not be distinguished from any other source.

2. That thermal energy due to friction which does not pass through the apparatus walls does not appear in the q term, but adds to the total energy of the fluid, and hence is included (or concealed) in the right-hand side of each Equation (3–46) or (3–48).

3-18. Adiabatic process

An adiabatic process is one in which no heat is added to or removed from the fluid mass; in other words $q = 0$. In some cases one can assume that the process is adiabatic and that the internal energy change is negligible. For such conditions, Equation (3–46) can be written in the form

$$W = \left(\frac{p_2}{\gamma_2} + \frac{V_2^2}{2g} + z_2\right) - \left(\frac{p_1}{\gamma_1} + \frac{V_1^2}{2g} + z_1\right) \tag{3-52}$$

If p were expressed in pounds per square foot, V in feet per second, z in feet, γ in pounds per cubic foot, and g in feet per second^2, then W, and each other term of Equation (3-52), would have the units of foot-pounds (of work or energy) per pound. The term p/γ is sometimes called a *pressure head*, and $V^2/2g$ a *velocity head*. z might be called an elevation or potential head. Each one of these terms has the net dimension of a length.

Example. In a centrifugal pump test the discharge gage reads 100 pounds per square inch, and the suction gage reads 5 pounds per square inch. Both gages indicate pressure above atmospheric. The gage centers are at the same level. The diameter of the suction pipe is 3 inches, and the diameter of the discharge pipe is 2 inches. Suction and discharge are at the same level. Oil (specific gravity $= 0.85$) is being pumped at the rate of 100 gallons per minute. Assuming an adiabatic process and no internal energy changes, what is the power delivered to the pump?

Let the suction be section 1 and the discharge be section 2. 1 gallon $= 231$ cubic inches. The energy equation for this case is

$$W = \left(\frac{p_2}{\gamma} + \frac{V_2^2}{2g}\right) - \left(\frac{p_1}{\gamma} + \frac{V_1^2}{2g}\right)$$

The rate of discharge Q is $100(231)/60(1728) = 0.223$ cubic feet per second. From the continuity equation $A_1 V_1 = A_2 V_2 = Q$,

$$V_1 = \frac{0.223}{\frac{\pi}{4}\left(\frac{3}{12}\right)^2} = 4.54 \text{ feet per second}$$

$$V_2 = \frac{0.223}{\frac{\pi}{4}\left(\frac{2}{12}\right)^2} = 10.22 \text{ feet per second}$$

$$W = \frac{144}{62.4(0.85)}(100 - 5) + \frac{1}{64.4}[(10.22)^2 - (4.54)^2]$$
$$= 259.3 \text{ foot-pounds per pound}$$

Since 1 horsepower $= 550$ foot-pounds per second,

$$\text{horsepower} = \frac{259.3(0.223)62.4(0.85)}{550} = 5.58$$

3-19. Values of internal energy and enthalpy

The internal energy and enthalpy of a fluid depends on the fluid and its state. Values of these properties for liquids and vapors can be found in different reference tables. The change in internal energy and enthalpy for a gas following the simple relation $pv = R_0 T$ can be found by the following relations:

$$u_2 - u_1 = c_v(T_2 - T_1) \tag{3-53}$$
$$h_2 - h_1 = c_p(T_2 - T_1) \tag{3-54}$$

where c_v is the specific heat at constant volume, and c_p is the specific heat at constant pressure. The specific heat of a substance is defined as the amount of heat required to change the temperature of unit amount of the substance one degree. In the American system the specific heat is sometimes taken as the number of Btu required to change the temperature of one pound of a substance one degree Fahrenheit. Table 3–1 gives some approximate values of c_p and c_v using these units; more precise values can be found in the reference literature.

TABLE 3–1

SPECIFIC HEATS FOR SEVERAL GASES

	c_p	c_v
Air...................	0.239	0.171
Carbon monoxide	0.248	0.177
Helium	1.251	0.754
Nitrogen	0.248	0.177
Oxygen	0.219	0.156

PROBLEMS

3-1. A pipe 12 inches in diameter reduces to a diameter of 6 inches, and then expands to a diameter of 10 inches. If the average velocity in the 6-inch pipe is 15 feet per second, what is the average velocity at the other sections for *any* incompressible fluid?

3-2. A pipe 8 inches in diameter expands to 14 inches diameter. Air flows through the pipe at a rate of 0.466 slug per minute. At one section, in the 8-inch pipe the gage pressure is 40 pounds per square inch, and the temperature is 120 degrees Fahrenheit. At another section, in the 14-inch pipe, the gage pressure is 25 pounds per square inch and the temperature is 80 degrees Fahrenheit. Barometric pressure is 30.0 inches of mercury. What is the mean velocity at each section?

3-3. There is two-dimensional flow between two parallel walls. The velocity distribution at a section is given by a parabola with the maximum velocity V_m at the center of the channel and zero velocity at each wall. What is the average velocity V in terms of V_m?

3-4. Water flows through a 2-inch diameter pipe at a rate of 6 pounds per second. What is the average velocity and the rate in gallons per minute?

3-5. At section A in a two-dimensional incompressible flow, the distance between streamlines is 1 inch and the average velocity is 34 feet per second. At another section B the distance between these streamlines is 1.7 inches. What is the average velocity at section B?

3-6. A vertical cylindrical tank, 14 inches in diameter, contains a liquid. At the bottom is a circular hole through which the liquid flows steadily with an average linear velocity of 10 feet per second. If the jet diameter is 1.2 inches, what is the volume rate of flow? What is the average velocity in the tank just upstream from the orifice?

3-7. Fluid flows through a circular pipe having an internal diameter of 12 inches. The velocity distribution is a paraboloid of revolution with the maximum velocity of 20 feet per second at the center of the pipe. What is the volume rate of flow through the pipe?

3-8. Water flows through a centrifugal pump impeller at 4 cubic feet per second. In a region close to the impeller exit, the impeller consists essentially of two flat circular disks which are parallel. Outside diameter of the impeller is 14 inches. At exit the effective distance between the parallel surfaces is 2 inches. What is the radial component of the average velocity leaving the impeller?

3-9. Air at 60 pounds per square inch absolute and 40 degrees Fahrenheit flows through a 12-inch diameter pipe at 0.5 slug per second. What is the average velocity of the air?

3-10. A jet of oil (specific gravity = 0.80) moving at 60 feet per second strikes a fixed flat plate normal to the stream. The dimensions of the jet are small in comparison with those of the plate. The rate of discharge is 0.20 cubic foot per second. What is the force on the plate?

3-11. A jet of lye solution (specific gravity = 1.10) 2 inches in diameter has an absolute velocity of 40 feet per second. It strikes a single flat plate which is moving away from the nozzle with an absolute velocity of 15 feet per second. The plate makes an angle of 60 degrees with the horizontal. Calculate the total force acting on the plate. Assume no friction force along the plate surface.

3-12. A jet of glycerine $1\frac{1}{2}$ inches in diameter is deflected through an angle of 80 degrees by a single vane. The jet velocity is 35 feet per second, and the vane moves at 10 feet per second away from the nozzle in the direction of the entering jet. Determine the total force acting on the vane. Assume no friction force along the vane surface.

3-13. At the end of a water pipe of 4-inch diameter is a nozzle which discharges a jet having a diameter of $1\frac{1}{2}$ inches into the open atmosphere. The pressure in the pipe is 60 pounds per square inch gage, and the rate of discharge is 0.88 cubic foot per second. What are the magnitude and direction of the force necessary to hold the nozzle stationary?

3-14. A 90-degree bend is in a water pipe of 10-inch diameter in which the pressure is 100 pounds per square inch gage. The rate of flow is 40 gallons per second. What is the magnitude and direction of the force necessary to "anchor" the bend?

3-15. The jet leaving a rocket has a diameter of 6 inches, an average velocity of 1500 feet per second relative to the rocket, and a gas density of 0.0015 slug per cubic foot. What is the thrust on the rocket?

3-16. A 1-inch diameter horizontal jet of water has a velocity of 20 feet per second toward the left. It strikes a fixed, large flat plate which tilts upward to the left at 45 degrees with the horizontal. Assume no friction. What volume rate of flow goes upward, and what volume rate goes downward?

3-17. A jet of incompressible fluid moving at 50 feet per second leaves a nozzle and strikes a flat plate normal to the stream. The dimensions of the jet are small in comparison with those of the plate. The plate moves at 20 feet per second away from the nozzle in the direction of the nozzle axis. What is the direction of the fluid after it leaves the plate with respect to the nozzle axis? Neglect friction.

3-18. A small diameter jet of water with a velocity of 60 feet per second strikes a large, fixed flat plate which is normal to the axis of the jet. The force of the water on the plate is 50 pounds. What force is exerted on the plate if the jet velocity is doubled while the jet diameter is held constant?

3-19. A fluid weighs 25.0 pounds per cubic foot at a location where the gravitational acceleration is 32.17 feet per second per second. What is its density? What is the specific weight of this same fluid at a different location where the gravitational acceleration is 32.13 feet per second per second?

3-20. Standard air enters an 8-inch diameter horizontal pipe with a velocity of 650 feet per second. The exit velocity is 80 feet per second. Neglect friction. What are the exit pressure and the exist temperature?

3-21. Air flows through a 6-inch diameter horizontal pipe. At inlet the pressure is 120 pounds per square inch absolute, the temperature is 100 degrees Fahrenheit, and the velocity is 220 feet per second. At exit the pressure is 80 pounds per square inch absolute and the temperature is 110 degrees Fahrenheit. What is the friction force opposing the motion?

3-22. A stream of grain and air leaves a combine at atmospheric pressure and enters a bend with an average velocity of 9 feet per second. The bend is of uniform cross section, 10 by 10 inches square. The bend turns the stream through an angle of 135 degrees into an open truck. The specific weight of the mixture of grain and air is 48.0 pounds per cubic foot. What force does the stream exert on the bend?

3-23. A particle moves in a curved path with a linear velocity V. The radius of curvature of the path is R. Show that the normal or centripetal acceleration is V^2/R.

3-24. A stream of water in the open air from a hose rises to a peak and then falls. At the peak the velocity is 20 feet per second in a horizontal direction. What is the radius of curvature at this peak point?

3-25. A person can swing a bucket of paint in a vertical plane without spilling if certain conditions exist. Consider the state as the bucket passes its highest point. Assume an effective radius of 26 inches. What is the minimum tangential velocity which would avoid spilling?

3-26. A rectangular open container is on a truck. The container is 6 feet long (in the direction of truck motion), 3 feet wide, and 2 feet high. With the truck at rest, the container is half full. What is the maximum horizontal acceleration of the truck without spilling any water?

3-27. A rectangular tank 4 feet wide, 9 feet long, and 5 feet deep contains water to a depth of 3 feet. It is accelerated horizontally at 8 feet per second per second in the direction of its length. Calculate the depth of water at each end of the tank, and the total force on each end of the tank.

3-28. A glass U-tube with vertical legs 20 inches apart is partly filled with water to be used as an accelerometer. It is installed in a car which accelerates from 10 to 45 miles per hour in 15 seconds. What is the difference in level of the two legs for this acceleration?

3-29. A rectangular open container is 6 feet long (in the direction of motion), 3 feet wide, and 2 feet high. When at rest this container is filled with water. Determine the volume spilled after the container has accelerated in a horizontal direction at 8 feet per second per second.

3-30. A glass with water 4 inches deep is on an elevator which is accelerating upward at 10 feet per second per second. What is the pressure at the bottom of the glass?

3-31. A vertical circular cylinder containing water is rotated about a vertical axis at a steady speed of 1700 revolutions per minute. Consider a point A at a radial distance of 5 inches from the axis of rotation. What is the ratio between the normal or centripetal acceleration at point A and the gravitational acceleration?

3-32. A cylindrical tank 1 foot in diameter and 3 feet high is half full of glycerine. The tank is rotated about its vertical axis. What speed of rotation will cause the liquid to reach the top? What will then be the maximum pressure?

3-33. A centrifugal water pump impeller 20 inches in diameter rotates at 1500 revolutions per minute. Assume no flow through the pump. What is the maximum pressure developed by the runner due to centrifugal action?

3-34. A simple centrifugal oil separator consists essentially of a cylindrical container rotating about its axis. In some cases the impurities inherent in an oil combine both chemically and mechanically with foreign impurities to form what is commonly called sludge. If the sludge is denser than the oil, at what points of the simple separator should pipes be located for removing the clean oil, and for removing the sludge?

3-35. In a water sprinkler two arms rotate in a horizontal plane about a vertical axis. Water flows through each arm at 0.082 cubic foot per second. Each exit nozzle has a diameter of 1 inch, moves with a tangential velocity of 5 feet per second, has a tangential discharge, and is at a radius of 11 inches. The water enters each arm with a relative velocity in a radial direction. What is the net torque due to the water flow in the two arms?

3-36. Water approaches the body shown in Fig. 3–11 with a velocity of 10 feet per second. If the velocity at point B is 20 feet per second, what is the pressure difference between points O and B?

3-37. An airship flies through stationary standard air at 80 miles per hour. What is the gage pressure at the stagnation point?

3-38. A propulsion system on a boat has a horizontal converging nozzle which is fastened to the boat. Water enters the nozzle at 8 pounds per square inch gage with a velocity of 7 feet per second. Inlet area is one-half square foot. Water leaves the nozzle at 1 pound per square inch gage. Neglect friction. What force, in the direction of the nozzle, does the nozzle transmit to the boat?

3-39. Water flows through a conical expanding channel at 2.62 cubic feet per second. At inlet the diameter is 4 inches and the pressure is 15 pounds per square inch gage. At exit the diameter is 10 inches. Assuming frictionless flow, what is the resultant axial force of the channel on the water?

3-40. Derive Equation (3–36) by starting with Equation (3–37).

3-41. In a free vortex, at a radius of 1.5 feet the velocity is 200 feet per second, and the pressure is 14.7 pounds per square inch absolute. What is the pressure at a radius of 4 feet? Fluid is standard air.

3-42. A right-angle bend is formed by two concentric circular arcs. Assume two-dimensional flow, a free vortex, and that the fluid is water. The inner radius is 2 feet, and the outer radius is 3 feet. The difference in pressure at the two walls is 0.80 pound per square inch. What is the velocity at the inner radius?

3-43. An incompressible fluid passes through a channel or bend formed by fixed walls of concentric circular arcs. Assume two-dimensional frictionless flow, and a depth of 12 inches perpendicular to the plane of flow. The inner radius is 4 feet, the outer radius is 20 feet, and the velocity at the middle of the bend is 13 feet per second. What is the volume rate of flow?

3-44. Water flows through a right-angle bend formed by two concentric circular arcs. Assume two-dimensional flow, depth one foot perpendicular to the plane of flow, and a free vortex. The inner radius is 1 foot; the outer radius is 2 feet. The difference in pressure at the two walls is 0.60 pound per square inch. What is the volume rate of flow through the channel?

3-45. Consider the whirlpool or vortex action which forms above an open water drain in a shallow basin as essentially a free vortex. Point *A* at a distance of 10 inches from the center has a tangential velocity of 4 inches per second. Point *B* is at a distance of 5 inches from the center. What is the change in surface elevation between points *A* and *B*?

3-46. Assume that a "cyclone" or circular movement in the atmosphere is essentially a free vortex. Assume standard air. At 60 miles from the cyclone center the velocity is 5 miles per hour and the barometer reads 30.00 inches of mercury. What would be the barometric reading 6 miles from the cyclone center?

3-47. Brine (specific gravity = 1.20) flows through a pump at 2000 gallons per minute. The pump inlet is 12 inches in diameter. At the inlet the vacuum is 6 inches of mercury. The pump outlet, 8 inches in diameter, is 4 feet above the inlet. The outlet pressure is 20 pounds per square inch gage. What power does the pump add to the fluid?

3-48. Air enters a horizontal pipe 14 inches in diameter at an absolute pressure of 14.8 pounds per square inch and a temperature of 60 degrees Fahrenheit. At exit the pressure is 14.5 pounds per square inch absolute. Entrance velocity is 35 feet per second; exit velocity is 50 feet per second. What is the heat added or abstracted?

3-49. A liquid is being heated in a vertical tube of uniform diameter, 50 feet long. The flow is upward. At entrance the average velocity is 3.5 feet per second, the absolute pressure is 50 pounds per square inch, and the specific volume is 0.017 cubic feet per pound. At exit the mixture of liquid and vapor has the absolute pressure of 48 pounds per square inch, and the specific volume is 0.90 cubic feet per pound. The increase in internal energy is 10 Btu per pound. Find the heat added.

3-50. Where would be a desirable place for locating ventilating openings along the length of a streetcar if (a) draft upon starting and stopping is to be avoided and (b) maximum exchange of air is desired?

3-51. Air flows through a 6-inch diameter horizontal pipe at a constant temperature of 70 degrees Fahrenheit. At inlet the pressure is 44 pounds per square inch gage, and the velocity is 100 feet per second. At outlet the pressure is 17.0 pounds per square inch absolute. One horsepower of mechanical shaft-work is added to the fluid between inlet and outlet. What is the heat taken away or added?

3-52. A Thomas flow meter is inserted in a section of horizontal pipe 6 inches in diameter. This meter consists of two thermometers and an electric heating element. Carbon dioxide flows through the thermally insulated pipe. At section 1 the temperature is 70 degrees Fahrenheit, and the absolute pressure is 30.2 inches of mercury. At section 2 the temperature is 80 degrees Fahrenheit, and the absolute pressure is 29.92 inches of mercury. Between sections the electric heating element adds a total of 100 Btu per minute to the gas. $c_p = 0.21$. Consider the fluid incompressible if the density variation is less than 4 per cent. Find the weight of gas flowing per minute.

3-53. A fluid flows adiabatically through a horizontal nozzle. The entrance velocity is 1000 feet per second. Vapor tables give 1200 Btu per pound for exit enthalpy and 1322.5 Btu per pound for entrance enthalpy. What is the exit velocity?

3-54. A compressor handles air at a rate of 1000 pounds per hour. Air at standard conditions is taken in through the inlet of 0.60 square foot cross-section area and is compressed to 100 pounds per square inch absolute and 180 degrees Fahrenheit. The discharge is through an area of 0.11 square foot. The heat taken from the air is 50,000 Btu per hour. The change in elevation is negligible. What is the work done on the air?

3-55. A rocket travels through air at 650 miles per hour. The temperature of the air at rest far from the rocket is 35 degrees Fahrenheit. Assume an adiabatic compression process and that potential energy changes can be neglected. What is the air temperature at the stagnation point of the rocket?

3-56. Starting with the dynamic equation, derive the relation for kinetic energy.

3-57. At a certain section in a pipe, water flows at a pressure of 80 pounds per square inch and with a linear velocity of 9 feet per second. What is the total flow-work for 1.5 cubic feet that passes this section?

3-58. What vacuum in the mouth is necessary to draw water a distance of 6 inches through a small diameter straw? Assume no friction.

3-59. At one section in a horizontal water pipe line the diameter is 5 inches, and the static pressure is 10 pounds per square inch gage. At another section the diameter is 10 inches, and the static pressure is 20 pounds per square inch gage. For a flow rate of 3 cubic feet per second, what is the direction of flow?

3-60. Air flows steadily through a horizontal channel at constant temperature. At one section the diameter is 2 inches, the pressure is 35.0 pounds per square inch gage, and the temperature is 90 degrees Fahrenheit. At another section the diameter is 3 inches, and the pressure is 37.0 pounds per square inch gage. If the friction loss is 1700 foot-pounds per pound, what is the flow rate?

3-61. Liquid ammonia (specific gravity is 0.96) flows steadily through a 180-degree bend formed of 4-inch diameter pipe with an average velocity of 11 feet per second. At inlet to the bend the pressure is 30 pounds per square inch gage. Through the bend the loss equals 3 feet of fluid. What is the total force of the fluid on the bend?

3-62. In a certain waterfall, water drops from an initial state of rest to a level 900 feet lower. At this lower level the water strikes a horizontal surface. Neglect friction and thermal effects. What is the temperature rise of the water?

3-63. A centrifugal air compressor handles 140 pounds per minute from 15 pounds per square inch absolute and 72 degrees Fahrenheit to 30 pounds per square inch absolute and 240 degrees Fahrenheit. Assume an adiabatic process and neglect changes in kinetic and potential energies. What is the power added to the fluid?

3-64. A pump draws water from a reservoir through a 6-inch diameter pipe. Point *A* in the pipe is 7 feet above the free surface in the reservoir. If the flow rate is 2 cubic feet per second, what is the pressure at point *A*? Neglect friction.

3-65. A liquid jet leaves a nozzle with an absolute velocity of 30 feet per second directed upward into the atmosphere at an angle of 45 degrees with the horizontal. Neglect friction and thermal effects. What is the initial radius of curvature of the jet?

3-66. A liquid jet leaves a nozzle with an absolute velocity *V* directed upward into the atmosphere at an angle *A* with the horizontal. The diameter of the jet is relatively small. Neglect friction and thermal effects. Show that the maximum elevation of the jet above the nozzle equals $(V \sin A)^2/2g$.

3-67. A jet of water leaving a nozzle is directed vertically upward. The jet leaves the nozzle with a speed of 30 feet per second. Neglect thermal and friction effects. How far above the nozzle will the water travel?

3-68. A circular jet of water leaves a nozzle in a vertical upward direction with a velocity of 20 feet per second; jet diameter is 1 inch. A large circular disk weighing 2 pounds is held in a horizontal position above the nozzle. Neglect thermal and friction effects. What is the distance between nozzle and disk?

3-69. A body weighing 20 pounds falls from rest. Assume no thermal effects and friction. What is the velocity of the body at a distance of 200 feet from the rest position?

3-70. Steam enters the blades of a turbine runner with a velocity of 3500 feet per second and leaves with a velocity of 1500 feet per second. Assume an adiabatic process and no change in enthalpy. What is the work transfer to the runner?

3-71. A locomotive pulls a water tender at a constant velocity of 50 feet per second. Water is scooped up from a trough between the tracks at 12 cubic feet per second. The pipe from the scoop has a constant diameter right angle bend. Assume no friction and no static pressure change in the bend. What power is required to move the scoop?

3-72. Air flows steadily through a machine. At inlet the area is one square foot, the pressure is 100 pounds per square inch absolute, the temperature is 90 degrees Fahrenheit, and the velocity is 190 feet per second. The exit is at the same level as the inlet. At exit the area is one square foot, the pressure is 80 pounds per square inch absolute, and the temperature is 81 degrees Fahrenheit. For an adiabatic process, what power does the machine add to or receive from the air?

3-73. Air enters a vacuum pump at 1 pound per square inch absolute, 40 degrees Fahrenheit with a velocity of 200 feet per second through a cross-sectional area of 0.10 square foot and is discharged at atmospheric pressure of 14.7 pounds per square inch absolute with negligible velocity. Power input to the pump is 20 horsepower, and cooling water removes 600 Btu per minute from the air passing through the pump. Pump outlet is at the same level as pump inlet. Calculate the air discharge temperature.

3-74. Atmospheric air at 14.7 pounds per square inch absolute and 60 degrees Fahrenheit enters a gas turbine power plant unit with a velocity of 350 feet per second through an area of 1 square foot. The air passes through a compressor, heat exchanger, and a turbine runner. The air leaves the power plant unit at 10 pounds per square inch gage and 320 degrees Fahrenheit. Exit area is 1 square foot and the unit delivers 410 horsepower at a rotating shaft. Exit is at the same height as inlet. What is the heat transfer to the air passing through the unit?

3-75. Air flows through a 10-inch diameter horizontal pipe. At inlet the pressure is 150 pounds per square inch absolute, the temperature is 80 degrees Fahrenheit, and the velocity is 160 feet per second. At outlet the pressure is 110 pounds per square inch absolute, and the velocity is 200 feet per second. What is the heat transfer?

3-76. Air flows through an 8-inch diameter horizontal pipe at a constant temperature of 90 degrees Fahrenheit. At inlet the pressure is 40 pounds per square inch gage, and the velocity is 200 feet per second. At outlet the pressure is 22 pounds per square inch absolute. Between inlet and outlet 6 Btu per pound are added to the fluid. What power is added to a machine between inlet and outlet?

3-77. An air compressor takes in 100 cubic feet of standard air per minute and discharges it at 80 pounds per square inch absolute and 280 degrees Fahrenheit through an area of 0.04 square foot. Neglect inlet velocity. Assume inlet at same level as outlet. Input to the air is 14.2 horsepower. What is the heat transfer?

3-78. Linseed oil flows through a pump at 2.3 cubic feet per second. The pump inlet is 8 inches in diameter, the pump outlet is 6 inches in diameter, and pump inlet is level with pump outlet. At inlet the vacuum is 5 inches of mercury.

The center of the discharge gage is 3 feet above the outlet. The outlet gage reads 36 pounds per square inch gage. What power does the pump add to the liquid?

3-79. Air at 60 pounds per square inch absolute and 140 degrees Fahrenheit enters a horizontal, insulated 12-inch diameter pipe at a velocity of 300 feet per second. What is the velocity at a point in the pipe where the air temperature is 100 degrees Fahrenheit?

3-80. Four pounds of helium enclosed in a tank at 50 pounds per square inch absolute are heated from 60 degrees Fahrenheit to 100 degrees Fahrenheit. What is the heat transfer?

3-81. In a nonflow process 5 pounds of air expand from 100 degrees Fahrenheit to 500 degrees Fahrenheit at constant pressure. What is the work?

3-82. A fan receiving 0.60 horsepower delivers 6000 cubic feet of standard air per minute. Fan inlet area equals fan outlet area. Fan inlet pressure is atmospheric. A manometer at the fan outlet shows a static pressure above atmospheric equivalent to 0.30 inch of water. Neglect changes in potential energy. What is the power loss between fan inlet and outlet?

PART II

Selected Topics
in Fluid Mechanics

Part I brought out certain basic relations which are available for tackling a problem. We have the equation of state (relation between pressure, temperature, and density), the equation of continuity, the dynamic or momentum equation, and the energy equation. In solving a particular problem we may use one or more of these equations. For organizing experimental data we have the tools of dimensional analysis and dynamic similarity.

Various cases will be treated in Part II. For example, one chapter deals with the flow of incompressible fluids in pipes. Another chapter deals with the flow of compressible fluids in pipes. One chapter discusses fluid meters; another chapter discusses fluid machinery. Consideration is given to the flow around bodies as well as to the flow through channels. Besides giving information about various cases of flow, Part II illustrates how the basic relations of Part I can be used. With Part I as a background, each chapter in Part II is essentially self-contained.

CHAPTER FOUR

Dimensional Analysis
and Dynamic Similarity

But then the rigorous logic of the matter is not plain! Well, what of that? Shall I refuse my dinner because I do not fully understand the process of digestion?

—OLIVER HEAVISIDE.[1]

Modern fluid mechanics is based upon a healthy combination of physical analysis and experimental observations. The general objective is to provide dependable practical results and a thorough understanding of fundamental flow features. Many flow phenomena are so complex that a purely mathematical solution is impractical, incomplete, or impossible, and it is necessary to give serious consideration to experimental results. The present chapter discusses dimensional analysis and dynamic similarity, two tools which have proved very useful in the organization, correlation, and interpretation of experimental data. The following chapters will take advantage of these tools.

DIMENSIONAL ANALYSIS

4–1. Dimensions

Dimensional analysis is a mathematical method useful in checking equations, changing units, determining a convenient arrangement of variables of a physical relation, and planning systematic experiments. The following discussion deals mainly with the problem of determining a convenient arrangement of variables.

Dimensional analysis results in a sound, orderly arrangement of the variable physical quantities involved in a problem. Reference to experimental data must be made in order to obtain the necessary constants or coefficients

---

[1] *Electromagnetic Theory*, vol. II. The Electrician Printing and Publishing Co., London, 1899, page 9.

for a complete numerical expression. It should be kept in mind that a study of dimensions by itself does not yield any information about the physical phenomena or the functional relation between the variables involved. A study of dimensions, however, frequently aids in making an easier and more convenient description of the phenomena.

The first step in treating a problem is to list all the variables involved. This listing may be the result of experience or judgment. The next step, the dimensional analysis, can be made by following a formal procedure. The final arrangement of variables in the equation which results from a dimensional analysis is no more accurate or complete than the original listing of variables.

Let the letter L be the dimensional symbol for length, M the dimensional symbol for mass, F for force, and T for time. If length is arbitrarily selected as the fundamental or primary dimension of space, units of area and volume can be devised as secondary or derived units. Area is represented by the dimensional symbol L^2; volume is represented by the dimensional symbol L^3. The volume dimension depends upon the third power of the primary dimension length, regardless of the dimensional units employed. If time is also taken as a primary dimension, then the unit of linear velocity can be arbitrarily taken as a secondary or derived unit. The dimensional symbol for linear velocity is LT^{-1}. Similarly, the dimensional symbol for linear acceleration becomes LT^{-2}.

The selection of force or mass as a primary dimension is solely a matter of convenience. Either one could be selected; neither is "absolute." If one is selected, the other becomes a secondary or derived dimension. The dimensional relation between these two quantities can be found from the relation

Force equals mass times acceleration

In dimensional symbols,

$$F = MLT^{-2} \quad \text{or} \quad M = FT^2L^{-1}$$

TABLE 4–1

DIMENSIONS OF SOME PHYSICAL QUANTITIES

Quantity	Dimensions	Quantity	Dimensions
Length	L	Angular velocity	T^{-1}
Mass	M	Torque or moment (FL)	ML^2T^{-2}
Time	T	Density	ML^{-3}
Force (F)	MLT^{-2}	Specific weight (FL^{-3})	$ML^{-2}T^{-2}$
Area	L^2	Dynamic viscosity	$ML^{-1}T^{-1}$
Volume	L^3	Kinematic viscosity	L^2T^{-1}
Linear velocity	LT^{-1}	Pressure or Stress (FL^{-3})	$ML^{-1}T^{-2}$
Linear acceleration	LT^{-2}	Work, energy, or heat	ML^2T^{-2}
Angular measure	None		

Table 4–1 gives the dimensions of some physical quantities using the M, L, and T system.[2]

4–2. Equations and concept of function

This book will follow the customary practice in engineering work of dealing solely with physical equations which are dimensionally homogeneous. All the terms of a dimensionally homogeneous equation have the same dimensions. We will not add or equate masses and times or masses and velocities; instead, we will add only masses and masses or add only velocities and velocities. Dissimilar quantities cannot be added or subtracted to form a true physical relation. A nonhomogeneous equation is

$$\text{Land} + \text{cows} = \text{farm}$$

which may have some meaning but is not the type of relation under consideration.

In order to illustrate the concept of a "function," let A_1, A_2, A_3, and A_4 represent all the physical quantities or variables involved in some physical action. For example, A_1 might be density, A_2 might be a length of some body, A_3 might be a pressure, and A_4 might be a velocity. The physical action can be described saying that A_1 is "some function of A_2, A_3, and A_4." This statement means that for certain values assigned to A_2, A_3, and A_4 there are certain corresponding values of A_1. This relation is written symbolically as

$$A_1 = f(A_2, A_3, A_4) \tag{4–1}$$

It is important to note that Equation (4–1) is a shorthand or symbolic form because the functional relation between the variables may be expressed by means of a table or graph and not necessarily by an analytical or mathematical equation. An alternate symbolic form is

$$\phi(A_1, A_2, A_3, A_4) = 0$$

meaning that some function of the four variables is equal to zero.

As an example of an analytical expression, if $A_1 = 4A_2^2 + 2 \sin A_3 + \log A_4$, for a value of A_2, a value of A_3, and a value of A_4, there is a corresponding value for A_1. The variables A_2, A_3, and A_4 might be called "independent" variables, and A_1 the "dependent" variable. This relation can also be written as

$$4A_2^2 + 2 \sin A_3 + \log A_4 - A_1 = 0$$

---

[2] In heat transfer the fundamental dimensions are usually taken as length, time, temperature, mass, and heat. Heat may be used as a fundamental dimension as well as mass, as long as no interchange of mechanical energy and heat takes place. If appreciable conversion of mechanical energy into heat takes place, then under no condition may heat be regarded as a fundamental dimension.

4–3. Dimensionless groups

Imagine a certain physical phenomenon which involves an object with one side x and another side y acted upon by a force P and a force R. The variables x and y each have the dimension of a length. Assume that these four variables are the only ones involved in the action. There is a question as to the physical relation or equation which describes the action. An inspection of the dimensions indicates that we can form two dimensionless coordinates, one the ratio x/y and the other P/R. As illustrated in Fig. 4–1, actual data can be plotted using these coordinates. From experimental data or some other study besides dimensional analysis we determine the actual curve or the functional relation between the coordinates: *from dimensional analysis alone we determine only a set of convenient coordinates.* From this analysis we can write only that P/R is some function of x/y, or

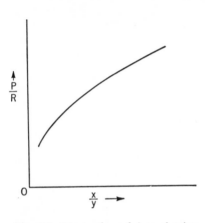

Fig. 4-1. Illustration of data plotting with dimensionless ratios.

$$\frac{P}{R} = f\left(\frac{x}{y}\right)$$

Without the data we do not know what the relation is. Suppose we make some experiments and discover that the relation is

$$\frac{P}{R} = \log\left(\frac{x}{y}\right) + 8\left(\frac{x}{y}\right)^3$$

Then we have the actual curve.

The entire problem of data organization and use involves two separate steps. This chapter discusses the first step, that of obtaining a convenient set of coordinates. The next step, that of actual measurements, will be treated in subsequent chapters of this book.

There are other possible arrangements of coordinates. For example, R/P could be plotted vertically with x/y horizontally, P/R could be plotted vertically with y/x horizontally, and R/P could be plotted vertically with y/x horizontally. Thus there are four ways of arranging the coordinates, each of which is correct from a dimensional point of view. Custom, however, may favor the use of one coordinate. For example, if R were a tangential friction force and P were a normal force, custom would favor the use of the ratio R/P with the familiar name of "friction coefficient."

In some problems it may be possible to discover convenient coordinates very easily. In many cases, however, the number of variables is large, and a general method of arranging variables efficiently is very desirable.

Let A_1, A_2, A_3, . . . , A_n be n physical quantities which are involved in some physical phenomenon. Examples of these physical quantities are velocity, viscosity, and density. The physical equation, or the functional relation between these quantities, can be written in the form

$$f(A_1, A_2, A_3. \ldots, A_n) = 0$$

Let m be the number of all the primary or fundamental units (such as length, mass, and time) involved in this group of physical quantities. Let k represent the maximum number of variables in a problem which can be combined without forming a dimensionless product; k cannot be greater than m. In other words, k is the number of A's whose units can be used to measure all the variables. Van Driest[3] proposed a rule which can be proved rigorously. The rule is that the foregoing physical equation can be written alternately as

$$\phi(\pi_1, \pi_2, \ldots, \pi_{n-k}) = 0$$

where each π is an independent dimensionless product of some of the A's. The use of the symbol π is conventional; it has no relation to the number 3.1416. The π term, in a theorem, was introduced by an early investigator E. Buckingham. Note that the number of terms in the physical equation has been reduced from n to $n - k$. In various cases k may equal m; however, this is not always so, and one should make a check with the basic rule.

In the preceding example (Fig. 4–1) $n = 4$, there are two primary dimensions: $m = 2$, and $k = 2$. Thus there are two dimensionless ratios.

Consider the case in which there are three variables, pressure p, density ρ, and velocity V. There are three variables $n = 3$, there are three primary dimensions $m = 3$, and the largest number of variables that will form a dimensionless product is 2, $k = 2$. Thus there is one dimensionless ratio, as $p/\rho V^2$.

In tackling a problem it may help to follow the procedure:

1. Select from among the list of variables a number of variables equal to the number of fundamental units and including all the fundamental units.

2. Set up dimensional equations combining the variables selected with each of the others in turn. When setting up the π terms, one of the variables is not included (a different one for each π term) to insure that the π terms will be independent.

The general method of approach can be explained by several examples.

---

[3] "On Dimensional Analysis and the Presentation of Data in Fluid-Flow Problems" by E. R. Van Driest, *Journal of Applied Mechanics*, Vol. 13, No. 1, 1946, page A-34.

4–4. Illustrations

Example 1. Consider a rotating shaft in a well-lubricated bearing. The problem is to determine a convenient set of coordinates for the equation giving the frictional resistance in terms of pertinent variables. The case is taken in which the following variables are involved:

Variable	Symbol	Dimensions of variable
Tangential friction force	R	MLT^{-2}
Force normal to shaft	P	MLT^{-2}
Shaft revolutions per unit time....	N	T^{-1}
Viscosity of lubricant	μ	$ML^{-1}T^{-1}$
Shaft diameter	D	L

Since there are five variables ($n = 5$) and three fundamental units ($m = 3$), and since $k = 3$, the physical equation has two dimensionless ratios. Let π_1 and π_2 represent these ratios. Dimensional equations are written combining the three variables P, N, and D [selected as suggested in (1) of Article 4–3] with each of the remaining variables in turn. Thus

$$\pi_1 = P^{x_1}N^{y_1}D^{z_1}R, \qquad \pi_2 = P^{x_2}N^{y_2}D^{z_2}\mu$$

The variable R and the variable μ each is taken to the first power for the sake of simplicity. Each of the resulting dimensionless ratios (as π_1, π_2) can be raised to some power greater than one and each still remain a dimensionless ratio. The six exponents are found from dimensional considerations. Substituting the dimensions for the symbols in the equation of π_1 gives

$$\left(\frac{ML}{T^2}\right)^{x_1}\left(\frac{1}{T}\right)^{y_1}L^{z_1}\frac{ML}{T^2} = L^0M^0T^0$$

where $L^0M^0T^0$ represents the fact that the π_1 ratio is dimensionless or has zero dimensions. Solving for each dimension separately gives

$$
\begin{array}{llll}
M & x_1 + 1 = 0 & & x_1 = -1 \\
L & x_1 + z_1 + 1 = 0 & & z_1 = 0 \\
T & -2x_1 - y_1 - 2 = 0 & & y_1 = 0
\end{array}
$$

Therefore

$$\pi_1 = \frac{R}{P}$$

which is easily checked. R/P is commonly called a friction coefficient f.

Substituting the dimensions for the symbols in the equation for π_2 gives

$$\left(\frac{ML}{T^2}\right)^{x_2}\left(\frac{1}{T}\right)^{y_2}L^{z_2}\frac{M}{LT} = L^0M^0T^0$$

$$
\begin{array}{llll}
M & x_2 + 1 = 0 & & x_2 = -1
\end{array}
$$

$$L \qquad x_2 + z_2 - 1 = 0 \qquad z_2 = 2$$
$$T \qquad -2x_2 - y_2 - 1 = 0 \qquad y_2 = 1$$

Thus

$$\pi_2 = \frac{N\mu D^2}{P}$$

The physical equation has the form

$$\pi_1 = \phi(\pi_2) \qquad \text{or} \qquad f = \phi\left(\frac{N\mu D^2}{P}\right)$$

where ϕ means "some function of"; that is, the friction coefficient f is some function of the dimensionless ratio $N\mu D^2/P$. The foregoing coordinates are used in lubrication studies. Experimental data are necessary in order to determine the exact nature of the functional relation; such data are available in current lubrication literature.

R, N, and D were selected to give *one* solution. Another set of three variables could have been selected to give *another* solution. For example, the dimensionless ratio P/R, from a dimensional point of view, is just as satisfactory and sound as its reciprocal. Custom, however, has arbitrarily established the ratio R/P as a friction coefficient.

Example 2. The problem is to find convenient coordinates for studying the thrust of a screw propeller completely immersed in a fluid. It is judged that the following variables are involved:

Variable	Symbol	Dimensions of variable
Thrust (axial force)	P	MLT^{-2}
Propeller diameter	D	L
Velocity of advance	V	LT^{-1}
Revolutions per unit time	N	T^{-1}
Gravitational acceleration	g	LT^{-2}
Density of fluid	ρ	ML^{-3}
Kinematic viscosity of fluid	v	$L^2 T^{-1}$

Since there are seven variables and three fundamental units, and since $k = 3$, there are four dimensionless ratios. Dimensional relations are written combining D, V, and ρ with each of the remaining variables in turn:

$$\pi_1 = D^{x_1}V^{y_1}\rho^{z_1}P$$
$$\pi_2 = D^{x_2}V^{y_2}\rho^{z_2}N$$
$$\pi_3 = D^{x_3}V^{y_3}\rho^{z_3}g$$
$$\pi_4 = D^{x_4}V^{y_4}\rho^{z_4}v$$

The twelve exponents in x, y, and z are determined such that each π function is dimensionless. The result is

$$\pi_1 = \frac{P}{D^2 V^2 \rho}, \qquad \pi_2 = \frac{DN}{V}, \qquad \pi_3 = \frac{Dg}{V^2}, \qquad \pi_4 = \frac{v}{DV}$$

This dimensional analysis gives *one* solution to the problem. There are *other* solutions each involving different dimensionless groupings of the variables. The above indicates that the form of the physical equation involves the four ratios π_1, π_2, π_3, and π_4. One solution of a variety can be written as

$$\pi_1 = \phi(\pi_2, \pi_3, \pi_4)$$

meaning that π_1 is some function of the other three π ratios, or

$$\frac{P}{D^2 V^2 \rho} = \phi\left(\frac{DN}{V}, \frac{Dg}{V^2}, \frac{v}{DV}\right)$$

Proper experiments would give the functional relation.

The last example includes one feature, in connection with the ratio $\pi_4 = v/DV$, which merits special attention. The reciprocal of this number is still a dimensionless ratio. The ratio DV/v occurs frequently in fluid flow problems and is significant in establishing criteria of flow. Its significance will be discussed more fully throughout the remainder of this book. This particular number is called Reynolds number.

DYNAMIC SIMILARITY

4–5. Mechanically similar flows

The important and useful concept of similarity can be illustrated by a simple geometrical case. Imagine one triangle with sides x, y, and z. Imagine another triangle with sides X, Y, and Z. The length x may be considerably different from the length X. The two triangles, however, are similar if the dimensionless ratio x/y equals the ratio X/Y, or if the ratio x/z equals the ratio X/Z. We can specify geometrical triangle similarity in a very simple and concise fashion.

The concept of similar force triangles is useful in dealing with force systems. As an example, consider a body No. 1 moving along a surface to the right, as shown in Fig. 4–2. If only a total of three forces act on the body, as P_1, R_1, and S_1, then for equilibrium these three forces must balance, and a simple force triangle can be formed. The vector sum of P_1 plus R_1 is balanced by S_1. Consider another body No. 2 which is geometrically similar to body No. 1 acted upon by the three forces P_2, R_2, and S_2. The force P_2 may be considerably different in magnitude from P_1. The ratio R_1/P_1 will be called a friction coefficient. If the friction coefficient for body 1 is the same as that for body 2, then there is a certain force or dynamic similarity between the two bodies. Assume that body 2 is larger than body 1 and that the frictional force R_2 is unknown. If the normal or vertical force P_2 could be calculated, as from the body weight or vertical component of S_2, then experiments upon a smaller body could be made to determine the friction coefficient. If true dynamic similarity is realized, then the measured friction coefficient for the smaller body could be applied directly to the larger body and the frictional force

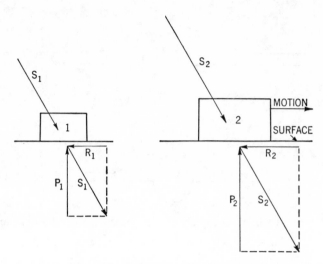

Fig. 4-2. Illustration of force similarity.

R_2 calculated. For dynamic similarity the two force triangles shown in Fig.4–2 are similar.

In order to obtain information regarding the flow phenomena in or around a structure or machine, called the original or *prototype*, it is often convenient, economical, and sound engineering to experiment with a copy or *model* of the prototype. The model may be geometrically smaller than, equal to, or larger than, the prototype in size. Model experiments for pumps, turbines, airplanes, ships, pipes, canals, and other structures and machines have resulted in savings which have more than justified the investments. Model tests provide an advantage in research, design, and performance-prediction work which cannot be obtained from theoretical calculations alone.

Certain laws of similarity must be observed in order to insure that the model-test data can be applied to the prototype. These laws, in turn, provide means for correlating and interpreting test data. For the flow around two bodies, as illustrated in Fig. 4–3, there is this question: Under what conditions is the flow around one body mechanically similar to the flow around the other? Mechanical similarity implies not only geometric similarity but also similarity with respect to the forces acting, or dynamic similarity. The flow around two bodies can be similar only if the body shapes are geometrically similar; this geometric similarity is a necessary condition, but it is not a sufficient condition.

The streamline pattern for one body must also be similar to the streamline pattern for the other body. For corresponding points with respect to the bodies, the velocity direction in one flow must be the same as the velocity

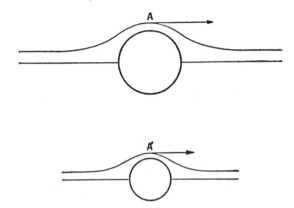

Fig. 4-3. Similar motions for a prototype and its model. A and A' are corresponding points.

direction in the other flow. The velocity direction at any point in the field of flow is determined completely by the *ratio of the forces* acting on a fluid particle at the point. Therefore, mechanical similarity is realized when the ratio of the forces acting on a fluid particle in one flow is the same as the ratio of the forces acting at a corresponding point in the other flow. Assume that the flow in Fig. 4–3 involves only three types of forces. For dynamic similarity, the force triangle at point A must be similar to the force triangle at A'.

Various laws of similitude could be devised, depending upon the type of forces acting. The present discussion will consider cases in which the following forces predominate:[4] (1) inertia forces, (2) viscous forces, (3) gravity forces, (4) pressure forces.

By "inertia force" is meant the resistance of an inert mass to acceleration. The *magnitude* of the inertia force equals the product of the particle mass and particle acceleration. The direction of this inertia force is opposite to the direction of the acceleration of the particle.

It is customary to consider separately two cases, or combinations, each with only three forces thus: (1) viscous, inertia, and pressure forces; (2) inertia, gravity, and pressure forces. In each case the pressure force or force due to fluid pressure is uniquely determined by the other two forces. In each case, specifying two of the forces automatically specifies the third force because the three forces are in equilibrium (recalling D'Alembert's principle). Therefore, in case (1) the significant pair of forces can be taken as viscous and inertia, whereas in case (2) the significant pair of forces can be taken as inertia and gravity. Each combination will be treated in the next two articles.

---

[4] Elastic forces resulting from the compressibility of the fluid are discussed in Chapter Six.

4–6. The Reynolds number

For completely enclosed flow (as in pipes, many flow meters, fans, pumps, and turbines), or for flow in which bodies are fully immersed in a fluid (as vehicles, submarines, aircraft, and structures) in such fashion that free surfaces do not enter into the consideration and gravity forces are balanced by buoyant forces, the inertia and viscous forces are the only ones which need to be taken into account. Mechanical similarity exists if, at points similarly located with respect to the bodies, the ratios of the inertia forces to the viscous forces are the same. The force triangle at a point in one flow must be similar to the force triangle at the corresponding point in the other flow.

Since the product of mass multiplied by acceleration is proportional to

$$\text{volume} \times \text{density} \times \frac{\text{velocity}}{\text{time}}$$

the inertia force is proportional to

$$\frac{l^3 \rho V}{t} = \frac{l^3 \rho V}{l/V} = l^2 \rho V^2$$

where V is some characteristic velocity (for example, the velocity of a body in a fluid or the average velocity over a fixed cross section of pipe), and l is some characteristic length (such as the diameter or length of a body, or the internal diameter of a pipe). The viscous force is proportional to τl^2, where τ is the viscous shear stress. Since $\tau = \mu(du/dy)$, the internal friction or viscous force is proportional to $\mu V l$. Then the dimensionless ratio inertia force/viscous force is proportional to

$$\frac{l^2 \rho V^2}{\mu V l} = \frac{\rho V l}{\mu} = \text{Reynolds number}$$

If viscous and inertia forces determine the flow for a prototype, then mechanical similarity between model and prototype is realized when the dimensionless Reynolds number for the model equals the Reynolds number for the prototype. In subsequent chapters, experimental data for various flow phenomena will be expressed as a function of the Reynolds number.

The nature of any particular flow of a real fluid may be judged to some extent from the corresponding Reynolds number. A small Reynolds number indicates that viscous forces predominate, whereas a large value of Reynolds number indicates that inertia forces predominate. The laws of motion of fluids and the laws of resistance to motion are very different for these two cases.

Example. Standard air flows through a pipe 32 inches in diameter with an average velocity of 6 feet per second. A model of this pipe 3 inches in diameter is constructed for water flow. What average velocity of water is necessary for the model flow to be dynamically similar to that of the prototype? Taking the diameter D of the pipe as a characteristic length,

$$\frac{V_1 D_1}{\nu_1}(\text{model}) = \frac{V_2 D_2}{\nu_2}(\text{prototype})$$

Using values given in Table 1–3, Chapter One, the velocity V_1 becomes

$$V_1 = V_2\left(\frac{D_2}{D_1}\right)\frac{\nu_1}{\nu_2} = 6\left(\frac{32}{3}\right)\left(\frac{1.233 \times 10}{1.57 \times 10}\right)$$

$$V_1 = 5.03 \text{ feet per second}$$

for the average velocity in the model pipe.

Frequently a model is tested in a fluid which is the same as that of the prototype flow. If the Reynolds number for each is the same and the kinematic viscosity of each fluid is the same, then similarity is given by the relation $V_1 l_1$ for the model equals $V_2 l_2$ for the prototype. If the prototype is larger in size than the model, then the model velocity will be higher than the prototype velocity.

4–7. The Froude number

Gravity forces are taken into account in cases in which a fluid free surface plays an essential role, as for example the surface waves produced by a ship or a seaplane hull, or the flow in an open channel. In the surface waves produced by a ship, the form of the liquid is displaced above and below the mean surface level, and hence a weight or gravity force is involved. A model of a ship will produce the same shape of surface waves as the full-size ship if the ratio of inertia force to gravity force is the same at similarly located points with respect to the model and the prototype. The force triangles must be similar for dynamic similarity. The gravity force for a fluid particle equals the particle mass times the gravitational acceleration g. The gravity force is proportional to $\rho l^3 g$, where l is some characteristic length, like the length of the ship. Then the dimensionless ratio inertia force/gravity force is proportional to

$$\frac{l^2 \rho V^2}{\rho l^3 g} = \frac{V^2}{lg}$$

The shapes of the water waves produced by the ship model and the prototype will be similar if the value of V^2/lg is the same for both. This dimensionless ratio, or usually its square root, is called the Froude number:

$$\text{Froude number} = \frac{V}{\sqrt{lg}}$$

Example. An ocean vessel 600 feet long is to travel 15 knots (nautical miles per hour). What would be the corresponding speed for a geometrically similar model 6 feet long towed in water for studies of wave resistance?

$$\frac{V_1}{\sqrt{l_1 g}}(\text{model}) = \frac{V_2}{\sqrt{l_2 g}}(\text{prototype})$$

$$V_1 = 15\sqrt{\frac{6}{600}} = 1.5 \text{ knots}$$

If the model is smaller than the prototype, dynamic similarity on the basis of the Froude number shows a model velocity lower than that of the prototype.

4–8. Surface tension

"Surface tension" is a molecular property of liquids which produces certain phenomena at liquid surfaces. By liquid surfaces is meant such surfaces of separation as that between water and air and that between two different immiscible liquids. Surface tension is the result of forces exerted on the surface molecules by molecules in the interior of the liquid and the result of forces exerted by the molecules of the substance in contact with the liquid. The force of surface tension is tangent to the surface of separation and has the same magnitude perpendicular to any imaginary line in the surface.

Surface tension can be expressed in such units as dynes per centimeter or pounds per foot. The surface tension for water in contact with air, at 68 degrees Fahrenheit, is about 0.0005 pound per foot.

Consider the case in which the forces involved are surface tension, inertia, and pressure. The inertia force is proportional to $l^2 \rho V^2$. Let σ represent surface tension. Then the surface tension force is proportional to σl. Then the dimensionless ratio inertia force/surface tension force is proportional to

$$\frac{\rho l^2 V^2}{\sigma l} = \frac{\rho l V^2}{\sigma}$$

The ratio $\rho l V^2/\sigma$, or the square root of this ratio, is sometimes called the Weber number.

4–9. Other numbers

Two foregoing articles illustrated dynamic similarity with the Reynolds number and the Froude number. Other force ratios could be devised. Chapter Six, dealing with high-velocity compressible flow, will treat elastic forces resulting from the compressibility of the fluid; in that chapter one other number, called the Mach number, will be discussed.

The force due to pressure, or the pressure force, was not explicitly stated in the discussion of the Reynolds number. In some applications it is essential to have the pressure in explicit form. For example, consider a body immersed in a stream, as shown in Fig. 4–4. The total force of the

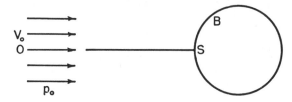

Fig. 4-4. Flow around a body.

fluid on the body depends upon the pressure acting at each point on the body surface or the pressure distribution over the body surface. In some cases models are constructed so that pressures can be measured at different points on the surface. Examples are models of fan blades and airfoil sections. There is a question as to the organization of test data in a form most suitable for general application. This question can be handled solely by dimensional analysis. Let us treat it, however, from a slightly different point of view in an attempt to give a physical picture.

Assume that the fluid is incompressible and approaches the body with a velocity V_0. The upstream pressure p_0 is constant, but the pressure at B varies over the surface of the body. Let the magnitude of the pressure difference between points B and O be represented by Δp. In practice it is common to define a dimensionless pressure coefficient in the form

$$\text{Pressure coefficient} = \frac{\Delta p}{\rho V_0^2/2}$$

If the pressure coefficient at point B has the magnitude 0.40, then the pressure difference between points B and O is 0.40 of upstream dynamic pressure.

For the case of a model and a prototype, the flow for one is similar to the flow for the other if the pressure coefficient for one is the same as the pressure coefficient for the other at corresponding points. If we wanted to know the pressure at various points along a large fan blade, we could experiment with a smaller blade and express the measurements in terms of pressure coefficients.

As another approach, let us form a force ratio in a manner similar to that followed in the two previous articles. The force due to pressure acting on some body is proportional to some pressure (or pressure difference) times an area. This force is proportional to $\Delta p l^2$ where l is some characteristic dimension of the body. The inertia force is proportional to $\rho V^2 l^2$. Therefore the ratio of pressure force/inertia force is proportional to

$$\frac{\Delta p l^2}{l^2 \rho V^2} = \frac{\Delta p}{\rho V^2}$$

Let $E = \Delta p / \rho V^2$. This ratio or number has the pressure explicitly stated. For dynamic similarity, E for the model must equal E for the prototype at corresponding points. Some writers prefer to call this number or its reciprocal the Euler number, whereas others prefer to call it the Newton number. Except for the numerical factor of $\frac{1}{2}$, this force ratio has the same form as the pressure coefficient.

The application of a special similarity law depends upon the forces determining the particular flow. Frequently the flow depends not on one ratio of forces only, but on two or possibly three ratios. An example is the problem of the total resistance of a ship moving on a surface of water. The total resistance consists mainly of frictional resistance due to the viscous forces and of surface wave resistance due to gravity force. The total resistance of the ship thus depends on both the Reynolds number and the Froude number. Consider a ship model, smaller in size than the prototype, tested in water. The Reynolds law indicates a model speed higher than that of the prototype, whereas the Froude law indicates a model speed lower than that of the prototype. Theoretically, it may be possible to combine both laws of similarity by using two different liquids. Actually, however, no two practical liquids are available which have sufficiently different kinematic viscosities. In practice, ship-model tests are based on the Froude number, and viscous forces are taken into account indirectly by computation.

PROBLEMS

4-1. The velocity of sound c in a gas depends upon the density ρ, pressure p, and dynamic viscosity μ. Find a convenient set of coordinates for organizing data.

4-2. Observations indicate that the resistance R which the air offers to an airplane wing depends mainly on some characteristic length l, the speed V of the wing, and the density and viscosity of the air. Find a convenient set of coordinates for organizing data.

4-3. Consider that the resistance R of a flat plate immersed in a fluid is dependent on the fluid density and viscosity, the velocity, and the width b and the height h of the plate. Find a convenient set of coordinates for organizing data.

4-4. It is judged that the performance of a lubricating oil ring depends upon the following variables: Q, quantity of oil delivered per unit time; D, inside diameter of the ring; N, shaft speed in revolutions per unit time; μ, oil viscosity; ρ, oil density; γ, specific weight of the oil; and σ, the surface tension in air (force per unit length). Find a convenient set of coordinates for organizing data.

4-5. For a certain type of installation, observations show that the gases in the wake of a smokestack flow downward a certain distance, called the *downwash*.

Assume that the pertinent variables involved in this phenomenon are depth of downwash l, diameter of stack D, wind velocity V_1, gas velocity V_2, and gas temperature θ. Find a convenient set of coordinates for organizing data.

4-6. The elevation h of a liquid in a capillary tube depends on the vertical component of surface tension (force per unit length) t, the radius of the tube R, density of liquid ρ, and the gravitational acceleration g. Find a convenient set of coordinates for organizing data.

4-7. Air at a certain pressure and temperature flows through a pipe 4 inches in diameter at an average velocity of 100 feet per second. For dynamically similar flows, what is the average velocity for air at the same pressure and temperature through a 10-inch diameter pipe?

4-8. An airship model is tested in standard air at a speed of 100 feet per second. Calculate the towing speed of the same model when tested submerged in water under similar dynamic conditions.

4-9. An automobile is to travel through standard air at 60 miles per hour. A model, one-fifteenth the length of the prototype, is considered for tests in water. What should be the model velocity?

4-10. A torpedo 19 feet long is to travel 40 knots submerged in fresh water. A model of this torpedo is 8 feet long. What would be the model speed if tested in fresh water? What would be the model speed if tested in standard air?

4-11. An airplane model has linear dimensions that are one-twentieth those of the full-sized plane. It is desired to test the model in a wind tunnel with an air speed equal to the flying speed of the full-size plane. What must be the pressure in the wind tunnel? Assume the same air temperature and dynamic viscosity in each case.

4-12. A centrifugal pump is to handle castor oil at 59 degrees Fahrenheit, at a speed of 1000 revolutions per minute. A model pump is exactly two times as large. Experiments are to be made with the model pump using standard air. Determine the speed of the air pump.

4-13. A submerged submarine is to move at 6 knots. A model is one-tenth as long as the prototype. What would be the model speed if the water were the same in both cases?

4-14. A propeller-type mixer, 9 feet in diameter, is to be used submerged in soybean oil at 30 degrees centigrade (specific weight 56.5 pounds per cubic foot, viscosity 0.085 slug per foot-second). Tests on a model, 6 inches in diameter in water at 59 degrees Fahrenheit, show that the best operating speed for the model is 18 revolutions per minute. Estimate the best angular speed for the prototype.

4-15. An ocean vessel 500 feet long is to travel at 10 knots. What would be the speed for a model 20 feet long towed in water for studies of surface wave resistance?

4-16. An airplane flies at 300 miles per hour through standard air at sea level. Determine the craft speed at 10,000 feet above sea level in an isothermal atmosphere for dynamically similar conditions.

4-17. Castor oil at 59 degrees Fahrenheit flows through a 3-inch diameter pipe at an average velocity of 15 feet per second. A model of this pipe, 2 inches in diameter, is arranged for flow with standard air. What average velocity of air is necessary for dynamically similar flows?

4-18. The Reynolds number for water flowing through a certain pipe is 100,000. What is the Reynolds number for the same weight rate of flow if the pipe diameter is doubled?

4-19. It is desired to obtain data for the design of an airship 230 feet long to travel at 130 miles per hour in standard air. It is proposed to test a geometrically similar model, 17 inches long, in water. What should be the model speed?

4-20. Consider the flow through the impeller of a centrifugal compressor. Let D represent the impeller diameter, ρ the density of the fluid, N the angular speed of the impeller, and V the exit flow velocity. What combination of variables or number is proportional to the ratio of total inertia force divided by the centrifugal force?

4-21. Consider that the performance of a centrifugal compressor depends upon the following factors: inlet pressure p, fluid density ρ, impeller diameter D, average inlet velocity V, and the angular speed of the shaft N. Find a convenient set of coordinates for organizing data.

4-22. Consider a flow in which there are three types of forces acting: viscous, inertia, and pressure. Derive a number proportional to the ratio of the viscous force divided by the pressure force.

4-23. Consider a flow in which there are three types of forces acting: inertia, gravity, and pressure. Derive a number proportional to the ratio of gravity force divided by pressure force.

4-24. Imagine flow approaching and passing along a flat plate. The so-called "boundary layer" is a layer of fluid near the plate in which there is a velocity variation. Let D be the thickness of the boundary layer at distance l from the leading edge of the plate, V the approach velocity, ρ the density, and μ the dynamic viscosity of the fluid. Find a set of convenient coordinates for organizing data.

4-25. Why is it that certain insects can walk on a water surface?

4-26. Imagine that the flow over a weir depends upon the following quantities: volume rate Q, head H, density ρ, dynamic viscosity μ, and surface tension σ. Find a convenient set of coordinates for organizing data.

4-27. The formation of drops by the breaking up of liquid jets is encountered in connection with sprays and with fuel injection in Diesel engines. It is judged that the following variables are involved: density ρ, velocity V, a length l, dynamic viscosity μ, and surface tension σ. Find a convenient set of coordinates for organizing data.

4-28. In studying the action of bubbles rising in a liquid, it is judged that the following variables are involved: surface tension σ, V velocity of bubble with respect to the liquid, d diameter of bubble, ρ density of liquid, ρ_1 density of bubble, and μ viscosity of the liquid. Find a convenient set of coordinates for organizing data.

4-29. The noise produced by a propeller in water involves the following variables; i sound intensity (power per unit area), ρ density of fluid, p static pressure, V tip velocity of propeller, D diameter of propeller, d distance from propeller, and blade width c. Find a convenient set of coordinates for organizing data.

4-30. In a certain flow the velocity V is a function of pressure p, density ρ, and a factor E which has the dimensions of force per unit area. Find a convenient set of coordinates for organizing data.

CHAPTER FIVE

Steady Motion
of Viscous Fluids

*There be three things which are too wonderful for me, yea, four which
I know not:
The way of an eagle in the air; the way of a serpent upon the rock;
the way of a ship in the midst of the sea; and the way of a man with a
maid.*

—PROVERBS.

FLOW OF INCOMPRESSIBLE FLUIDS IN PIPES

A study of pipe flow is useful in many ways besides giving information
about a common case. Such a study serves to illustrate laminar flow and
turbulent flow, how such flows can be correlated with a simple parameter,
and how a problem with a large number of variables can be reduced to one
with a convenient set of coordinates.

5–1. Laminar and turbulent flow

There are two different types of flow: laminar and turbulent. In laminar
flow the fluid moves in layers, or laminas. In turbulent flow secondary irregular motions and velocity fluctuations are superimposed on the principal or
average flow. As pointed out by H. L. Dryden, a common occurrence showing
these types of flow is found in the rising column of smoke from a cigarette
lying on an ashtray in a quiet room. For some distance the smoke rises in
smooth filaments which may wave around but do not lose their identity;
this flow is laminar. The filaments suddenly break up into a confused eddying
motion at some distance above the cigarette; this flow is turbulent. The transition between laminar and turbulent flows moves closer to the cigarette when
the air in the room is disturbed. Turbulent flow is very common; it is found
in countless engineering cases, as for example, the flow through pipes, through
other channels, and in different machines. Laminar flow is found in numerous
cases where the flow channel is relatively small, the velocity relatively low,

105

and the viscosity relatively high; examples are the flow of oil lubricated bearings, the flow of liquids in small channels in the human body and in instruments for measuring fluid properties.

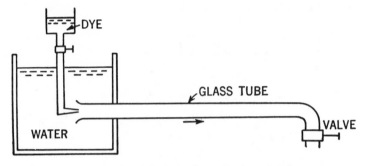

Fig. 5–1. Apparatus for investigating pipe flow.

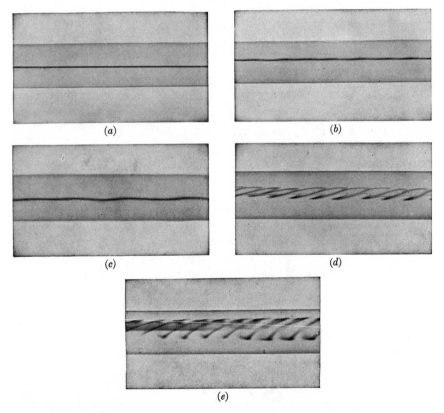

Fig. 5–2. Transition from laminar to turbulent flow for water in a glass tube, with increasing values of the Reynolds number.

The type of apparatus shown diagrammatically in Fig. 5–1 can be used to illustrate various types of flow. Colored liquid entering the mouth of the glass tube moves along with the water, to indicate the nature of the flow. At low velocities the colored filament is straight and stable. At velocities above a critical value, the dye shows irregular patterns. The pictures in Fig. 5–2 are in the order of increasing velocity. Figure 5–2(a), with the lowest velocity, definitely shows laminar flow, whereas Figures 5–2(d) and 5–2(e) show turbulent flow.

5–2. Critical Reynolds numbers

It would be helpful to have a correlation between the type or nature of the flow and some quantitative measurements. Let ρ represent the density of the fluid passing through the pipe, D the internal diameter of the pipe, μ the dynamic viscosity of the fluid, and V the average velocity in the pipe. Dimensional analysis shows that we can arrange these variables in a dimensionless grouping $\mathrm{Re} = \rho V D / \mu$ which is commonly called the *Reynolds number*.

Experiments show that the stable form of motion for pipe flow is normally laminar for Re less than 2000. Below $\mathrm{Re} = 2000$ an initially turbulent flow, with its typical irregular mixing or eddying motion, cannot be maintained indefinitely. Thus a *lower critical velocity* V_c may be established in which

$$V_c = 2000 \frac{\mu}{\rho D} \qquad (5\text{–}1)$$

If the actual average velocity V is below V_c, the stable flow will be laminar.

For usual conditions the pipe flow is turbulent for values of Re above about 3000. In the transition range $\mathrm{Re} = 2000$ to $\mathrm{Re} = 3000$ there are various possible conditions, depending upon the initial disturbances, the pipe entrance, and the pipe roughness. By very careful manipulation of the apparatus, laminar flow has been obtained with values of Re considerably above 3000; such flow, however, is not inherently stable, for any disturbance, once started, tends to break up the laminar pattern. Laminar flow at values of Re considerably above 3000 is a phenomenon which is somewhat similar to supersaturation or to undercooling. Normally, laminar flow does not occur for values of Re above about 3000.

The discussion on dynamic similarity in the previous chapter showed that the Reynolds number is proportional to the ratio inertia force/viscous force. A relatively small Re indicates that viscous forces predominate in laminar flow, whereas a relatively large Re indicates that inertia forces predominate in turbulent flow.

It is to be noted that the foregoing critical values of the Reynolds number apply *only* to pipe flow. Critical values of the Reynolds number for other cases, as for flow around various bodies immersed in a stream, are different from those given in the preceding discussion. Further, it is difficult to specify exactly one universal critical Reynolds number or one specific value which will provide a very sharp and general distinction for all pipe installations. Turbulent flow depends upon the pipe roughness, initial disturbances, entrance conditions, and other factors. Hence, the several critical values given in the foregoing discussion have been qualified.

5–3. Energy relations for pipe flow

Using the notation shown in Fig. 5–3, the case is taken in which the fluid is incompressible and there is no pump, turbine, or similar machine between sections 1 and 2. The energy equation then becomes

$$(q + u_1 - u_2) = \frac{p_2 - p_1}{\gamma} + \frac{V_2^2 - V_1^2}{2g} + (z_2 - z_1) \qquad (5\text{–}2)$$

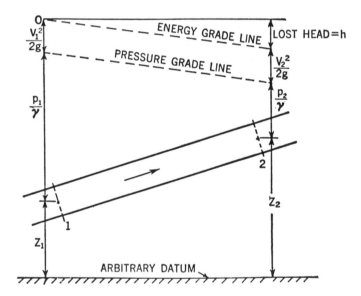

Fig. 5–3. Notation for pipe flow.

As the fluid passes through the pipe, some mechanical energy is degraded into unavailable energy; there is a so-called "friction loss" or a "head loss" due to friction or viscosity of the fluid and the turbulent motion. No energy is actually destroyed. Some energy, however, is transformed into a form which is not available for maintaining the flow; thus, from the point of view of the flow, it is "lost." Friction opposes flow. Let *h* represent this lost head. Then the general energy equation can be written as

$$h = \frac{p_1 - p_2}{\gamma} + \frac{V_1^2 - V_2^2}{2g} + (z_1 - z_2) \qquad (5\text{–}3)$$

Each term in Equation (5–3) can be expressed in units of mechanical energy per unit weight of fluid flowing. The lost energy h, for example, can be stated in terms of foot-pounds per pound of fluid, or simply feet, or some other net unit of length.

The terms of the preceding energy equation can be represented by vertical ordinates, as is done in Fig. 5–3. The *pressure grade line* is an imaginary line giving the sum of $(z + p/\gamma)$ at any point along the pipe; it is the line which a series of manometers would indicate. Plotting the *total head*

$$\left(z + \frac{p}{\gamma} + \frac{V^2}{2g} \right)$$

at each point gives the *energy grade line*. The energy grade line slopes down from the horizontal in the direction of flow. The vertical distance between the energy grade and a horizontal line through 0 represents the lost head. If the area at section 1 equals the area at section 2, the equation of continuity specifies that $V_1 = V_2$. In this case the pressure grade line is parallel to the energy grade line, or

$$h = \left(\frac{p_1}{\gamma} + z_1 \right) - \left(\frac{p_2}{\gamma} + z_2 \right) \qquad (5\text{–}4)$$

If the pipe is horizontal, then $z_2 = z_1$, and the head loss h equals $(p_1 - p_2)/\gamma$.

5–4. Smooth and rough pipes

Let h represent the head loss in the pipe length l. It is customary to express the lost head in the form

$$h = f \frac{l}{D} \cdot \frac{V^2}{2g} \qquad (5\text{–}5)$$

The ratio l/D is dimensionless. The factor $V^2/2g$ is frequently called a *velocity head;* it has the same dimensions as h, namely, energy per unit weight or the net dimension of a length. Thus the friction factor or coefficient f is dimensionless.

For ordinary velocities a great many data have established the fact that the friction coefficient f depends on two factors: the Reynolds number of the flow and the geometry of the conduit surface. For steady flow through a conduit of circular cross section, the geometry of the conduit can be expressed in terms of a parameter called the *relative roughness* of the pipe surface. Dimensional analysis can be used to show that the friction factor is a function of the Reynolds number and the relative roughness.

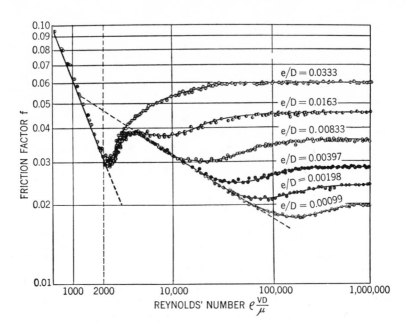

Fig. 5–4. Plot of Nikuradse's data on flow in artificially roughened pipes.

Figure 5–4 illustrates the usual method of plotting pipe friction data. The friction factor is plotted as a function of the Reynolds number using logarithmic scales. The points shown are the results of tests by Nikuradse[1] in which the pipe walls were coated with closely packed, uniform sand grains. This technique provided an artificial but measurable roughened surface. The mean diameter of the sand grain e is arbitrarily called the *absolute roughness;* it is a length and can be expressed in such a unit as the foot. The dimensionless ratio e/D is arbitrarily called the *relative roughness.*

For a Reynolds number below about 2000, these tests and others show that the friction factor is given by the simple relation $f = 64/Re$ and that the friction factor is independent of the pipe surface. For laminar flow, then, pipes of different surface roughness have the same friction factor for the same Reynolds number.

For turbulent flow, if the surface roughness is very slight, as for glass tubing or "smooth pipes," the friction factor is a function of the Reynolds number only. Work by Kármán,[2] Nikuradse, and Prandtl resulted in the

[1] "Strömungsgesetze in rauhen Rohren," by J. Nikuradse. Forschungsheft 361, *Beilage zu Forschung auf dem Gebiete des Ingenieurwesens*, VDI-Verlag, Berlin, 1933.

[2] "Turbulence and Skin Friction," by T. v. Kármán. *Journal of the Aeronautical Sciences*, vol. 1, no. 1, January, 1934, page 1.

"Mechanische Ähnlichkeit und Turbulenz," by T. v. Kármán. *Proceedings of the Third Congress for Applied Mechanics*, Stockholm, 1930, vol. 1, page 85.

following analytical relation for flow above about Re = 4000 for smooth pipes:

$$\frac{1}{\sqrt{f}} = 2\log_{10}(\text{Re}\sqrt{f}) - 0.8 \tag{5-6}$$

This relation has a complicated but rational background, and checks the different data very closely.

Refer to Fig. 5–4 and consider the region above a Reynolds number of about 10,000. In this region, for the same Reynolds number, the friction factor increases as the relative roughness increases. For each relative roughness value, particularly at the higher values, there is a range of the Reynolds number in which the friction factor is practically constant. For convenience in reference, let us call this the "rough-pipe" zone. In this zone, the friction is not a function of the Reynolds number but a function of relative roughness only. In this zone, apparently, the surface irregularities or the surface absolute roughness is large in comparison with the laminar sublayer; complete turbulence is established throughout the flow. Viscous forces then become extremely small in comparison with inertia forces, and the friction factor curve is horizontal for a constant relative roughness. Kármán has provided an analytical function for this rough-pipe zone. The numerical constants have been determined by Nikuradse; the result is as follows:

$$\frac{1}{\sqrt{f}} = 1.74 - 2\log\left(\frac{2e}{D}\right) \tag{5-7}$$

Above a Reynolds number of about 3000 there are two regions, the "transition" zone and the rough-pipe zone. The transition zone lies between the rough-pipe zone and the lower Reynolds numbers. Colebrook[3] has developed an analytical relation for this transition region which checks the data very well for commercial pipes:

$$\frac{1}{\sqrt{f}} = -2\log\left(\frac{e/D}{3.7} + \frac{2.51}{\text{Re}\sqrt{f}}\right) \tag{5-8}$$

5–5. Friction losses in circular pipes

Figure 5–5 shows an organization of data due to Moody. This chart is limited to steady flow through circular pipes, and only to new and clean piping. The effect of age on pipe roughness is frequently very difficult to predict. One cannot expect a high degree of accuracy in determining f. For smooth tubing actual data may show a variation or scatter of ±5 per cent in f. For commercial steel and wrought iron pipes, actual data may show a variation within ±10 per cent in f.

[3] "Turbulent Flow in Pipes, with Special Reference to the Transition Region Between the Smooth and Rough Pipe Laws," by C. F. Colebrook. *Journal of the Institution of Civil Engineers*, vol. 11, 1938–1939, page 133.

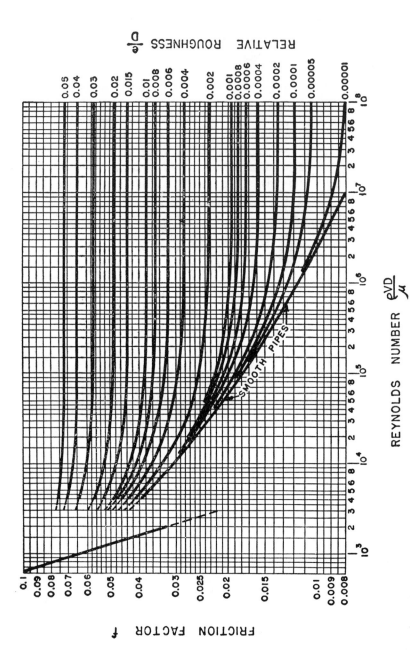

Fig. 5–5. Friction factors for flow in circular pipes.[4]

[4] Adapted from "Friction Factors for Pipe Flow," by L. F. Moody. *A.S.M.E. Transactions*, vol. 66, no. 8. November, 1944, 671.

Table 5–1 shows absolute roughness values e (feet) for different commercial pipes. This roughness scale is arbitrary, but it is the best that is available. Figure 5–6 provides a chart for determining relative roughness. One common and important type of problem is the determination of pressure drop for a given length, diameter, dynamic viscosity, density, and average velocity (or rate of discharge). In solving such a problem the first suggested step is to calculate Re. If the flow is turbulent, the next step is to determine the relative roughness. Reference to the particular curve in Fig. 5–5 gives the friction factor. Use of Equation (5–5) gives the head loss, and use of the energy equation gives the pressure drop.

TABLE 5–1

ABSOLUTE ROUGHNESS VALUES FOR
COMMERCIAL PIPE

Pipe	Absolute roughness e (feet)
Commercial steel or wrought iron	0.00015
Asphalted cast iron	0.0004
Galvanized iron	0.0005
Cast iron	0.00085
Wood stave	0.0006 to 0.003
Concrete	0.001 to 0.01
Riveted steel	0.003 to 0.03

Example. Benzene at 50 degrees Fahrenheit (specific gravity = 0.90) flows through a horizontal commercial steel pipe 6 inches in diameter with an average velocity of 11.0 feet per second. Calculate the pressure drop in 200 feet of pipe.

Solution. Figure 1–4 shows that the dynamic viscosity of the benzene is 1.6×10^{-5} slug per foot-second.

$$\text{Re} = \frac{0.90(62.4)11.0(6)}{32.2(12)1.6 \times 10^{-5}} = 6.0 \times 10^5$$

The absolute roughness factor e for this pipe is 0.00015 foot. Calculation or reference to Fig. 5–6 shows that the relative roughness e/D is 0.0003. On Fig. 5–5 there are only curves of relative roughness for $e/D = 0.0002$ and $e/D = 0.0004$. Therefore it is necessary to interpolate between these two curves. Figure 5–5 thus gives a friction factor of $f = 0.016$. Then the lost head is

$$h = 0.016 \left(\frac{200}{0.5}\right) \frac{(11.0)^2}{2(32.2)} = 12.02 \text{ feet (or foot-pounds per pound)}$$

$$\Delta p = \gamma h = 0.90(62.4)\frac{12.02}{144} = 4.69 \text{ pounds per square inch}$$

Between Reynolds numbers of 2000 and 4000 the behavior of the fluid, and thus the friction factor, is somewhat difficult to predict. There are various

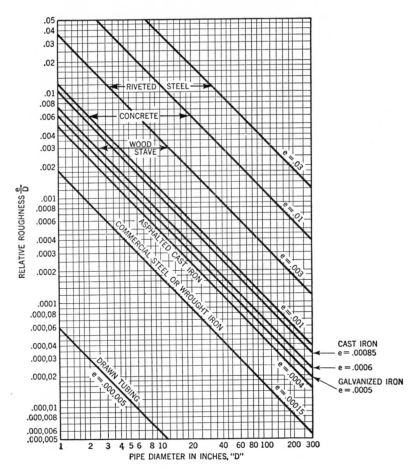

Fig. 5-6. Chart for determining relative roughness of pipes.[5]

possible conditions, depending upon the initial disturbances, the pipe entrance, and the roughness. Nikuradse's plot shows a certain curve in this
region, but this type of variation in this region is not generally found with
commercial pipes.

Subsequent chapters, such as those dealing with flow meters and with
the resistance of moving immersed bodies, will bring out the facts that other
flow coefficients are correlated with the Reynolds number, and that the
method of calculation in many of these other problems is somewhat similar
to that used in pipe flow problems.

[5] Adapted from same source as that for Fig. 5-5.

5–6. Hagen-Poiseuille law for laminar flow
in circular pipes

Careful experiments have shown that for fully developed *laminar* flow in all circular pipes $f = 64/\text{Re}$. For this case the pressure drop in a horizontal pipe becomes

$$p_1 - p_2 = \Delta p = \gamma h = \gamma \frac{64}{\text{Re}} \cdot \frac{l}{D} \cdot \frac{V^2}{2g} = \frac{32\mu l V}{D^2} \qquad (5\text{-}9)$$

The rate of discharge $Q = (\pi D^2/4)V$. Then

$$\Delta p = \frac{128\mu l Q}{\pi D^4} \qquad (5\text{-}10)$$

Equations (5–9) and (5–10) are forms of the so-called Hagen-Poiseuille law (after G. Hagen and L. J. M. Poiseuille). This law can be derived on an analytical basis, as will be done in the next paragraph, to show that the mathematical analysis of laminar flow checks very accurately with experimental results.

In Fig. 5–7 the pressure difference $p_1 - p_2$ exerts the force $(p_1 - p_2)\pi y^2$

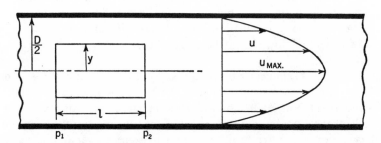

Fig. 5–7. Laminar flow in a horizontal circular pipe, with a parabolic velocity distribution.

on a cylindrical mass of fluid of radius y; p_1 acts on one circular flat surface and p_2 acts on the other circular flat surface. The opposing force consists of friction on the curved cylindrical surface; this total force is $2\pi y l \tau$, where τ is the shear stress acting on the curved surface. Equating these two forces for steady, nonaccelerated flow gives the relation for shear stress:

$$\tau = \frac{p_1 - p_2}{l} \cdot \frac{y}{2} = \frac{\Delta p}{l} \cdot \frac{y}{2} \qquad (5\text{-}11)$$

For laminar flow, $\tau = \mu(du/dy)$. With the coordinate system employed and the highest velocity at the pipe center, du is a decrement. Then

$$\frac{du}{dy} = \frac{-\Delta p}{2\mu l} y$$

Integrating, and determining the integration constant by the fact that $u = 0$ at $y = D/2$ (fluid adheres to the pipe wall), there results

$$u = \frac{\Delta p}{4\mu l}\left(\frac{D^2}{4} - y^2\right) \tag{5-12}$$

Equation (5-12) indicates a parabolic velocity distribution, as represented in Fig. 5-7. The maximum velocity occurs at the center where $y = 0$. Thus

$$u_{\text{max.}} = \frac{\Delta p D^2}{16\mu l} \tag{5-13}$$

Q can be determined by integration.

$$Q = \int_0^{D/2} 2\pi y\, dy\, u = \frac{\pi}{\mu l}\Delta p \frac{D^4}{128} \tag{5-14}$$

which checks Equation (5-10). The Hagen-Poiseuille law provides a good way for determining the dynamic viscosity of a fluid; and it will be discussed further in this respect in the chapter on Flow Measurement. Inspection of Equations (5-9) and (5-13) shows that $V = u_{\text{max.}}/2$ and that the average velocity is one-half the maximum velocity at the center. It is to be emphasized that the Hagen-Poiseuille law applies only for *laminar* flow in circular pipes. Before the results of a calculation based on the Hagen-Poiseuille law are judged, the type of flow should be established, as by a calculation of the Reynolds number.

Equation (5-11) indicates that the shear-stress distribution across a pipe section for laminar flow is a linear function of the radial distance from the pipe center. Equation (5-9) shows that the pressure drop for laminar flow is directly proportional to the first power of the velocity. For turbulent flow, the pressure drop is directly proportional to some power of the velocity between 1 and 2. The value of the power can be calculated by referring to the experimental data given in Fig. 5-5.

5-7. Velocity distributions in circular pipes

The velocity distribution at a pipe section depends upon various factors, such as the length of pipe preceding the section. At the rounded entrance in Fig. 5-8 the velocity profile is approximately uniform except for a thin film at the walls. For laminar flow, a short distance downstream the velocity profile becomes a combination of a nearly parabolic distribution at the walls and a core in which the velocity is nearly uniform. Fully developed laminar flow, with the parabolic velocity profile, is reached further downstream. The distance for complete transition depends upon Re.

For fully developed flow, the velocity profile becomes flatter as Re increases. Figure 5-9 shows a comparison of three velocity profiles at differ-

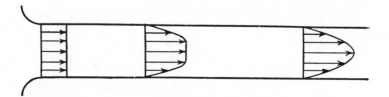

Fig. 5–8. Velocity profiles along a pipe for laminar flow.

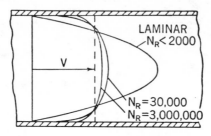

Fig. 5–9. Comparison of velocity profiles in a smooth pipe for the same average velocity V and different Reynolds numbers.

ent values of Re, but each for the same average velocity. The ratio V/u_m is sometimes called the *pipe coefficient or pipe factor*, where u_m is the velocity at the center of the pipe. Figure 5–10 shows a plot of V/u_m against the Reynolds number based upon data presented by Folsom and Iversen.[6] Figure 5–10 is for fully developed flow. Each curve in Fig. 5–10 is for a certain constant relative roughness; the relative roughness values e/D are indicated along the right-hand edge. For laminar flow the ratio is 0.50, whereas for turbulent flow the ratio is greater than 0.50.

Kármán and Prandtl have shown that the velocity distributions for turbulent flow may be expressed empirically by power laws over a wide range. Let u be the velocity at a distance y from the wall, and let r be the pipe radius. Then the velocity distribution in the main stream can be represented by the relation

$$u = u_m \left(\frac{y}{r}\right)^n \qquad (5\text{–}15)$$

The value of n is $\frac{1}{7}$ for turbulent flow in smooth tubes up to values of Re of about 100,000. If the pipe factor (Fig. 5–10) is taken as 0.82, then $u_m = 1.22V$. For this case,

$$u = 1.22V \left(\frac{y}{r}\right)^{1/7} \qquad (5\text{–}16)$$

which relation agrees fairly well with experimental results, For higher values

[6] "Pipe Factors for Quantity Rate Flow Measurements with Pitot Tubes," by R. G. Folsom and H. W. Iversen. *American Society of Mechanical Engineers Paper* 48-A-35.

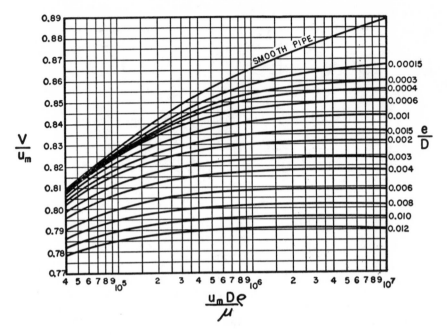

Fig. 5-10. Pipe factors as a function of the Reynolds number and relative roughness (e/D). u_m is the velocity at the center of the pipe.

of Re, above about 100,000 to about 400,000, the velocity distribution for smooth tubes is better represented with $n = \frac{1}{8}$. Higher powers, such as $n = \frac{1}{5}$, are found with rough pipes.

Differentiation of Equation (5-16) shows that the velocity gradient du/dy becomes infinitely large as y approaches zero. This result would indicate that the shear stress becomes infinitely large as y approaches zero. The impossibility can be explained by noting that the seventh-root relation for turbulent velocity distribution ceases to be valid at the pipe wall. The seventh-root relation only holds for the main stream up to a very small distance from the wall. In this short distance from the wall there is a thin layer in which the flow is laminar; sometimes this thin layer is called a *laminar sublayer*. The velocity gradient for laminar flow becomes finite at the wall, as can be seen by examining the parabolic velocity distribution for such flow.

5-8. Friction losses in noncircular pipes

The foregoing discussions in this chapter have been confined to the flow of fluids in pipes of circular cross section. In some applications it is necessary to make flow calculations for noncircular and annular shapes. A parameter called the *profile radius, hydraulic mean depth,* or *hydraulic radius* has been devised to handle these problems. This term is defined as

$$\text{profile radius} = R = \frac{\text{cross-sectional area}}{\text{wetted perimeter}}$$

For a circular pipe flowing full, the profile radius R equals $D/4$. If $4R$ is substituted for D, the Reynolds number becomes

$$\text{Re} = \frac{4VR\rho}{\mu} \tag{5–17}$$

and Equation (5–5) can be written as

$$h = \frac{f}{4} \cdot \frac{l}{R} \cdot \frac{V^2}{2g} \tag{5–18}$$

Use of Equations (5–17) and (5–18) in connection with Fig. 5–5 for ordinary shapes gives results which are in fair agreement with experimental data if the flow is turbulent. Use of the profile radius for laminar flow may give very inaccurate results. Lamb[7] presents some theoretical results for laminar flow in sections other than circular.

5–9. Loss due to a sudden enlargement

The foregoing articles in this chapter have been confined to the flow in straight pipes with constant cross section. In many applications this type of flow is the major feature. In some applications, however, it is necessary to give attention to the flow in channels in which there are changes in sections, bends, fittings, and valves. Figure 5–11 illustrates a channel with an upstream

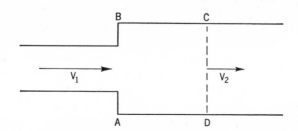

Fig. 5–11. Sudden enlargement.

area A_1 suddenly enlarging to a channel with area A_2. Just before the enlargement the pressure is p_1 and the average velocity is V_1. After the fluid passes the section AB it breaks away from the wall, or "separates," and eddying

[7] *Hydrodynamics*, by H. Lamb, 6th ed. Cambridge University Press, London, 1932, page 586.

flow takes place. At some distance downstream the flow follows the wall; let this section be represented by CD. At section CD the average velocity is V_2 and the pressure is p_2. Neglecting change in potential evergy, the general energy equation gives the lost head h_0 in the form

$$h_0 = \left(\frac{p_1}{\gamma} + \frac{V_1^2}{2g}\right) - \left(\frac{p_2}{\gamma} + \frac{V_2^2}{2g}\right) \tag{5-19}$$

Another equation, the dynamic relation, is helpful in eliminating some of the variables and obtaining a convenient form for the loss. Consider the forces acting on the fluid between sections AB and CD. Neglect shear forces along the wall, and assume that p_1 acts directly inside the enlarged area, that is, over the area AB in the fluid. Then the net force in the direction of flow decelerating the fluid is $(p_2 - p_1)A_2$. Force equals mass times acceleration. The mass rate of flow is $A_2V_2\gamma/g$, where γ is the specific weight of the fluid. For each unit of time, the mass $A_2V_2\gamma/g$ has its velocity reduced from V_1 to V_2. Therefore the dynamic equation shows that

$$(p_2 - p_1)A_2 = A_2V_2\frac{\gamma}{g}(V_1 - V_2) \tag{5-20}$$

Eliminating the pressure difference from Equations (5–19) and (5–20) gives the final form:

$$h_0 = \frac{1}{2g}(V_1 - V_2)^2 \tag{5-21}$$

Experimental results show fair agreement with Equation (5–21); for practical purposes it appears to be a fair estimate of the loss.

5–10. Efficiency of diffusers

The word "nozzle" refers to a channel in which the velocity of the fluid is increased and the pressure is reduced. For an incompressible fluid a nozzle is formed by a converging channel; the fluid is accelerated in the converging passage, and some pressure head is converted into velocity head. In general, this conversion is a stable process, and can be made with low losses. The word "diffuser" refers to a channel in which the velocity of the fluid is decreased and the pressure is increased. For an incompressible fluid a diffuser is formed by a diverging channel. The sudden enlargement in the previous article is a special case of a diffuser.

Flow in a diffuser is one of the most common, most important, and most troublesome problems in practice. Diffusers are found in all sorts of piping systems and fluid machines; examples are water and gas piping systems, the passage through fan and pump blades, wind tunnels for testing models, and some fluid meters.

Figure 5–12 illustrates a diffuser; the vectors shown are velocity vectors.

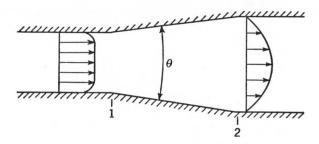

Fig. 5-12. Diverging channel or diffuser.

The flow involves a conversion from velocity head to pressure head. This process is much more difficult and troublesome than the reverse process of flow in a converging nozzle. In the diverging channel some of the kinetic energy is dissipated into unavailable energy because of the viscosity of fluid.

As an example, assume that the fluid enters the diffuser with a uniform velocity distribution across the entire channel. As fluid moves downstream, energy is lost as adjacent layers of fluid near the wall slide by each other. Also, the fluid may not fill the channel completely; the fluid may break away from the walls or "separate," with the result that eddies are formed and energy is dissipated by turbulent mixing. Thus flow in a diffuser may be an unstable, inefficient process, accompanied by large losses and appreciable eddying regions.

In evaluating the efficiency of or loss in a diffuser it is necessary to consider the velocity variation across the channel. Refer to section 1 at the diffuser inlet of Fig. 5-12. Let u represent the variable velocity at any point in the section. Assume that the flow is essentially in the direction of the axis of the diffuser; that is, there is no spiral motion. For the sake of illustration, picture a circular cross section, and let y represent the radial distance from the center of the channel. An infinitesimal area dA is formed by an annular ring of length $2\pi y$ and radial infinitesimal distance dy. The volume of fluid per unit time passing through this annular area is $u2\pi y dy$, or udA. The mass rate of flow through this area is $\rho u dA$ (as slugs per second). The kinetic energy of this fluid per unit time is $\frac{1}{2}\rho u dA u^2$ or $\frac{1}{2}\rho u^3 dA$; this term has the dimensions of power. The total kinetic energy per unit time passing section 1 is the integration over the entire area and is represented by the expression

$$\frac{1}{2}\rho \int_{A_1} u^3 dA$$

where the subscript A_1 indicates integration over section 1. In a similar fashion, the kinetic energy per unit time for section 2 is

$$\frac{1}{2}\rho \int_{A_2} u^3 dA$$

As the fluid passes through the diffuser, the actual reduction in kinetic energy per unit time, or the power available for transformation in the diffuser, is the difference

$$\frac{1}{2}\rho \int_{A_1} u^3 dA - \frac{1}{2}\rho \int_{A_2} u^3 dA \qquad (5\text{-}22)$$

Not all of this available energy is transformed into useful work. It may be helpful to imagine the diffuser as a "pump"; the "pump" raises the pressure of the fluid entering. Equation (5-22) might be regarded as the power supplied or input to the "pump." At section 1 the flow work is pv (as foot-pounds per pound of fluid flowing). The weight rate of flow through the infinitesimal area dA is udA/v (as pounds per second). The flow work per unit time is $pvudA/v$ or $pudA$. Thus the actual power transformation or the actual increase in flow work per unit time is

$$\int_{A_2} pu\, dA - \int_{A_1} pu\, dA \qquad (5\text{-}23)$$

where the subscript in A_2 or A_1 refers to an integration over the entire particular section. Equation (5-23) gives the actual power the "pump" adds to the fluid.

The purpose of the diffuser is to convert kinetic energy into useful pressure energy or flow work. The efficiency E of the diffuser is defined by the relation

$$E = \frac{\displaystyle\int_{A_2} pu\, dA - \int_{A_1} pu\, dA}{\dfrac{1}{2}\rho \displaystyle\int_{A_1} u^3 dA - \dfrac{1}{2}\rho \int_{A_2} u^3 dA} \qquad (5\text{-}24)$$

Let V_1 represent the average velocity at section 1, and V_2 the average velocity at section 2. Let

$$\frac{1}{2}\rho \int_{A_1} u^3 dA = B\left(\frac{1}{2}\rho V_1^3 A_1\right) \qquad \frac{1}{2}\rho \int_{A_2} u^3 dA = C\left(\frac{1}{2}\rho V_2^3 A_2\right)$$

where B and C are numerical, dimensionless constants which are determined from an integration over the section. Then the efficiency takes the final form

$$E = \frac{p_2 - p_1}{\dfrac{1}{2}\rho V_1^2\left[B - C\left(\dfrac{A_1}{A_2}\right)^2\right]} \qquad (5\text{-}25)$$

If the velocity is uniform over the initial section and the final section, then $B = 1$ and $C = 1$.

Unfortunately, the available experimental data on the efficiency of diffusers are very meager. Some data for conical diffusers are shown in Fig. 5-13 as an illustration. The diffuser efficiency is plotted as a function of the total included angle θ (see Fig. 5-12). Each of the curves refers to a particular expansion ratio, or ratio of final area to initial area. Note that the efficiency is high in the region between 0 and 20 degrees. The diffuser efficiency may be

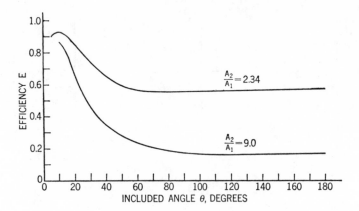

Fig. 5–13. Efficiency E of conical diffusers.

80 to 90 per cent. Beyond 20 degrees the efficiency decreases with increase in included angle. The efficiency may be very low at high angles of divergence. In practice, channel divergence angles of 7 to 8 degrees are found. The value of these choices is illustrated by the high efficiency for these angles.

5–11. Losses in fittings and valves

A lost head h_0 is frequently expressed as equal to $K(V^2/2g)$, where K is a dimensionless loss coefficient and V is some characteristic velocity. Reliable loss coefficients for many cases have not been fully established at the present time. Real difficulties have been encountered in trying to correlate experimental data, particularly measurements with different fluids. The values given in this particular article are to be regarded as approximations, because

TABLE 5–2

VALUES OF K_1 FOR A SUDDEN
CONTRACTION

$\dfrac{D_2}{D_1}$	0.1	0.3	0.5	0.7	0.9
K_1	0.45	0.39	0.33	0.22	0.06

they are based on somewhat limited experimental results. Table 5–2 gives values of K_1 for the sudden contraction shown in Fig. 5–14. Values of losses for pipe entrances are shown in Fig. 5–15.

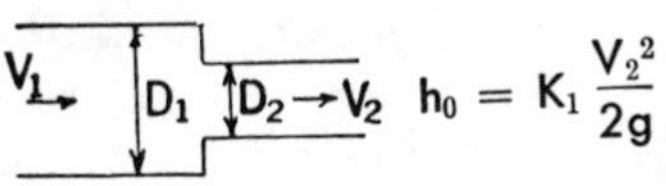

$$h_0 = K_1 \frac{V_2^2}{2g}$$

Fig. 5–14. Sudden contraction.

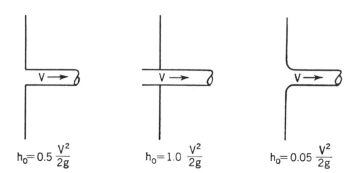

Fig. 5-15. Pipe entrance losses.

The head loss in pipe fittings and valves may also be expressed as equal to $K_2(V^2/2g)$, where V is the velocity in a pipe of the nominal size of the fitting. Some values of K_2 are listed in Table 5-3.

TABLE 5-3

LOSS COEFFICIENTS FOR VALVES AND FITTINGS

Valve or Fitting	K_2	Valve or Fitting	K_2
Globe valve, wide open	10	Return bend	2.2
Angle valve, wide open	5	Standard tee	1.8
Gate valve, wide open	0.19	Standard elbow	0.9
Gate valve, $\frac{1}{4}$ closed	1.15	Medium sweep elbow	0.75
Gate valve, $\frac{1}{2}$ closed	5.6	Long sweep elbow	0.60
Gate valve, $\frac{3}{4}$ closed	24.0	45-degree elbow	0.42

5-12. Direct computation of pipe-line flow

There are various practical pipe flow problems which cannot be solved directly by Fig. 5-5 alone. For example, there are the following cases:

(1) Given ρ, μ, D, and the pressure drop Δp in a length l, what is the velocity V or the discharge Q?

(2) Given ρ, μ, Q, Δp, and l, what is D?

In either case the Reynolds number cannot be calculated; in one case the velocity is not known; in the other case the diameter is not known. At first thought it might appear that a trial-and-error method is necessary. A direct solution, however, can be obtained by dimensional analysis.

For pipes of the same relative roughness, there is a definite functional relation between the variables ρ, μ, D, Δp, l, and V (or Q). Viewing the general problem of pipe flow, the Reynolds number is only one convenient dimensionless ratio; there are other dimensionless ratios which would be helpful in solving cases (1) and (2). Specifically, for case (1) one ratio should not include the velocity; for case (2) one ratio should not include the diameter. Dimensional analysis yields two suitable dimensionless ratios; one will be called an S number and the other a T number for convenience in reference.

$$S = \frac{(p'D^3\rho)^{1/2}}{\mu} \qquad T = \frac{(Q^3 p' \rho^4)^{1/5}}{\mu}$$

where p' is the pressure gradient $\Delta p/l$. S does not include V, and T does not include D. Any set of consistent units could be used for each ratio.

Figure 5–16 illustrates a method of using the S coordinate on a plot of friction factor versus the Reynolds number. The S number equals $\mathrm{Re}(f/2)^{1/2}$ Thus lines of constant S number can be plotted using the relation

$$\mathrm{Re}\left(\frac{f}{2}\right)^{1/2} = S = \text{constant}$$

For case (1) the procedure is to compute the S number and follow the S coordinate to the pipe curve applying. Then f or Re is determined from which V or Q can be calculated.

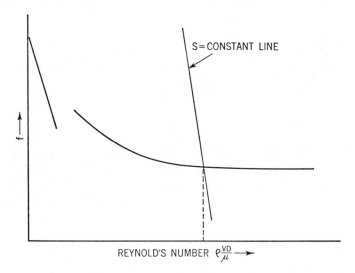

Fig. 5–16. Example of plot of friction versus Reynolds number with the additional coordinate S.

Figure 5–17 illustrates a method of using the T coordinate on a plot of friction factor versus Reynolds number. The T number equals

$$\frac{\mathrm{Re}}{4}(8\pi^3 f)^{1/5}$$

Thus lines of constant T number can be calculated and plotted from the relation

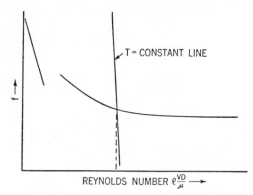

REYNOLDS NUMBER $\varrho\frac{VD}{\mu}$ →

Fig. 5–17. Example of plot of friction factor versus Reynolds number with the additional coordinate T.

$$\frac{\text{Re}}{4}(8\pi^3 f)^{1/5} = T = \text{constant}$$

The procedure is to compute the T number and follow the T coordinate to the pipe curve applying. Thus f or Re is determined, from which D can be calculated. The relative roughness may not be known exactly, and hence some trial may be necessary. The T coordinate, however, reduces the amount of trial and error which would exist if a diagram like Fig. 5–5 were used alone.

The friction coefficient f can be written as

$$f = \frac{2Dp'}{\rho V^2} = \frac{\pi^2}{8} \cdot \frac{D^5 p'}{\rho Q^2}$$

There is another type of problem in which the Reynolds number cannot be calculated in the first step. In some problems such as those involving the determination of the temperature (and consequently viscosity) at which it is economical to pump an oil a long distance, D, p', ρ, V (or Q) are given, and it is required to determine an allowable value of the dynamic viscosity. For such a problem f can be calculated from one of the preceding forms. Following the horizontal value of f to the curve applying gives the Re, from which the dynamic viscosity, and hence the temperature, can be determined.

5–13. Parallel pipes

Various piping arrangements are found in practice. For incompressible fluids in general these arrangements can be analyzed by the basic relations and concepts presented.

Some discussion of one type of piping system will be given; it is indicated in Fig. 5–18. Let Q represent the volume rate of flow in pipe A; this equals the amount flowing in pipe B. Let Q_1 represent the amount flowing in branch

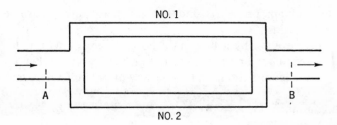

Fig. 5–18. Flow in parallel pipes.

No. 1 and Q_2 the amount flowing in branch No. 2. Let h_1 represent the head loss across pipe No. 1 and h_2 the head loss across pipe No. 2. A simple accounting of mass or weight gives the relation $Q = Q_1 + Q_2$. If pipe junction losses are neglected, then $h_1 = h_2$. If Q and h_1 are given, the problem is to determine Q_1 and Q_2. This problem can be solved directly as case (1) in the previous article.

FORCES ON IMMERSED BODIES

The following discusses the forces acting on a body completely immersed in a relatively large expanse of fluid. Let V be the uniform, undisturbed velocity some distance ahead of the body at rest in Fig. 5–19. The fluid exerts a

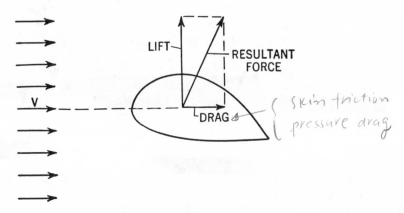

Fig. 5–19. Lift and drag forces.

resultant force on the body; it is common practice to resolve this resultant force into two components. One component, along the line of V, is called the resistance or *drag*. The other component, at right angles to V, is called the *lift*. Drag and lift studies are important in connection with fluid machines (such as fans, pumps, propellers, and turbines), aeronautics, marine engineering, and other fields.

The force exerted by a fluid on a body depends only on the relative velocity between body and fluid, and not on the absolute velocity of either fluid or body, Figure 5–19 indicates one way of obtaining a certain relative motion. The same relative motion could be realized if the body were moving with the constant velocity V through a mass of fluid at rest some distance away from the body.

5–14. Deformation resistance at very low Reynolds numbers

The flow, and consequently the drag, for a completely immersed body is determined by the viscous and inertia forces acting. It was pointed out that the Reynolds number $\rho Vl/\mu$ is proportional to the ratio of inertia forces and viscous forces. l is some characteristic length or dimension of the body.

The Reynolds number can become very small if the fluid is very viscous, if the velocity V is very low, or if the body dimensions are very small. The motion at a very small Reynolds number, in which viscous forces predominate, is sometimes called a *creeping* motion. The corresponding resistance is sometimes called a *deformation* resistance. The resistance is due primarily to the deformation of the fluid particles; this deformation extends to large distances from the body.

5–15. Fluid resistance at high Reynolds numbers

The Reynolds number, on the other hand, is high in many flows of practical importance. A high Reynolds number, however, should not imply that the effect of viscosity on drag can be neglected. A fluid may have a very low viscosity; this small viscosity can have an appreciable effect, directly and indirectly, on the flow. The total resistance of a body at high Reynolds numbers may be divided into separate components, one the *skin-friction* drag, and the other the *pressure* drag.

Figure 5–20 indicates a very thin, smooth plate parallel to the approach-

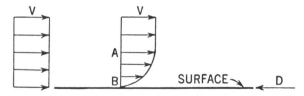

Fig. 5–20. Flow along a flat plate.

ing flow; the velocity distribution is uniform ahead of the leading edge of the plate. If the fluid were nonviscous, the fluid would simply slip over the surface with the velocity V; at all points along the surface the velocity distribution would be uniform and identical with that ahead of the leading edge. No drag would result if the fluid were frictionless. A drag results, however, with real, viscous fluids. A very thin film of fluid adheres to the surface, whereas some distance normal to the plate the fluid has the velocity V. There is a velocity gradient from zero to V in the *boundary layer* (as AB in Fig. 5–20). The boundary layer exists because the fluid is viscous; no boundary layer would exist if the fluid were nonviscous. Tangential or shear stresses are developed in the boundary layer because adjacent layers of fluid move with different velocities. A force D is necessary to hold the plate in the stream to overcome the *skin-friction* drag.

Pressure drag is the contribution due to pressure differences over the surface of the body. Viscosity effects can change the flow pattern, as to cause eddies, and thus influence the pressure distribution. As an example, the flow closes in smoothly behind a circular cylinder at very low velocities. If the flow were symmetrical as shown in Fig. 5–21, and if the fluid were

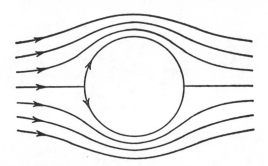

Fig. 5–21. Two-dimensional symmetrical flow around a circular cylinder.

frictionless, then the drag would be zero. The Reynolds number for the circular cylinder will be taken as equal to VD/v, where D is the diameter of the cylinder, V is the approaching velocity, and v is the kinematic viscosity. For real fluids, the flow in Fig. 5–22(a) is obtained if the velocity is increased, or if the Reynolds number is increased beyond 1. The fluid "separates" from the body, and an eddying wake is formed behind the body. Figure 5–22(a), 5–22(b), and 5–22(c), are photographs of the flow of air, a fluid which has a small dynamic viscosity. Figure 5–22(a) shows a difference between the streamlines upstream and downstream from the body. This difference indicates a variation in the pressure distribution around the body. The resultant of the pressure forces, over the body, in the line of undisturbed motion is the *pressure* drag force. The total drag of the body equals the sum of the skin

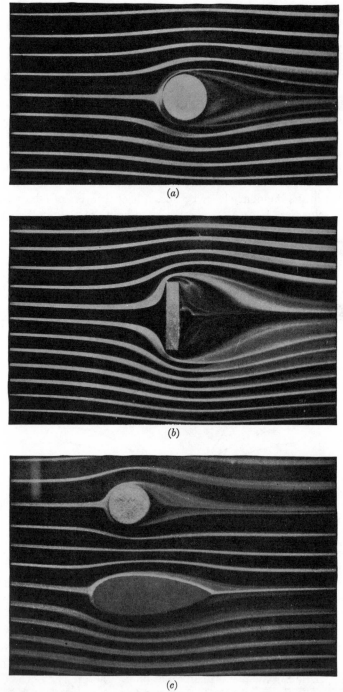

(a)

(b)

(c)

Fig. 5–22. Two-dimensional flow around various objects. Flow is from left to right.

friction drag and the pressure drag. The point of separation, and therefore the drag, may be different for different Reynolds numbers.

Figure 5–22(*b*), for the flow around a flat plate, shows an appreciable wake behind the body; the pressure drag is relatively large. Figure 5–22(*c*) shows the flow around a cylinder and around a streamlined strut in the same stream. The streamlined section has practically no wake; the pressure drag is very small (as compared, for example, with that of a flat plate). The examples in Fig. 5–22 show that the phenomena giving rise to resistance are markedly affected by the *rear* of the body as well as by the front of the body.

5–16. Drag coefficients

It is customary to express the total drag D as

$$D = C_D \rho \frac{V^2}{2} (\text{area}) \tag{5-26}$$

where C_D is the drag coefficient, a dimensionless ratio. The product $\rho V^2/2$ is the dynamic pressure. In some cases the area is taken as the projected area normal to the stream (as for spheres and cylinders). In other cases the area is taken as the largest projected area of the object; this area may be parallel or nearly parallel to V (as for airfoils and flat plates tilted slightly with respect to V).

For a completely immersed body, dynamic similarity and dimensional analysis show that C_D is a function of the Reynolds number $\text{Re} = \rho V l / \mu$. The following articles will give experimental values of C_D against Re for various objects. For a particular problem, C_D can be determined once Re is established, and the drag then computed by Equation (5–26). Drag coefficients are correlated in a manner similar to that used in correlating pipe friction factors.

5–17. Drag of sphere, disk, and bodies of revolution

Figure 5–23 shows a plot of C_D against Reynolds number for spheres and circular disks. For the sphere, the point of separation, and hence the drag coefficient, is different for different Reynolds numbers.

For Re below about 0.4, experiments show that the drag coefficient for a sphere is $C_D = 24/\text{Re}$. This relation, for laminar or viscous flow only, gives a drag value

$$D = C_D \frac{\rho V^2}{2} (\text{area}) = \frac{24}{\rho V d / \mu} \cdot \frac{\rho V^2}{2} \cdot \frac{\pi d^2}{4} \tag{5-27}$$

$$D = 3\mu V \pi d$$

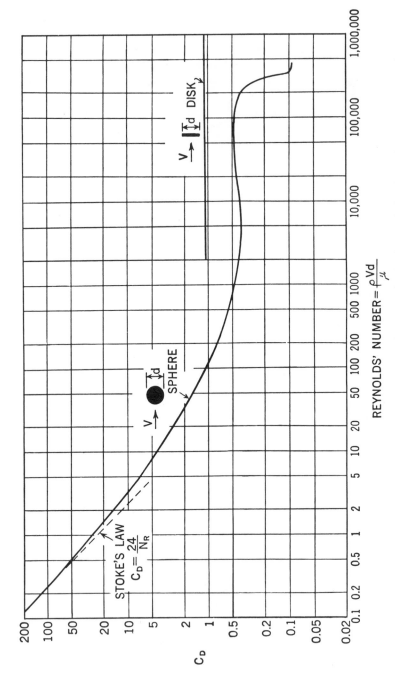

Fig. 5–23. Drag coefficients for sphere and circular disk. Area in the drag relation is projected area normal to the stream. (Data adapted from "Das Widerstandsproblem," by F. Eisner, *Proc. Third Int. Cong. App. Mech.*, Stockholm, 1931.)

Equation (5–27) is commonly known as Stokes' law. Stokes' law is sometimes applied to problems dealing with the rise or fall of solid particles, drops, or bubbles through a fluid. The transportation of flue dust in the atmosphere, the transportation of silt in streams, the action in some oredressing processes, the measurement of dynamic viscosity by the "falling-ball" method, and the operation of air-lift pumps are examples. A solid particle in a fluid will rise or fall depending on whether the particle is lighter or heavier than the fluid. For steady vertical flow, the magnitude of the fluid resistance equals the difference between particle weight and the buoyant force. The limit of application of Stokes' law should be carefully noted.

Figure 5–24 shows values of C_D against Re for two ellipsoids of revolu-

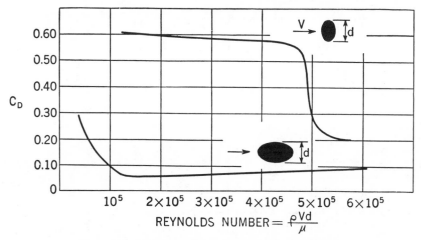

Fig. 5–24. Drag coefficients for ellipsoids of revolution.[8]

tion. The upper curve applies to a body in which the ratio of major axis to minor axis is $\frac{4}{3}$; the axis of revolution is parallel to the flow direction. The lower curve applies to a body in which this ratio is $1:8$; the axis of revolution is perpendicular to the flow direction. In Fig. 5–24 C_D is based on the area of the largest cross section normal to the undisturbed flow.

5–18. Drag of cylinder, flat plate, and streamlined sections

Figure 5–25 gives values of C_D against Re for the two-dimensional flow around a cylinder and a flat plate. Two-dimensional flow can be obtained with an infinite length, or approached with a finite length with a flat plate at

---

[8] Data adapted from *Ergebnisse der aerodynamischen Versuchanstalt zu Göttingen* by L. Prandtl, vol. II. R. Oldenbourg, 1923.

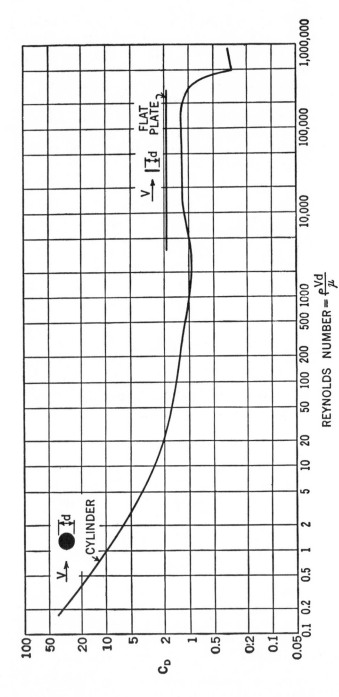

Fig. 5-25. Drag coefficient for two-dimensional flow around a cylinder and a flat plate. Area in the drag relation is projected area normal to the stream. (Data adapted from "Das Widerstandsproblem," by F. Eisner, *Proc. Third Int. Cong. App. Mech.,* Stockholm, 1931.)

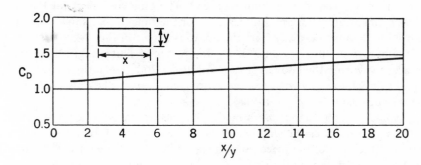

Fig. 5-26. Drag coefficients for a flat plate of finite length normal to flow.

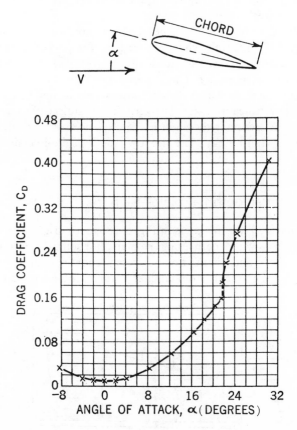

Fig. 5-27. Drag coefficient plotted against angle of attack for N.A.C.A. airfoil 0018. Reynolds number = 2,970,000. (N.A.C.A. *Technical Report*, No.669, 1939, page 523.)

each end to prevent flow around each end. The drag coefficient for the flat plate is practically constant and nearly 2 for the range of Re shown. In this range the fluid breaks away from the body, as shown in Fig. 5–22(*b*).

Figure 5–26 shows values of C_D for plates of finite length; the fluid breaks away from the body as shown in Fig. 5–22(*b*). The ratio x/y is sometimes called the *aspect ratio*. As the ratio x/y increases, the drag coefficient approaches the value given in Fig. 5–25.

Sometimes the statement is made that the resistance of a body varies as the square of the speed, and that the power required to overcome resistance varies as the cube of the speed. Such a statement is accurate only if the drag coefficient is constant in the range of speeds under consideration. Inspection of Figures 5–23, 5–24, and 5–25 shows that in some ranges the drag coefficient is practically constant, particularly at high Reynolds numbers. In other ranges, however, the drag may vary approximately with the first power of the velocity; Stokes' law is an example.

Streamlined sections of the general shape shown in the lower portion of Fig. 5–22(*c*) are desirable when low drag is of importance, as for airfoil, fan, and propeller sections. The gradually tapering tail serves to minimize the pressure or eddying drag. There are applications, however, as in heat-exchange equipment, in which it may be desirable to induce eddying flow in order to promote heat transfer. The variation of drag coefficient with angle of attack for one section is shown in Fig. 5–27. The area in the drag relation is the projected chord area. Note the low values of drag coefficient at low angles of attack.

5–19. Drag of smooth flat plate—skin friction

The boundary-layer thickness in Fig. 5–28 is zero just before the leading edge, and increases with length along the plate. For some distance the boundary-layer flow is laminar, with viscous forces predominating. The boundary-layer flow changes gradually from laminar to turbulent in a transition region.

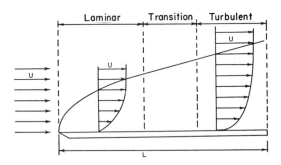

Fig. 5–28. Boundary layers along a flat plate.

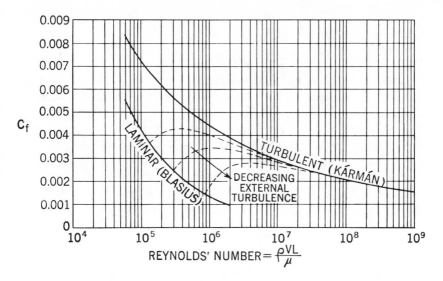

Fig. 5-29. Skin friction for smooth flat plates.[9]

Figure 5-29 gives values of the skin-friction drag coefficient C_f against Reynolds numbers for smooth flat plates. The total skin friction drag D is expressed as

$$D = C_f \frac{\rho V^2}{2} S \tag{5-28}$$

where S is the wetted area. The L in the Reynolds number refers to the total length of the plate in the direction of flow. If the nature of the boundary layer is known, C_f can be found after Re is determined, and the drag calculated by Equation (5-28). The dotted curves in Fig. 5-29 are for different transition conditions.

It is very difficult to formulate perfectly general rules as to the nature of the boundary layer for all cases. The transition from laminar to turbulent flow is not sharp. The boundary layer depends upon the initial turbulence or fluctuations in the stream ahead of the plate, the shape of the leading edge, and the roughness of the plate. Turbulent boundary layers at low Reynolds numbers have been obtained with relatively blunt noses. Information about the particular application involved may be helpful in deciding whether the boundary layer is completely laminar or turbulent. For example, in calculations of skin friction for marine vessel hulls, it is common practice to regard the boundary layer as completely turbulent for usual conditions of operation.

[9] "Turbulence and Skin Friction," by T. v. Kármán. *Journal of the Aeronautical Sciences*, 1, No. 1 (January, 1934).

5–20. Resistance of ships

The total resistance of a partially immersed or floating body, such as a marine vessel, involves wave resistance in addition to the skin friction and the drag due to eddying flow. Both the Froude law and the Reynolds law of similarity are involved. Consider a ship model, smaller in size than the prototype, to be tested in water. The Reynolds law indicates a model speed higher than that of the prototype, whereas the Froude law indicates a model speed lower than that of the prototype. Both laws might be satisfied by the use of different liquids, but to do this is not practical.

The difficulty can be avoided by the following procedure. The model test is made on the basis of the Froude law, and the total drag measured. A calculated skin-friction drag is subtracted from the total model drag, to leave a *residuary resistance*. This residuary resistance is extended to the full-size ship. A calculated skin-friction drag is then added to the residuary resistance to give a value of total drag for the prototype ship. Further details can be found in references on naval architecture.

5–21. Direct determination of the velocity of a body

In the foregoing articles the drag coefficient is plotted as a function of the Reynolds number. This is a common and sound method of organizing data. If ρ, l, V, and μ are given, the Reynolds number can be determined, the drag coefficient found from the proper plot, and the drag force calculated. There are cases, however, in which ρ, l, μ, and the drag are known, and it is desired to calculate the speed of the body. At first thought it might appear that a trial-and-error method of computation is necessary; a possible procedure is to assume first a velocity (or Re), next to determine the corresponding drag coefficient, and then to check, or to repeat until the calculated velocity (or Re) agrees with the assumed value. Such a trial-and-error method, on the other hand, can be eliminated by using dimensional analysis.

The method of approach will be illustrated by specific reference to the velocity of rise or fall of spheres in fluids. Imagine that there is given a certain sphere with a given size and weight immersed in a certain fluid; what is the velocity of rise or fall? For steady vertical motion the resistance to relative motion D equals the difference between the buoyant force and the weight of the sphere. Let A represent the normal projected area of the sphere, and d the diameter of the sphere. Then

$$D = C_D \frac{\rho V^2}{2} A, \qquad \mathrm{Re} = \frac{\rho V d}{\mu}$$

These relations can be arranged in terms of a parameter which does not include velocity:

$$\frac{2D}{\rho A} = C_D V^2 \frac{C_D \mathrm{Re}^2 \mu^2}{\rho^2 d^2}$$

Then

$$\frac{2D\rho d^2}{A\mu^2} = C_D \mathrm{Re}^2 \tag{5-29}$$

In practical problems it is convenient to take the square root of each side of Equation (5–29) to give

$$B = \left(\frac{2D\rho d^2}{\mu^2 A}\right)^{1/2} = \mathrm{Re}\,C_D^{1/2} \tag{5-30}$$

where the left side will be designated as a B number. Note that B is a dimensionless ratio. For the problem under consideration, B is known, but Re and C_D are not known. Re and C_D, however, are coordinates of the common plot. If the plot employs logarithmic scales, then a constant B line is a straight line and requires only two points for plotting, because

$$\log B = \log \mathrm{Re} + \tfrac{1}{2} \log C_D$$

Figure 5–30 shows a plot of C_D versus Re to which B lines have been

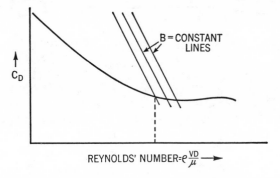

Fig. 5–30. Plot of drag coefficient versus Reynolds number for spheres, with additional B coordinates.

added. In making a velocity calculation, the procedure is to calculate B, follow along a B line to the curve, determine the Reynolds number, and then compute the velocity. The procedure is direct and does not require any trial and error.

The foregoing discussion outlines a procedure which has general application. B lines can be easily constructed on other plots of drag coefficient versus Reynolds number.

5–22. Development of lift

The "curve" of a baseball, a tennis ball, or a golf ball is a familiar example of the development of a dynamic lift; a twirl or spin of the ball is necessary in order to produce the curve.

Figures 5–31 and 5–32(*a*) show that the flow around a stationary cylinder

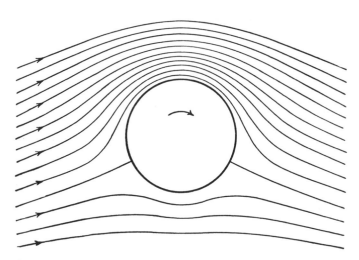

Fig. 5–31. Two-dimensional unsymmetrical flow around a circular cylinder.

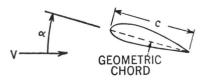

Fig. 5–32. Notation for a lifting vane.

is symmetrical with respect to a line through the cylinder center in the direction of the undisturbed flow. The pressure distribution over the cylinder on one side of this line is the same as that on the other side; therefore there is no lift. If the cylinder in Fig. 5–31 were rotated clockwise in a viscous fluid, for example, the velocities above the cylinder would be higher than those below the cylinder. Application of the dynamic equation shows that the pressures on the lower surface would be greater than those on the upper surface—there would be a force at right angles to the undisturbed flow.

This effect of lift generation due to spinning is commonly called the Magnus effect, after its discoverer Magnus. The Flettner rotor,[10] which employs the Magnus effect, has been applied to the propulsion of marine vessels. Vertical cylinders are extended some distance above the deck. Each

---

[10] *The Story of the Rotor*, by A. Flettner. F. O. Willhoft, New York, 1926.

cylinder is rotated about its axis by a small motor, and an air force is produced for moving the craft. It is to be noted that the unsymmetrical flow around the rotating cylinder is caused by viscosity effects. If the fluid were nonviscous, the rotating cylinder itself would not generate a lift.

The behavior of a frictionless fluid, however, is helpful in explaining certain features of the lift action. The two-dimensional unsymmetrical flow of a frictionless fluid shown in Fig. 5–31 can be obtained by the superposition of two separate flows: (1) a translatory flow such as shown in Fig. 5–31; and (2) a circulatory flow, or a free vortex whose origin is at the center of the cylinder. The resulting streamlines in Fig. 5–31 have been so drawn that the volumetric rate of flow is the same in each space between the streamlines. Thus the spacing between streamlines gives an indication of the magnitude of the velocity. The streamlines are crowded (velocity is relatively high) over the top surface, whereas the streamlines are spread apart (velocity is relatively low) over the lower surface.

Figure 5–32 shows some conventions frequently employed in discussing lifting vanes or sections. The geometric chord of length c is an arbitrary line usually established by the designer in laying out the section. The angle α between the chord and the line of the undisturbed velocity V is called the *angle of attack*.

Figure 5–33 shows a photograph of the two-dimensional flow around a

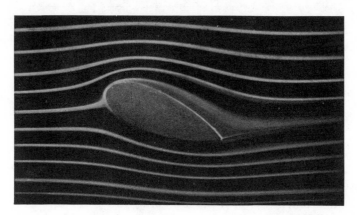

Fig. 5–33. Two-dimensional flow around a lifting vane.

section placed at an angle of attack with the undisturbed flow. The flow is from left to right, and can be regarded as incompressible. The streamlines were equally spaced some distance ahead of the nose. Consider first the flow over the top surface. The velocity is increased in a region near the nose (the streamlines are crowded). The velocity decreases in a region near the trailing edge; the flow resembles that in a diverging channel. Consider next the flow under the lower surface. The velocity is low below the nose. In a region near the trailing edge the velocity increases; the flow resembles that in a converging channel. The velocity and pressure distribution is not uniform over each surface. The net effect, however, is a lift at right angles to the approaching flow.

Let V represent the velocity and p_0 the pressure some distance upstream from a vane. Let p represent the pressure at any point on the surface of the vane. The upstream velocity or dynamic pressure is $\frac{1}{2}\rho V^2$. It is common practice to express pressure distribution measurements in terms of a pressure coefficient P defined by the dimensionless ratio

$$P = \frac{p - p_0}{\frac{1}{2}\rho V^2}$$

The upstream pressure p_0 is constant. The pressure p may be above or below p_0.

Figure 5–34 illustrates a pressure distribution measurement. The pressure

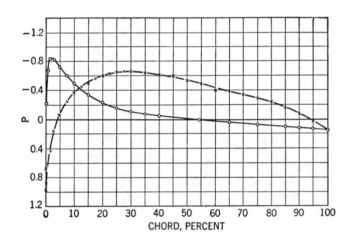

Fig. 5–34. Pressure distribution for N.A.C.A. 4412 airfoil. (From N.A.C.A. *Technical Report* No. 646.)

coefficient is plotted with the negative values up and the positive values down for different positions along the surface. Each position of measurement is expressed in terms of per cent of the chord from the leading edge of the vane. The curve marked with crosses refers to the upper surface, and the curve marked with circles refers to the lower surface (see Fig. 5–33 for notation). For example, if $P = -0.60$ at a particular point, the pressure p at that point is $0.60\rho V^2/2$ below the pressure p_0. Considering the region between 20 and 100 per cent chord, the pressure on the top surface is below the pressure on the lower surface; there is evidence of a lift force. The net area between the two curves is approximately proportional to the lift (the pressures are perpendicular to the surface of the vane and not exactly in the direction of the lift).

5–23. Lift coefficient

It is common to express the lift in the form

$$L = \text{lift} = C_L \frac{\rho V^2}{2} (\text{area}) \tag{5-31}$$

where C_L is a dimensionless lift coefficient. The area in Equation (5-31) is commonly taken as the projected chord area.

Numerous lifting-vane sections, particulary symmetrical sections, show a behavior which in many respects approaches that of a flat plate. The lift coefficient for a flat plate might be regarded as a theoretical maximum or ideal. Kutta[11] gave the following theoretical value for the lift coefficient of a thin flat plate in two-dimensional flow:

$$C_L = 2\pi \sin \alpha \tag{5-32}$$

Experimental results for modern airfoil sections, in a normal range of operation, show lift coefficient values about 90 per cent of the foregoing theoretical value.

Measured values of C_L for different sections can be found in the reference literature. Some typical features are illustrated by the sample shown in Fig. 5–35. The lift coefficient increases, reaches a maximum, and then drops as the angle of attack increases. The *stall* is the condition of a lifting vane (such as a pump vane, propeller element, airfoil, or airplane) at which it is operating at an angle of attack greater than the angle of attack at maximum lift. At the stall the fluid separates from the vane and forms a marked eddying wake. Figure 5–35 shows that the drag coefficient increases considerably as the stall is reached.

BOUNDARY LAYER FLOW

Various cases of flow are possible; some cases are simple, whereas other cases are complex. There is a question as to a suitable general method of approach in studying these various flow phenomena, particularly for the complex patterns. The flow may be around a body or through a channel. For the sake of illustration, picture the flow around a body. It is convenient and helpful to divide the entire flow field into two regions. In one region, close to the surface of the body, the viscosity of the fluid exerts a major effect. In the other region, relatively far from the body surface, the effect of viscosity may become minor; in many cases we can regard the fluid as frictionless, for practical purposes, in this outer region. The following focuses attention on the region close to a surface in which viscosity plays a predominant role.

[11] "General Aerodynamic Theory—Perfect Fluids," by T. v. Kármán, in vol. II, Division E, of *Aerodynamic Theory*, edited by W. F. Durand. Julius Springer, Berlin, 1935.

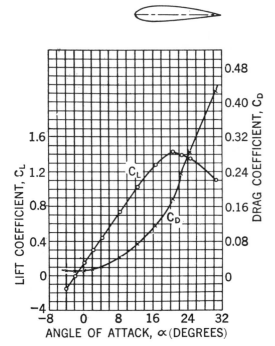

Fig. 5–35. Lift and drag coefficients plotted against angle of attack for N.A.C.A. 2418 airfoil. Reynolds number = 3,060,000. (From N.A.C.A. *Technical Report* No. 669.)

5–24. Boundary layer

As illustrated in Fig. 5–36, picture a fluid flowing along a surface or solid boundary. There is a very thin layer of fluid which adheres to the surface, and thus has a zero velocity with respect to that surface. Some distance away from the surface, in a region normal to the surface (or in the y-direction), the fluid velicoty has a constant value u_1.

There is a region, normal to the wall, in which there is a velocity gradient, or the velocity u varies from zero to u_1. This region in which there is a velocity variation is called a "boundary layer." The concept of the boundary layer was introduced by L. Prandtl in 1904.

Picture the state of flow at a distance y from the wall in the boundary layer where the indicated velocity is u. In "laminar flow," the fluid moves in layers; one layer tends to slide with respect to its neighbor. The shear stress τ is

$$\tau = \mu \frac{du}{dy} \qquad (5\text{--}33)$$

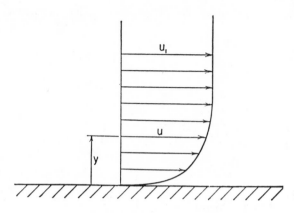

Fig. 5–36. Boundary layer.

where du/dy is the velocity gradient, and μ is the dynamic viscosity. This shear in laminar flow is caused by internal friction; the interaction between adjacent layers is molecular. There is no transfer of fluid masses between adjacent layers.

In laminar flow, the momentum transfer between adjacent layers is molecular. If the flow is turbulent, however, there is molecular friction plus an additional interaction due to momentum transfer of fluid masses between adjacent layers. The scientific term "turbulence" implies "irregular fluctuations," not a regular motion (laminar motion in a straight or curved path is regular motion). For the case shown in Fig. 5–36, if the flow were turbulent, u would be an average velocity at that particular point; velocity fluctuations would be superimposed on this average velocity. There would be a time fluctuation of velocity in the direction of u and a time fluctuation of velocity in a direction perpendicular to u. Fluid masses would be transported between layers. Referring to any point in the turbulent flow, it may be helpful to picture a certain "mixing length" in which the fluid masses move back and forth and are mixed in a direction perpendicular to the mean flow.

The region close to the wall or surface may be classed as a "turbulent boundary layer." For example, for turbulent flow in a pipe we might call the layer in which there is a velocity variation a "turbulent boundary layer." Close to the pipe wall, however, there is a region in which the velocity approaches zero, and the action between layers is solely molecular rather than turbulent. Thus, although the whole layer may be classed as "turbulent," the small laminar region close to the wall is frequently termed a "laminar (or viscous) sublayer." This laminar sublayer may have an appreciable thickness, or it may be so thin that the molecular viscous action is extremely small in comparison with the turbulent mixing in the central core of the pipe.

If the streamlines in the boundary layer do not curve much, the static pressure in the boundary layer is constant along a line perpendicular to the wall.

5–25. Separation

If p is the static pressure at a point and x is a distance measured positive in the direction of flow, then the "pressure gradient" is dp/dx. The passage between streamlines can be such that the velocity of the fluid is increased in the direction of flow, and the pressure is reduced. The pressure gradient is negative. Such streamlines may be formed in the flow around a body, as a turbine blade or some craft. Such streamlines may be formed by a solid wall; in such a case the unit is called a "nozzle." For an incompressible fluid, a nozzle-type channel is formed by converging streamlines. The fluid is accelerated in the converging passage, and some of the pressure head is converted into velocity head. Or we might say that some flow work is converted into kinetic energy. In general, this conversion is a stable process and can be made with low losses.

The passage between streamlines can be such that the velocity of the fluid is decreased in the direction of flow, and the pressure is increased. The pressure gradient is positive; sometimes this gradient is called "adverse." Such streamlines may be formed in the flow around a body, such as a compressor blade or some craft. Such streamlines may be formed by a solid wall; in such a case the unit is called a "diffuser." For an incompressible fluid a diffuser-type channel is formed by diverging streamlines.

Figure 5–37 illustrates the boundary layers at successive points along the wall in a diffuser. At section A there is a certain velocity distribution in the boundary layer. A boundary layer with flow toward the right could be maintained if all the kinetic energy were available for conversion. Some kinetic energy is dissipated into unavailable thermal energy, however, with the result that the velocity in the boundary layer is less than what it would be without friction. At point B the flow velocity is reduced. At C the velocity distribution curve is normal to the wall. C is the "point of separation." Separation of the main flow from a boundary is accompanied by a reversal in direction of the flow very close to the boundary behind the separation point and by the formation of a wake in which the velocity is much reduced. At D the fluid stream has separated from the wall, and an eddy is formed. The formation of eddies represents further dissipation of available into unavailable energy.

Thus flow in a diffuser may be an unstable, inefficient process, accompanied by large losses and appreciable eddying regions. If the streamlines around a body form a diffuser channel there may be a marked separation and an eddying wake. For lifting vanes, separation is associated with the "stall."

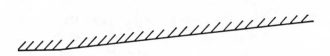

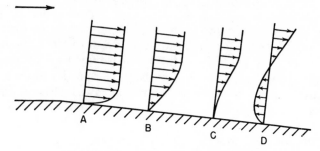

Fig. 5–37. Boundary layer in a diffuser.

The stall represents a drop in lift and an increase in drag as compared to the case in which there is no stall.

5–26. Transition between laminar and turbulent flow

"Transition" refers to the change from laminar to turbulent flow occurring in some limited region of the field of flow. As illustrated in Fig. 5–38, the boundary-layer thickness is zero at the leading edge, and it increases with

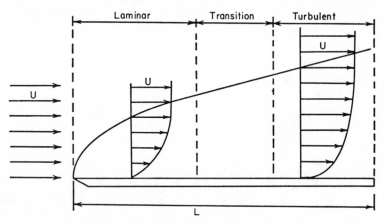

Fig. 5–38. Boundary layers along a flat plate.

length along the plate. For some distance the boundary-layer flow is laminar, with viscous forces predominating. The boundary-layer flow changes gradually from laminar to turbulent in the transition region. In the turbulent boundary layer, beyond the transition, the velocity profile is more blunt than the laminar profile.

5–27. Instability of laminar boundary layer

A boundary layer is formed as any real fluid flows along a wall or surface. In many cases a laminar layer will form on the upstream portion of a body, and a turbulent layer will form on the downstream portion of the body. In some cases a turbulent layer may cover the greater portion of the surface or body. It must be noted, however, that a turbulent layer is always formed from a laminar layer through a breakdown of laminar flow resulting from an unstable condition.

Small disturbances may be applied to the laminar boundary layer by surface roughness, by noise, or by turbulence in the stream surrounding the boundary layer. Oscillations observed at a fixed point suggest the presence of a traveling wave in the boundary layer. The superposition of small disturbance velocities on a uniform velocity in one direction provides to a fixed observer the appearance of a wave motion. The disturbance may be random and result in oscillations of many frequencies. Whatever small disturbances are initially imposed on the laminar layer are selectively damped and amplified until large sinusoidal oscillations are present. These large oscillations, however, do not constitute turbulent flow. The growth or decay of a disturbance depends on whether energy is transferred to the disturbance by absorption of energy from the main boundary-layer flow or extracted from the disturbance by viscous damping action. This flow of energy thus determines the stability of the boundary layer under the action of small disturbances.

Laminar boundary-layer oscillations are the velocity fluctuations that result from a wave traveling downstream through the boundary layer. Such a wave will cause transition when the amplitude is sufficiently large. The regular waves first grow in amplitude, then become very distorted, and then bursts of high frequency take place. There is a feeling that the small-scale eddies of turbulent motion arise from some dynamic instability, like the breaking up of water waves.

5–28. Mechanism of turbulent flow

The exact nature or mechanism of turbulent flow has not been completely established. Some concepts, however, have been developed; these are useful in many cases.

The dynamic viscosity μ is defined for laminar flow as the coefficient in the following expression for shear stress:

$$\tau = \mu \frac{du}{dy}$$

For turbulent flow the total shear stress can be written as

$$\tau = (\mu + \epsilon)\frac{du}{dy} \tag{5-34}$$

where ϵ (Greek letter epsilon) has the dimensions of dynamic viscosity; ϵ has been called the "exchange coefficient," "mechanical viscosity," or "eddy viscosity." ϵ is not a physical characteristic constant of the fluid like μ, but depends upon the Reynolds number and other parameters.

It would be highly desirable to have information about the laws which actually govern the turbulent motion, so that velocity and pressure distributions, energy transfer, and energy loss can be computed without any, or with a minimum of, empirical elements.

In experimental work it is common to measure motion subject to fluctuations in space and time. It is common to resolve the motion into a mean value which is taken constant with respect to time and oscillations about this mean. A "temporal mean" refers to values averaged for a single point of space over a long period of time. As illustrated in Fig. 5–39 picture a flow in the x-direc-

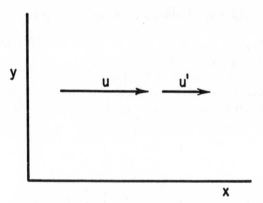

Fig. 5–39. Notation for velocity fluctuations.

tion. Let u represent the "temporal" or "mean" velocity with respect to time and u' the instantaneous fluctuation in the x-direction. Let t represent time, t_0 the start of a time interval, and $t_0 + T$ the end of a very long time interval. At any instant the velocity is $u + u'$. The temporal, or mean, value is

$$u = \frac{1}{T}\int_{t_0}^{t_0+T}(u + u')\,dt = u\frac{1}{T}\int_{t_0}^{t_0+T}u'\,dt \tag{5-35}$$

By definition, the integral of $u'\,dt$ over the long time interval is zero.

Imagine a flow which is two-dimensional in the x-y plane. Assume that the mean motion is parallel to the x-axis, with a variation in the y-direction, as illustrated in Fig. 5–40. Let u represent the temporal mean, u' the velocity fluctuation in the x-direction, and v' the velocity fluctuation in the y-direction. There is a transfer of fluid masses between adjacent layers, such as across the imaginary plane marked A-A in Fig. 5–40.

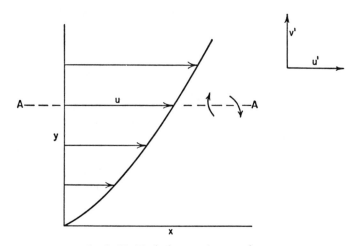

Fig. 5–40. Turbulent exchange of momentum.

Figure 5–41 gives a magnified view of part of Fig. 5–40. In Fig. 5–41 the fluid just below the plane A-A has the mean velocity u. Let us focus attention first on the motion of a single mass. Say that a fluid mass M (large in comparison with a molecule) just below A-A moves with the velocity v' in the positive y-direction through the distance Δy. After traveling this distance, the mass

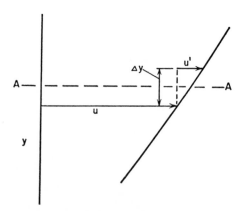

Fig. 5–41. Turbulent exchange of momentum.

M is in a region where the average velocity is $u + (du/dy)\Delta y$; in the new region the velocity increase

$$\left(\frac{du}{dy}\right)\Delta y$$

equals u'. The mass M which had an average velocity u is now in a region where the flow velocity is $u + u'$; the mass M tends to reduce from $u + u'$ to u, to make the reduction u'. The mass rate of flow through a unit area (normal to the y-direction) is v'. The velocity change is u'. The basic relation, force equals mass rate times velocity change, gives the magnitude of the shear stress (force per unit area) as $\rho u'v'$. We call this stress "negative" since it tends to decelerate the motion. Thus the shear stress due to turbulent fluctuation is $\rho u'v'$. This notation follows that used in defining viscosity, as in Equation (5–36).

The combination of $+v'$ and $-u'$ represents a positive shear force. The combination of $+u'$ and $-v'$ represents a positive shear force. Because of the foregoing relation between u' and v', the momentum transfer per unit time in the interval from time $t = t_0$ to $t = t_0 + T$ becomes

$$\frac{1}{T}\int_{t_0}^{t_0+T} (-\rho u'v')\, dt = -\overline{\rho u'v'}$$

where the bar over the product of the two components represents a time mean value of the product. This quantity is a tangential force per unit area. The apparent shear stress or the "turbulent friction" is thus

$$\tau = -\overline{\rho u'v'} \qquad (5\text{--}36)$$

For laminar flow there is only molecular momentum transfer. For turbulent flow there is also a momentum transfer of fluid masses. Thus the total amount of internal friction due to viscosity and turbulence can be written as

$$\tau = \mu\frac{du}{dy} - \overline{\rho u'v'} \qquad (5\text{--}37)$$

where the first term on the right side indicates laminar friction, and the second term indicates turbulent friction.

Equation (5–37) shows several features. At a solid wall the turbulent friction is zero because the velocity-fluctuation components vanish. In the immediate vicinity of the wall there is a laminar sublayer. For highly turbulent flow, the laminar term may be negligibly small in comparison with the turbulent term. If the fluctuations u' and v' were perfectly random, or completely independent, then the mean value $\overline{u'v'}$ would be zero. Thus the turbulent friction is different from zero only when there is a statistical correlation between u' and v'.

For molecular motion there is defined a term "mean free path" of the molecule, which means the average distance a molecule travels between collisions. "Mixing length" is analogous to "mean free path." The concept states that in the turbulent exchange the fluid masses displaced perpendicular to the mean flow carry momentum over a certain length perpendicular to the mean flow. The mixing length is defined as the average distance perpendicular to the mean flow covered by the mixing particle.

The turbulent shear stress τ is taken to be a function of density ρ, velocity gradient du/dy, and the mixing length L. The ratio τ/ρ has the dimension of a velocity squared. Prandtl proposed the following dimensionally correct relation for turbulent shear stress:

$$\tau = \rho L^2 \left| \frac{du}{dy} \right| \frac{du}{dy} \qquad (5\text{-}38)$$

where $|du/dy|$ is the absolute magnitude of the velocity gradient. The factor L or L^2 may be regarded as a correlation factor; all the unknown factors are concentrated into the mixing length L. The mixing length is determined indirectly from observations of the distribution of mean velocity. The concept of mixing length is useful in trying to gain an understanding of the fundamental laws of turbulent flow.

Kármán introduced (in 1930) a further concept by means of the principle of similarity. He felt that Equation (5-38) had a real physical sense only if the correlation is the same in all points of the fluid considered. Alternately, the mechanism of the turbulent friction can be fully characterized by a single length L only in the case where the flow pattern in the neighborhood of every point is similar, differing only in scale both as far as time and length are concerned. This assumption appears to be justified in all cases in which the characteristic length of the turbulent exchange is small in comparison with the dimensions of the cross section.

Kármán presented the following expression for L in terms of the mean flow

$$L = k \left| \frac{du/dy}{d^2u/dy^2} \right| \qquad (5\text{-}39)$$

where k is a universal numerical constant. The terms du/dy and d^2u/dy^2 are characteristic features of the velocity distribution and comprise a major factor in the mechanism of turbulence.

Various analytical extensions of these theoretical studies have been made. The results have been of great value in organizing data for practical engineering purposes. One of these extensions is presented in Section 5-32.

5-29. Types of flow

The "boundary layer" is found at a surface or wall; we might say it is "attached" to the wall. There is another type of layer called a "free layer," which is similar to a boundary layer. A free layer occurs in the flow, away from a surface or wall, and is recognized by the presence of velocity gradients much greater than elsewhere in the flow, A free layer is usually the continuation of a boundary layer attached to a surface. A free layer may be pictured as a boundary layer which has left or has become detached from a surface.

There are certain characteristic flow phenomena which are found in many fluid machines, the flow around different craft and bodies, and the flow in different channels. The flow pattern may be simple or complex. A study of these characteristic phenomena is very useful in gaining a general understanding of flow. We list these characteristic phenomena as: (1) formation of boundary layers and free layers; (2) laminar flow; (3) turbulent flow; (4) transition from laminar to turbulent flow; and (5) separation.

Various combinations of these characterstic flow phenomena or types of flow may be found in practice. As an illustration, we will study the two-dimensional flow of an incompressible fluid around a circular cylinder. Let D represent the cylinder diameter, U the approach velocity, ρ density, ν kinematic viscosity, and R the drag or resistance per unit length of cylinder. The drag coefficient C_D is defined as the dimensionless ratio

$$C_D = \frac{R}{\frac{1}{2}\rho U^2 D}$$

The drag coefficient is a function of the Reynolds number $\text{Re} = UD/\nu$.

By "inertia force" is meant the resistance of an inert mass to acceleration or velocity change. Let t represent time. Since force equals mass times acceleration, the magnitude of the inertia force is proportional to

$$\frac{D^2 \rho U}{t} = \frac{D^3 \rho U}{D/U} = D^2 \rho U^2$$

The viscous force is proportional to τD^2 where τ is the viscous shear stress. Since $\tau = \mu\, du/dy$, the viscous force is proportional to ρUD. Then the dimensionless ratio of the inertia force to the viscous force is proportional to

$$\frac{D^2 \rho U^2}{\mu UD} = \frac{UD}{\nu} = \text{Re} = \text{Reynolds number}$$

Figure 5–42 shows a plot of drag coefficient over a range of Reynolds numbers for a circular cylinder. Six types of flow, designated by A, B, C, D, E, and F, illustrate different combinations of characterstic flow phenomena.

Type A is at a Reynolds number less than 1. The inertia forces are small in comparison with the viscous forces. As illustrated in Fig. 5–43, the flow closes in smoothly behind the cylinder.

As the Reynolds number increases, the drag coefficient decreases. The inertia forces become larger, and the viscous forces become smaller; the character of the flow changes. Type B is at a Reynolds number of 20. This flow

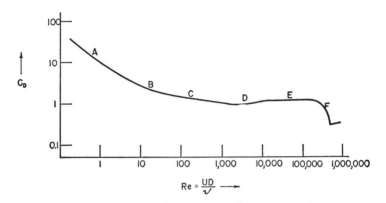

Fig. 5–42. Drag coefficient C_D for two-dimensional flow around a circular cylinder.

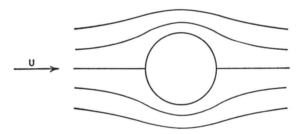

Fig. 5–43. Type A flow.

is laminar along the upstream surface of the cylinder. S in Fig. 5–44 is the separation point of the laminar boundary layer. On the downstream side of the cylinder there is a region in which there are two stationary eddies.

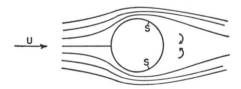

Fig. 5–44. Type B flow. S indicates separation point of laminar boundary layer.

As the Reynolds number increases above 20, the eddies become unstable, first vibrating irregularly and then breaking away alternately from the two sides. Figure 5–45 illustrates type C flow at an instant for a Reynolds number of about 170. The eddies break off alternately on either side in a periodic fashion; this is the so-called "Kármán vortex trail." Behind the

Fig. 5–45. Type *C* flow. Kármán vortex trail.

cylinder is a staggered stable arrangement or trail of vortices. The alternate shedding produces a periodic force acting on the cylinder normal to the undisturbed flow. In type *C* flow the boundary layer on the upstream side of the cylinder is laminar. This boundary layer separates behind the cylinder. In the wake behind the cylinder there are laminar free layers. Thus, the attached boundary layer and free layers are laminar.

At Reynolds numbers between 5000 and 15,000, the flow separates from the cylinder to form a symmetrical wake with free layers which are turbulent. The action for type *D* flow is illustrated in Fig. 5–46 for a Reynolds number of 5000. The cross-hatched areas indicate regions of turbulent free layers.

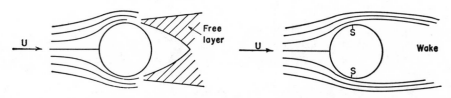

Fig. 5–46. Type *D* flow.   **Fig. 5–47.** Type *E* flow. *S* represents point
of separation of laminar boundary layer.

In the range of Reynolds numbers from 50,000 to 200,000 the drag coefficient is constant. In this range the point of transition in the free layer reaches the separation point. This type *E* flow is illustrated in Fig. 5–47. Point *S* is the separation point of the laminar boundary layer; transition occurs simultaneously with separation.

In a region with a Reynolds number above 200,000, transition begins in the laminar boundary layer ahead of the separation point. This type *F* flow is illustrated in Fig. 5–48. In the turbulent boundary layer there is an intense mixing of fluid portions and a more blunt velocity profile; thus separation is delayed. Behind the cylin-

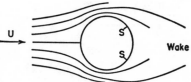

Fig. 5–48. Type *F* flow. *S* indicates point of separation of turbulent boundary layer.

der the turbulent boundary layer separates to form a turbulent wake.

The foregoing discussion has been simplified somewhat in not taking into account possible variations of turbulence in the approaching stream. Such variations may shift the distinguishing Reynolds number slightly, but the types of flow are still the same.

5-30. Momentum study of boundary layer

Different methods of approach could be followed in analyzing the boundary layer. Some approaches involve rather complicated mathematics. One approach, which is very useful in many applications, is to make a momentum study; this approach is outlined in the following paragraphs.

Figure 5-49 illustrates two-dimensional flow along a surface; the flow is identical in parallel planes. Upstream from the leading edge of the plate the uniform velocity is U. The indicated force F parallel to U is called a "skin-friction" force. If the fluid were frictionless, then F would be zero. With real fluids, however, F is definitely not zero.

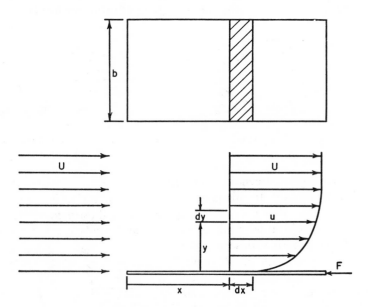

Fig. 5-49. Flow along a surface.

Let b represent the width of the plate perpendicular to the plane of flow. Picture the flow through the area $b \, dy$. The mass rate of flow through this area is $pubdy$. Ahead of the plate the fluid velocity is U; at the section a distance x from the leading edge the fluid velocity is u. The case will be taken in which there is no change in static pressure in the direction of flow. If p is static pressure, then the pressure gradient dp/dx is zero for this case. Thus the force

dF acting on the fluid equals the mass rate of flow times the velocity change, or

$$dF = \rho u b \, dy \, (U - u) \tag{5-40}$$

At section x there is a velocity variation normal to the plate. At $x = 0$, the velocity distribution is uniform. The total force of the plate on the fluid is then

$$F = b \int_0^\infty \rho u (U - u) \, dy \tag{5-41}$$

Theoretically speaking, the integral should be taken from 0 to infinity. For practical purposes, however, it is accurate enough to integrate from zero to some finite distance δ; δ is so selected that the velocity defect $(U - u)$ is very small for values of y greater than δ. δ is called the "disturbance thickness" of the boundary layer. It is the region in which the major part of the velocity variation occurs. Frequently, in organizing data, δ is taken as that value of y at which $u = 0.99U$. Then Equation (5-41) can be written as

$$F = b \int_0^\delta \rho u (U - u) \, dy \tag{5-42}$$

If we know the velocity distribution u as a function of y at section x, this functional relation can be put in Equation (5-42), the integration carried out, and the friction force F calculated. Thus, there are two ways of determining F: by a direct force measurement and by a measurement of the velocity distribution in the boundary layer.

The friction force dF on the strip $b\,dx$ can be expressed in terms of a local friction force per unit area or a wall shear stress τ_0 in the form $b\tau_0 \, dx$. Thus

$$\tau_0 = \frac{1}{b} \frac{dF}{dx} = \frac{d}{dx} \left[\int_0^\delta \rho u (U - u) \, dy \right] \tag{5-43}$$

It is convenient to organize the foregoing and other relations in terms of dimensionless ratios. Let the dimensionless ratio y/δ be denoted by n. Then u/U is some function of n, or

$$\frac{u}{U} = f\left(\frac{y}{\delta}\right) = f(n) \tag{5-44}$$

This function $f(n)$ equals zero when $y = n = 0$ and is very close to one when $y = \delta$ or $n = 1$. Then the momentum integral can be replaced by

$$F = \rho b U^2 \delta \int_0^1 f(n)[1 - f(n)] \, dn$$
$$F = \rho b U^2 \, \delta a \tag{5-45}$$

where the dimensionless ratio A is the value of the integral

$$A = \int_0^1 f(n)[1 - f(n)] \, dn \tag{5-46}$$

The wall shear stress can be arranged in the form

$$\tau_0 = \frac{dF}{b\,dx} = \rho U^2 A \frac{d\delta}{dx} \tag{5-47}$$

Another expression for shear stress can be developed which introduces viscosity. The wall shear stress equals the dynamic viscosity times the velocity gradient at the wall $(du/dy)_0$, where the subscript 0 means the slope at $y = 0$;

$$\tau_0 = \mu \left(\frac{du}{dy}\right)_0 = \mu \left(\frac{U}{\delta}\right)\left(\frac{df}{dn}\right)_0 \tag{5-48}$$

Let us represent $(df/dn)_0$ by the symbol B. Then, combining Equations (5-47) and (5-48), we get

$$\rho U^2 A \frac{d\delta}{dx} = \mu \frac{UB}{\delta}$$

$$\delta\,d\delta = \frac{\mu B}{\rho U A} dx \tag{5-49}$$

Integration of Equation (5-49) gives

$$\delta = \sqrt{\frac{2\mu Bx}{\rho U A}} \tag{5-50}$$

since $\delta = 0$ at $x = 0$. Putting this value of δ in Equation (5-42) gives

$$\tau_0 = \sqrt{\frac{AB\rho\mu U^3}{2x}} \tag{5-51}$$

We can put F in the form

$$F = b\int_0^x \tau_0\,dx = b\sqrt{2AB\rho\mu U^3 x} \tag{5-52}$$

We define a dimensionless skin-friction drag coefficient C_f as

$$C_f = \frac{F}{\frac{1}{2}\rho U^2 bx} \tag{5-53}$$

This coefficient can be arranged in an alternate form

$$C_f = \sqrt{\frac{8AB\mu}{\rho Ux}} \tag{5-54}$$

where $\rho Ux/\mu$ is a Reynolds number for the plate of length x.

As an illustration of the use of the foregoing relations, let us assume that the velocity distribution is given by the simple relation

$$\frac{u}{U} = \frac{3}{2}\left(\frac{y}{\delta}\right) - \frac{1}{2}\left(\frac{y}{\delta}\right)^2 \tag{5-55}$$

We then get values for A and B as

$$A = \frac{39}{280} \qquad B = \frac{3}{2}$$

The following relations result for this case:

$$\delta = 4.64\sqrt{\frac{\mu x}{\rho U}} \tag{5-56}$$

$$\tau_0 = 0.323\sqrt{\frac{\rho\mu U^3}{x}} \tag{5-57}$$

$$C_f = \frac{1.292}{\sqrt{\rho U x / \mu}} \tag{5-58}$$

The boundary-layer thickness varies as the square root of x, the shear stress varies inversely as the square root of x, and the skin-friction coefficient varies inversely as the square root of the Reynolds number for the plate.

5-31. Some velocity distributions for two-dimensional flow

The foregoing article points up the value of information as to velocity distribution. If we know the velocity distribution in the boundary layer, then we can calculate the resistance or drag, shear stress, and boundary layer thickness. Other problems, as certain problems in heat transfer, can be solved if the velocity distribution is known. Thus considerable attention has been given to this problem of velocity profiles.

Considering two-dimensional flow, one of the simplest velocity distributions is a linear relation, that is,

$$f(n) = n \qquad \frac{u}{U} = \frac{y}{\delta} \tag{5-59}$$

This linear relation satisfies the conditions that $u = 0$ when $y = 0$, and $u = U$ when $y = \delta$.

Another possible velocity distribution is the cubic relation

$$f(n) = \frac{3}{2}n - \frac{n^3}{2}$$

$$\frac{u}{U} = \frac{3}{2}\left(\frac{y}{\delta}\right) - \frac{1}{2}\left(\frac{y}{\delta}\right)^3 \tag{5-60}$$

This relation satisfies the various conditions that $u = 0$ at $y = 0$, $u = U$ at $y = \delta$, $du/dy = 0$ at $y = \delta$, and $d^2u/dy^2 = 0$ at $y = 0$.

Various other relations are possible, such as sine and other polynominals. In some cases these simple relations are convenient and accurate enough for practical purposes. The accuracy of a particular assumed velocity distribution can be checked by comparing theoretical results with experimental results.

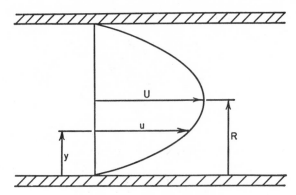

Fig. 5-50. Notation for pipe flow.

5-32. Logarithmic velocity distribution

Kármán has applied his turbulence concepts (Article 5–28) to various cases. An introduction will be given to his extension to the case of incompressible turbulent flow in pipes having smooth walls.

As illustrated in Fig. 5–50, let U represent the velocity in the center of the pipe, y the variable distance from the pipe wall, R the pipe radius, u the variable velocity at y. Let τ_0 represent the wall shear stress and μ the dynamic viscosity.

We divide the flow into two regions: (1) Outside of the laminar sublayer, we assume turbulent flow over the cross section and neglect viscosity effects; and (2) in the laminar sublayer the shear stress equals the velocity gradient times the dynamic viscosity.

The difference $(U - u)$ will be called a "velocity defect." The ratio τ_0/ρ has the dimensions of a velocity squared.

Referring to region 1, experimental data show that the dimensionless ratio $(U - u)/(\tau_0/\rho)^{1/2}$ is a general function of the dimensionless ratio y/R. Thus

$$\frac{U-u}{\sqrt{\tau_0/\rho}} = m\left(\frac{y}{R}\right) \qquad (5\text{–}61)$$

where $m(y/R)$ represents a universal function of the ratio y/R.

For smooth pipes, in region 2, in the laminar sublayer the velocity u is a function of τ_0, y, ρ, and μ. From dimensional analysis, we see that the relation can be expressed in the form

$$\frac{u}{\sqrt{\tau_0/\rho}} = g\left[\left(\frac{\tau_0}{\rho}\right)^{1/2}\frac{y\rho}{\mu}\right] \qquad (5\text{–}62)$$

where g means "some function" of the terms in the bracket. The terms in the brackets form a Reynolds number, y is a length, μ/ρ is a kinematic viscosity, and $(\tau_0/\rho)^{1/2}$ has the dimensions of a velocity.

The next step is to combine Equations (5-61) and (5-62) for the entire flow in the pipe. Kármán put the mixing length L as equal to ky, where k is the universal constant. Thus

$$L = k \left| \frac{du/dy}{d^2u/dy^2} \right| = ky \qquad (5\text{-}63)$$

$$\tau_0 = k^2 \rho \left(\frac{du}{dy}\right)^2 y^2$$

Integration gives

$$\frac{u}{(\tau_0/\rho)^{1/2}} = \text{constant} + \frac{1}{k} \log_e y \qquad (5\text{-}64)$$

for the form of the velocity-distribution curve. The foregoing relation is based upon the turbulent exchange action. This equation can be adapted to satisfy Equation (5-64) also if we introduce τ_0, ρ, and μ in the form

$$\frac{u}{(\tau_0/\rho)^{1/2}} = \text{constant} + \frac{1}{k} \log_e \left[\left(\frac{\tau_0}{\rho}\right)^{1/2} \frac{y\rho}{\mu} \right] \qquad (5\text{-}65)$$

A comparison of velocity distribution measurements shows that the following equation fits the data very well:

$$\frac{u}{(\tau_0/\rho)^{1/2}} = 5.5 + 2.5 \log_e \left[\left(\frac{\tau_0}{\rho}\right)^{1/2} \frac{y\rho}{\mu} \right] \qquad (5\text{-}66)$$

The factor 2.5 corresponds to a numerical value of the universal constant $k = 0.40$.

In some problems it is convenient to deal with a "wall shear stress." Picture a horizontal pipe with p_1, the static pressure at inlet, and p_2, the static pressure at outlet. The force pushing on the fluid in the direction of motion is $(p_1 - p_2)\pi D^2/4$. In terms of head loss, the pressure difference is

$$p_1 - p_2 = \gamma h = \gamma f \left(\frac{l}{D}\right) \frac{V^2}{2g} \qquad (5\text{-}67)$$

We can picture a shear stress τ_0 at the inner surface of the pipe; this shear stress acts on a cylindrical surface of fluid to oppose the motion. For equilibrium, in steady constant-velocity flow, the shear force just balances the pressure force. Thus

$$(p_1 - p_2)\frac{\pi D^2}{4} = \gamma f \left(\frac{l}{D}\right)\frac{V^2}{2g}\left(\frac{\pi D^2}{4}\right) = \tau_0 \pi D l$$

$$\tau_0 = f \frac{\rho V^2}{8} \qquad (5\text{-}68)$$

It is common to express the head loss h for pipe flow in the form

$$h = f \left(\frac{l}{D}\right)\frac{V^2}{2g} \qquad (5\text{-}69)$$

where V is the average velocity, and the friction factor f is a function of the Reynolds number $\text{Re} = \rho V D/\mu$.

Equation (5–66) can be employed to obtain an expression for the volume rate of flow through the pipe and thus an expression for V. This expression for V and Equation (5–66) can be used to arrange Equation (5–66) in terms of the Reynolds number and friction factor. The result is

$$\frac{1}{\sqrt{f}} = -0.8 + 2\log_{10}\text{Re}\sqrt{f} \qquad (5\text{–}70)$$

The foregoing analysis is not a pure rational theory in a strict sense because two constants need to be determined by experiment. The analysis however, does have a rational background and provides a framework which minimizes empirical relations.

PROBLEMS

5-1. 740 gallons of gasoline (specific gravity $= 0.75$) per minute flow through a horizontal pipe 8 inches in diameter and 8 miles long. The kinematic viscosity is 0.01 (centimeter)^2 per second, and the relative roughness of the pipe is 0.0004. What is the pressure drop?

5-2. Air flows through a horizontal galvanized iron pipe 3 inches in diameter. At one section the pressure is 150 pounds per square inch gage, the temperature is 90 degrees Fahrenheit, and the average velocity is 45 feet per second. Assume that the density change is negligible and that the dynamic viscosity is a function of temperature only. What is the pressure drop in 200 feet of pipe?

5-3. Carbon dioxide flows through a horizontal commercial steel pipe 4 inches in diameter. At inlet the pressure is 120 pounds per square inch gage, the temperature is 100 degrees Fahrenheit, and the average velocity is 40 feet per second. Assume that the density change is negligible and that the dynamic viscosity is a function of temperature only. What is the pressure drop in 150 feet of pipe?

5-4. 1500 gallons of benzine at 50 degrees Fahrenheit (specific gravity $= 0.90$) per hour are to be pumped through a $1\frac{1}{4}$-inch standard galvanized iron pipe (1.38 inches actual inside diameter). The total length of horizontal pipe is 1000 feet. The overall efficiency of the pump is 60 per cent. Find the horsepower input to the pump.

5-5. An oil weighs 58 pounds per cubic foot and has a kinematic viscosity of 0.09×10^{-3} foot squared per second. This oil flows through a horizontal 6-inch diameter smooth pipe with an average velocity of 6 feet per second. What is the pressure drop in 450 feet?

5-6. Castor oil at 59 degrees Fahrenheit flows through a horizontal wrought iron pipe 1 inch in diameter and 5 feet long at a rate of 0.090 pound per second. What is the pressure drop?

5-7. Water at 59 degrees Fahrenheit flows through a horizontal 6-inch diameter asphalted cast iron pipe with an average velocity of 4.0 feet per second. What is the pressure drop in a length of 120 feet?

5-8. Water at 59 degrees Fahrenheit flows through a vertical 6-inch diameter asphalted cast iron pipe with an average velocity of 4.0 feet per second. The flow is up. What is the pressure drop in a length of 120 feet?

5-9. Water at 59 degrees Fahrenheit flows through a 6-inch diameter smooth pipe 600 feet long. The velocity at the center of the pipe is 5.9 feet per second. The pipe makes an angle of 45 degrees with the horizontal. The friction loss is 105 feet of fluid, and the flow is up. What power must be added to the water to move it?

5-10. Calculate the power which must be added to the fluid to pump 22 cubic feet per minute of ethylene bromide at 20 degrees centigrade through a horizontal smooth pipe 900 feet long and 2 inches in diameter. The dynamic viscosity is 3.60×10^{-5} slug per foot-second; the specific gravity is 2.17.

5-11. 1500 gallons of water per hour flow through a smooth rectangular pipe whose cross section is 1.0 inch by 1.5 inches. The total length of pipe is 1000 feet, the pipe makes an angle of 60 degrees with the horizontal, and the flow is up. What is the pressure drop?

5-12. Castor oil at 59 degrees Fahrenheit flows through a vertical galvanized iron pipe 2 inches in diameter and 24 feet long at a rate of 0.20 pound per second. The flow is down. At inlet the pressure is 18 pounds per square inch absolute. What is the pressure at outlet?

5-13. Air flows through a duct of rectangular cross section 1 foot by 3 feet at 150 cubic feet per second. The duct is made of galvanized iron. As an alternate, consider a circular cross section using the same amount of sheet metal. Which cross section would give the lower pressure drop per unit length?

5-14. For laminar flow in a pipe, at what radial distance from the pipe center is the actual velocity equal to the average velocity? Express this radial distance as a fraction of the pipe diameter.

5-15. Consider turbulent flow in the apparatus shown in Fig. 5-1. Would a dye filament at a certain instant represent a streamline?

5-16. Consider laminar flow in a tube. At what radial position is the maximum energy dissipation or energy degradation?

5-17. When the faucet on a drinking fountain is opened, frequently the jet initially rises higher than its final level. Explain this action.

5-18. Picture laminar flow in a horizontal pipe. Take the rate of energy loss or dissipation per unit volume at a point as the product of shear stress and the velocity gradient. Starting with this relation, integrate over the pipe section, and show that the total power loss in a length of pipe equals the product of volume rate through the pipe and the pressure drop in the length of pipe.

5-19. Water flows through a 2-inch diameter cast iron pipe 500 feet long with a velocity of 10 feet per second. The pipe outlet is 100 feet above the pipe inlet. What is the pressure difference between inlet and outlet?

5-20. Water flows through a 6-inch diameter horizontal galvanized pipe 1000 feet long with an average velocity of 8 feet per second. What is the pressure drop?

5-21. Water flows through a 12-inch diameter galvanized iron pipe 800 feet long with a velocity of 9 feet per second. The pipe outlet is 50 feet above the pipe inlet. What is the pressure difference between inlet and outlet?

5-22. Water flows through a 10-inch diameter galvanized iron pipe 800 feet long with a velocity of 11 feet per second. The outlet is 40 feet above the inlet. What power is lost due to friction?

5-23. Water flows upward in an inclined pipe whose diameter is 10 inches with a flow rate of 0.0372 cubic foot per second. The sine of the angle of inclination is 0.2 and the static pressure gages mounted on the pipe centerline with 10 feet between their respective taps gives equivalent pressure readings of 4 and 8 inches of mercury. What is the friction factor?

5-24. Water flows through a smooth square pipe with an average velocity of 10 feet per second. The pipe cross section is 6 inches by 6 inches. The pipe makes an angle of 50 degrees with the horizontal. The flow is up. What is the pressure drop in a length of 800 feet?

5-25. Water flows through a 12-inch diameter galvanized iron pipe 500 feet long with an average velocity of 11 feet per second. The pipe outlet is 40 feet below the pipe inlet. What is the pressure difference between inlet and outlet?

5-26. Castor oil at 59 degrees Fahrenheit flows through a vertical wrought iron pipe 1 inch in diameter and 5 feet long at a rate of 0.090 pound per second. The flow is up. What is the pressure drop?

5-27. Walker, Lewis, and McAdams, in their book, *Principles of Chemical Engineering*, use a Reynolds number involving the following terms: diameter in inches, velocity in feet per second, specific gravity, and viscosity in centipoises. Determine the factor by which this number must be multiplied in order to give the Reynolds number in any consistent units.

5-28. Water flows through an 8-inch diameter pipe with an average velocity of 12 feet per second. There is a sudden enlargement to a 16-inch diameter pipe. What is the power loss due to the sudden enlargement?

5-29. Apply the general expression for diffuser efficiency to a sudden enlargement and get E in terms of only A_1 and A_2.

5-30. Standard air enters a conical diffuser which expands from an initial section of 1-foot diameter to a final section of 2-foot diameter. Assume incompressible flow. At inlet the velocity is 200 feet per second, and the pressure is atmospheric. The pressure rise is equivalent to 2 inches of water. At exit the velocity is given by the relation $u = 2V_2(1 - y^2/R^2)$, where V_2 is the average velocity, y is the radial distance to any point, and R is the pipe radius. What is the diffuser efficiency?

5-31. 117.9 cubic feet of water per minute flow through a smooth horizontal drawn tube. The pressure drop in 1000 feet is 18.2 pounds per square inch. What is the pipe diameter?

5-32. Air flows through a clean galvanized horizontal pipe 12 inches in diameter. The pressure drop in 100 feet is 0.040 pound per square inch. For standard conditions what is the rate of discharge?

5-33. A truck having a projected area of 68 square feet, traveling at 50 miles per hour, has a total resistance of 410 pounds. Of this amount, 25 per cent is due to rolling friction and the remainder is due to wind resistance. What is the drag coefficient?

5-34. A spherical piece of quartz (specific gravity $= 2.65$) falls through a body of water. If the diameter is 0.0030 inch, what is the velocity of settling for steady viscous flow?

5-35. A steel ball (specific gravity $= 7.85$) 0.060 inch in diameter falls 0.16 foot per second through a mass of oil (specific gravity $= 0.91$). What is the dynamic viscosity of the oil? First assume steady viscous flow, and then check.

5-36. A spherical drop of water 0.002 inch in diameter exists in standard air. Assume steady laminar motion. Which way would the water drop move if it were in a vertically rising air current having a speed of (a) 1.0 foot per second, (b) 0.6 foot per second, and (c) 0.2 foot per second?

5-37. A long wire $\frac{3}{8}$ inch in diameter is exposed to a stream of carbon dioxide at a velocity of 100 feet per second. Undisturbed gas pressure is 14.7 pounds per square inch absolute; temperature is 90 degrees Fahrenheit. What is the resistance per foot length of wire?

5-38. What is the total wind force on a rectangular sign board 1 foot by 4 feet, if the velocity is 20 miles per hour, the air is standard, and the flow is normal to the sign?

5-39. Find the resistance of a flat plate, 3 feet square, moving normal to itself at 20 feet per second, at 59 degrees Fahrenheit and atmospheric pressure, (a) through air and (b) through water.

5-40. What horsepower is required to move a vertical automobile windshield, 4 feet wide and 2 feet high, at a speed of 60 miles per hour? What horsepower is required if the speed is reduced to 30 miles per hour?

5-41. What is the wind force on a building 120 feet high and 60 feet wide, if the air is standard and the velocity is 60 miles per hour? Assume that the drag coefficient is the same as that for a rectangular plate.

5-42. A streamlined train 400 feet long travels at 85 miles per hour through standard air. Consider the sides and top of the train as a smooth flat plate 30 feet wide. If the boundary layer is turbulent, what horsepower must be expended to overcome the skin-friction drag of the sides and top?

5-43. The main portion of a torpedo consists of a cylinder 21 inches in diameter and 18 feet long. Consider the skin friction the same as that of a flat plate of the same area. For a turbulent boundary layer, what power is required to overcome the skin friction if the torpedo moves 50 miles per hour through salt water (specific weight $= 64.0$ pounds per cubic foot)?

5-44. A horizontal wind at standard conditions flows over a horizontal flat square area 500 feet on each side. The air enters across one edge of the area with a uniform velocity of 30 miles per hour. What is the frictional force on this area if the boundary layer is turbulent?

5-45. A boat 90 feet long has a total wetted surface of 4000 square feet. Calculate the skin-friction drag at 10 knots in fresh water for a turbulent boundary layer. Treat the wetted surface as a flat of the same length as the boat.

5-46. A metal ball (specific gravity is 7.2) is towed under water by means of a cable fastened to a boat moving steadily in a horizontal direction at 4 miles per hour. Neglect effect of water on cable. Diameter of the ball is 3 inches. The angle between the cable and vertical is 14 degrees. What power is required to pull the ball?

5-47. From an air flow point of view, why is "streamlining" or "fairing" of automobiles of less practical importance than the streamlining of aircraft?

5-48. Consider the steady flow of a fluid having no viscosity around a body. Explain why the drag is zero. Consider unsteady flow of a nonviscous fluid around a body; can there be a drag force in this case?

5-49. A rubber balloon 6 inches in diameter is filled with hydrogen. The balloon is in standard air. The hydrogen is at the same pressure and temperature as the air. The balloon is held by a single string. There is a steady horizontal wind of 10 miles per hour. Neglecting the effect of the wind on the string, what is the angle between the string and the vertical at the balloon?

5-50. The United Nations Building has a vertical rectangular surface 287 feet wide and 505 feet high. Assume standard air, a normal wind of 30 miles per hour, and a drag coefficient of 1.2. What is the maximum bending moment due to the air flow?

5-51. A balloon, 5 inches in diameter, is held by a string in a steady horizontal wind of 24 miles per hour. Assume standard air. What is the drag?

5-52. An aerial on an automobile is essentially a vertical circular cylinder $\frac{3}{8}$ inch in diameter and 3 feet high. What power is required to move this aerial through standard air at 60 miles per hour?

5-53. A cylindrical smokestack is 4 feet in diameter and 30 feet high. Assume two-dimensional flow and a wind of 5 miles per hour. What is the maximum bending moment on the stack due to the air flow?

5-54. A flat plate, 3 feet square, is held in standard air moving at 20 feet per second. In one position the plate is held normal to the stream. In another position the plate is held parallel to the flow. What is the ratio of the drag for the parallel position divided by the drag for the normal position? Assume a turbulent boundary layer.

5-55. Consider a kite weighing 3 pounds as essentially a flat plate with an area of 10 square feet. It is flown in standard air moving horizontally at 20 miles per hour. The kite makes an angle of 8 degrees with the horizontal. Assume the lift the same as that of the theoretical maximum for two-dimensional flow. If the string is at an angle of 45 degrees with the horizontal, what is the string pull?

5-56. A bird rests on a perch in a cage. The cage is placed on weighing scales. The bird leaves the perch and flies horizontally in the cage. For steady flight, is there any change in the scale reading?

5-57. A board having an area of 6 square feet is towed under water at a speed of 7 miles per hour. The angle of attack is 3 degrees. What is the theoretical or ideal lift?

5-58. A kite weighing 2.5 pounds is essentially a flat plate with an area of 8 square feet. It is flown in standard air moving horizontally at 18 miles per hour. The string holding the kite makes an angle of 60 degrees with the horizontal. The string pull is 2 pounds. What are the lift coefficient and the drag coefficient?

5-59. In some cases high winds have tended to lift roofs from buildings rather than to push them inward. Explain this action.

5-60. Why does an airplane take off and land against the wind?

5-61. Consider steady horizontal flight of an airplane over a certain ground area. Assuming proper equipment, as pressure gages, were available, explain how the airplane could be "weighed" by instruments on the ground as the plane passed aloft over the ground area.

CHAPTER SIX

Gas Dynamics, Compressible Flow in Channels

. . . The fundamental distinction between drag at supersonic and subsonic speeds has been pointed out by quite a number of physicists and ballisticians dealing with the problem; namely, that the energy loss corresponding to the drag at supersonic speeds is dissipated in the waves accompanying the projectile, in particular in the head wave emanating from the nose of the bullet, whereas the energy loss at subsonic velocities is mainly dissipated in eddies produced in the rear of the body.

—T. VON KÁRMÁN.[1]

GAS DYNAMICS

In some studies the fluid is regarded as incompressible, with relatively small pressure changes and a relatively low velocity. In this chapter we will discuss the steady flow of gases; we will consider the fluid as compressible, and study flow features over a wide range of velocities and with large pressure changes. Gas dynamics is involved in various cases, as in the motion of craft and missiles, and the flow through different structures and machines.

6–1. Some relations for gases

The "ideal" or "perfect" gas law is employed for many gases; this equation of state is

$$p = \rho RT \tag{6-1}$$

where p is absolute pressure, ρ is density, T is absolute temperature, and R

---

[1] "Problems of Flow in Compressible Fluids," in the book *Fluid Mechanics and Statistical Methods in Engineering.* University of Pennsylvania Press, Philadelphia, 1941.

is a gas factor; some average values of R are given in Table 1–2. As an example, if p is expressed in pounds per square foot, T in degrees Rankine ($t + 460$, approximately, where t is the Fahrenheit reading), and ρ in slugs per cubic foot, then, for air, R is about 1716 (feet)²/(second)² per degrees Rankine.

The specific heat of a substance is the amount of heat required to change the temperature of a unit quantity of the substance one degree. For a gas following the relation $p = \rho RT$, a parameter of importance is the ratio

$$k = \frac{\text{specific heat at constant pressure}}{\text{specific heat at constant volume}} = \frac{c_p}{c_v}$$

Some average values of k are listed in Table 6–1.

TABLE 6-1

	k
Air	1.40
Carbon dioxide	1.30
Carbon monoxide	1.40
Helium	1.40
Hydrogen	1.40
Methane	1.31
Oxygen	1.40

In making an energy balance in gas flow problems, there is the question of a convenient evaluation of change in internal energy or stored thermal energy u. The change in internal energy ($u_2 - u_1$) depends solely on the initial and final states, and not on how these states were reached. Internal energy, like temperature and height, is a "point" or "position" function. Any path or process could be employed to evaluate $u_2 - u_1$. A nonflow process will be selected because it is convenient.

Imagine the nonflow process for a gas confined in a cylinder with a movable piston, as shown in Fig. 6–1. The heat added to the gas equals the gain in internal energy plus the work done by the gas on the piston, that is,

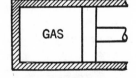

GAS

$$q = u_2 - u_1 + \int_1^2 p\,dv \qquad (6\text{–}2)$$

Fig. 6–1. Movable piston in a cylinder.

where q is heat transfer, p is pressure, v is volume, the amount of gas under consideration is a unit mass, and the units are consistent. The subscripts 1 and 2 refer to the initial and final states respectively. The work done is zero ($dv = 0$) for a constant-volume process. The heat added can be expressed as the product of specific heat and the change in temperature. Thus, for a constant-volume process,

$$q = u_2 - u_1 = c_v(T_2 - T_1) \qquad (6\text{–}3)$$

Since internal energy depends only on temperature, the internal energy change for *any* process equals $c_v(T_2 - T_1)$.

For a constant-pressure process, $q = c_p(T_2 - T_1)$, and the nonflow energy Equation (6–2) becomes

$$c_p(T_2 - T_1) = c_v(T_2 - T_1) + R(T_2 - T_1)$$

Therefore

$$c_p = c_v + R \qquad (6\text{–}4)$$

where each term is expressed in consistent units. For example, for air, the numerical values would be

$$c_p = 6006$$
$$c_v = 4290$$
$$R = 1716$$

in which each value is expressed in (feet)^2/(second)^2 per degree Rankine. For these values $k = 1.4$.

Using the definition of enthalpy and Equation (6–4), we can express the enthalpy change for *any* process in the form

$$h_2 - h_1 = c_p(T_2 - T_1) \qquad (6\text{–}5)$$

Equation (6–5) and the relation $k = c_p/c_v$ can be combined to give

$$c_v = \frac{R}{k - 1} \qquad (6\text{–}6)$$

$$c_p = \frac{kR}{k - 1} \qquad (6\text{–}7)$$

An adiabatic process is one in which no heat is added to or removed from the fluid mass. Consider a nonflow process for a unit mass of volume v. Equation (6–2) can be put in the differential form

$$dq = du + p\,dv = du + p\,d\left(\frac{1}{\rho}\right) \qquad (6\text{–}8)$$

Imagine a frictionless adiabatic process. Then $dq = 0$, and the basic relation is

$$c_v dT + p\,d\left(\frac{1}{\rho}\right) = 0 \qquad (6\text{–}9)$$

If we use the equation of state and Equation (6–6), Equation (6–9) gives the relation

$$\frac{dp}{p} = k\frac{d\rho}{\rho} \qquad (6\text{–}10)$$

Integration gives the final equation

$$\frac{p}{\rho^k} = \text{constant} \qquad (6\text{–}11)$$

If we consider states 1 and 2, then for the frictionless adiabatic process we can form the relations

$$\frac{p_1}{p_2} = \left(\frac{\rho_1}{\rho_2}\right)^k = \left(\frac{T_1}{T_2}\right)^{\frac{k}{k-1}} \tag{6-12}$$

6-2. Bulk modulus

If a volume of fluid v under a pressure p were subjected to an increase in pressure dp, there would be a decrease in volume dv. The *bulk modulus E* is defined as the ratio

$$\text{bulk modulus} = E = -\frac{dp}{dv/v} \tag{6-13}$$

The dimensionless ratio dv/v is a volumetric strain. The bulk modulus has the dimension of force per unit area. The negative sign in Equation (6-13) signifies a decrease in volume for a positive increment in pressure. E is about 300,000 pounds per square inch for water at ordinary conditions; water is about 100 times as compressible as mild steel. The bulk modulus for a gas depends upon the particular pressure-volume relation followed during the compression or expansion process. For an isothermal process, the equation of state for an ideal gas shows that $pv = $ constant, and

$$E = -\frac{dp}{dv/v} = \frac{dp}{dp/p} = p \tag{6-14}$$

The bulk modulus equals the pressure. For a frictionless adiabatic process, $pv^k = $ constant, and

$$E = -\frac{dp}{dv/v} = \frac{dp}{dp/kp} = kp \tag{6-15}$$

The bulk modulus equals the product of the pressure and the ratio of the specific heats. Most liquids have a relatively high bulk modulus, whereas the bulk modulus for a gas is relatively low.

6-3. Energy equation in convenient form

The energy Equation (3-46) has the form

$$q + \frac{p_1}{\rho_1} - \frac{p_2}{\rho_2} + W = u_2 - u_1 + \frac{V_2^2 - V_1^2}{2} + g(z_2 - z_1) \tag{6-16}$$

For many cases it is sufficiently accurate to regard the process as adiabatic. Consider the case in which $q = 0$, work $W = 0$, and $z_2 - z_1 = 0$. Then

$$\frac{p_1}{\rho_1} - \frac{p_2}{\rho_2} = u_2 - u_1 + \frac{V_2^2 - V_1^2}{2} \tag{6-17}$$

Equation (6-17) applies to *any* fluid; the fluid may be a liquid, gas, or vapor.

If the fluid is a gas following the relation $p/\rho = RT$, then the energy equation becomes

$$\frac{V_2^2 - V_1^2}{2} = \frac{p_1}{\rho_1} - \frac{p_2}{\rho_2} + c_v(T_1 - T_2) \qquad (6\text{--}18)$$

Using Equation (6–6) we can write Equation (6–18) in the forms

$$\frac{V_2^2 - V_1^2}{2} = \frac{k}{k-1}\left(\frac{p_1}{\rho_1} - \frac{p_2}{\rho_2}\right) \qquad (6\text{--}19)$$

$$\frac{V_2^2 - V_1^2}{2} = \frac{kR}{k-1}(T_1 - T_2) = c_p(T_1 - T_2) \qquad (6\text{--}20)$$

Equations (6–19) and (6–20) are convenient forms of the energy equation for gas flow. Note that the process has not yet been fully specified. If the process is a frictionless adiabatic, then

$$\frac{p_2}{p_1} = \left(\frac{\rho_2}{\rho_1}\right)^k$$

and the energy Equation (6–15) becomes

$$\frac{V_2^2 - V_1^2}{2} = \left(\frac{k}{k-1}\right)\frac{p_1}{\rho_1}\left[1 - \left(\frac{p_2}{p_1}\right)^{\frac{k-1}{k}}\right] \qquad (6\text{--}21)$$

6–4. Velocity of pressure propagation

In a perfectly rigid and incompressible medium an impulse is transmitted instantaneously from one element to another. In an elastic or compressible medium, however, the transmission of an impulse is retarded by the inertia of the displaced elements. The time of travel of an impulse may be short or difficult to observe if the distance of travel is short; on the other hand the time of travel may be noticeable if the distance is large. Thunder, for example, is heard some time after the lightning is seen if the observer is some distance from the storm; a finite time is required for the sound wave to travel a certain distance through the air. There are numerous cases, such as one method for measuring depth in the ocean, the sound detection of a submarine, and the location of a gun by sound ranging, which depend upon the fact that a pressure wave in a fluid requires time to travel a certain distance. The velocity of this wave travel is an important factor in high-velocity compressible flow studies.

Imagine a cylindrical tube with rigid walls, filled with a gas initially at rest, as illustrated in Fig. 6–2(a). It will be assumed that the motion is one-dimensional, with no friction and no heat transfer. Consider the pressure wave caused by the motion of the piston to the right in the tube. The piston moves with the linear velocity V_2; just ahead of the piston the gas pressure is p_2. Some distance ahead of the piston, to the right, the gas is at rest, the velocity $V_1 = 0$, and the pressure is p_1.

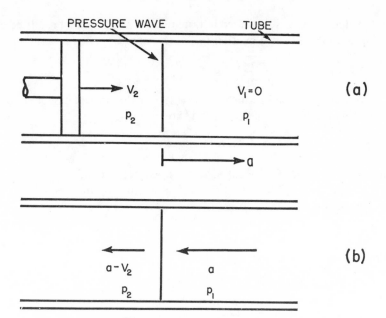

Fig. 6-2. Notation for pressure wave study.

Because of the motion of the piston there is a pressure wave traveling to the right with the velocity a. Across this pressure wave there is a sudden change in pressure from p_2 to p_1. Experimental work shows that the assumptions of no friction and no heat transfer are accurate. Figure 6-2(a) shows the pressure wave at a certain position at a certain instant. Let us imagine ourselves as observers moving along with the wave. Figure 6-2(b) illustrates this state. We can get this same relative motion by superimposing the velocity $-a$ on the flow illustrated in Fig. 6-2(a).

In Fig. 6-2(b) the fluid approaches the pressure wave with a velocity a to the left, a pressure p_1, and a density ρ_1. The velocity after the wave is $(a - V_2)$, the pressure is p_2, and the density is ρ_2. Let the cross-sectional area of the tube be A. The equation of continuity gives

$$Aa\rho_1 = A\rho_2(a - V_2) \tag{6-22}$$

The momentum equation gives

$$(p_1 - p_2)A = A\rho_1 a[(a - V_2) - a] \tag{6-23}$$

Application of the energy Equation (6-19) gives

$$\frac{(a - V_2)^2 - a^2}{2} = \frac{k}{k - 1}\left(\frac{p_1}{\rho_1} - \frac{p_2}{\rho_2}\right) \tag{6-24}$$

Equation (6-22) can be used to get an expression for ρ_2 that can be substituted in Equation (6-24). Equation (6-23) can be used to get an expression for V_2 which can be substituted in Equation (6–24). The final result of eliminating ρ_2 and V_2 from Equation (6–24) is

$$a = \sqrt{\frac{kp_1}{\rho_1}}\left(\frac{k-1}{2} + \frac{p_2}{p_1}\cdot\frac{k+1}{2}\right)^{1/2}\frac{1}{\sqrt{k}} \qquad (6\text{–}25)$$

Consider the case in which $p_2/p_1 = 1$. For this special case let a be represented by c_1. Then Equation (6–25) shows that

$$c_1 = \sqrt{\frac{kp_1}{\rho_1}} \qquad (6\text{–}26)$$

The velocity c_1 is variously called the "velocity of sound," "sonic velocity," or "acoustic velocity" for the gas with the pressure p_1 and density ρ_1. Since $E = kp$ we could write

$$c_1 = \sqrt{\frac{E_1}{\rho_1}} \qquad (6\text{–}27)$$

Using the relation $p/\rho = RT$ gives

$$c_1 = \sqrt{kRT_1} \qquad (6\text{–}28)$$

The foregoing equations show good agreement with measured values. c_1 is about 1120 feet per second for air at standard conditions.

Across a pressure wave moving with the acoustic velocity there is no large pressure change; we could say that the pressure change is infinitesimal. If the pressure change $p_2 - p_1$ is finite, greater than zero, however, the velocity of the compression wave is *higher* than the acoustic. This is illustrated

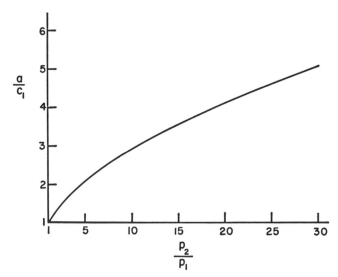

Fig. 6–3. Wave velocity ratio versus pressure ratio, for $k = 1.40$.

by the plot in Fig. 6–3 showing the ratio a/c_1 as a function of p_2/p_1. Note that at the higher pressure ratios the velocity a is considerably *higher* than the acoustic velocity c_1.

6–5. Dynamic similarity for compressible flow— Mach number

Various laws of similitude could be devised, depending upon the type of forces acting. For example, the Reynolds number is proportional to the ratio of inertia force/viscous force. The Reynolds number is useful in problems of incompressible flow, in establishing dynamic similarity, as a criterion for type of flow, and as a dimensionless parameter in correlating data. Two significant dimensionless parameters for compressible flow are the ratio of specific heats k and the so-called Mach number.

Imagine the flow of a compressible fluid around two geometrically similar bodies (or through two geometrically similar channels) in which the predominating forces are inertia, pressure, and elastic. If only three forces are involved, specifying two of the forces automatically specifies the third force because the three forces are in equilibrium (recalling d'Alembert's principle). A significant ratio is the ratio inertia force/elastic force. The inertia force is proportional to $\rho l^3 (V/t)$ or $\rho l^2 V^2$, where l is some characteristic length or dimension. The elastic or compressibility force is proportional to $E l^2$. Then inertia force/elastic force is proportional to

$$\frac{\rho l^2 V^2}{E l^2} = \frac{\rho V^2}{E} = \frac{V^2}{c^2}$$

If inertia and elastic forces determine the flow for a prototype, then mechanical similarity between model and prototype is realized when the ratio V^2/c^2 for the model equals the corresponding ratio V^2/c^2 for the prototype. Sometimes the ratio $\rho V^2/E$ is called Cauchy's number.

For purposes of dynamic similarity, the ratio V/c could be used just as well as the ratio V^2/c^2. The ratio $M = V/c$ is commonly called the Mach number, in honor of E. Mach, a Viennese physicist and philosopher.

For an ideal gas, the Mach number squared can be written in the form

$$M^2 = \frac{V^2}{kRT} \tag{6–29}$$

According to the kinetic-molecular theory of gases, the absolute temperature is directly proportional to the average kinetic energy of the molecular motion. The quantity V^2 is directly proportional to the kinetic energy of the flowing fluid. Thus the factor M^2 is proportional to the ratio of flow kinetic energy divided by molecular energy or internal energy. Subsequent discussions will bring out some of the important features characterized by the Mach number. A critical value of the Mach number is unity; this value marks a distinction between two types of flow.

6–6. Pressure at a stagnation point

The determination of the pressure at a stagnation point illustrates one application of the foregoing relations. Imagine the flow of a frictionless fluid around a body, as indicated in Fig. 6–4. The pressure is p_0 and the velocity is V_0 in the undisturbed stream to the left. At the "stagnation point" S the fluid velocity is zero, and the "stagnation pressure" is p_s. The dynamic equation will be applied to the small stream tube along the streamline O to S. The following analysis applies only for flow in which there is no friction and no sharp impact or shock.

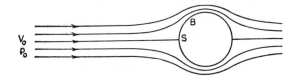

Fig. 6–4. Body in a stream of fluid.

For incompressible flow, use of the dynamic Equation (3–21) gives the familiar relation

$$p_s = p_0 + \tfrac{1}{2}\rho V_0^2 \tag{6–30}$$

For compressible adiabatic flow, Equation (6–21) gives the relation

$$\frac{p_s}{p_0} = \left[1 + \left(\frac{k-1}{2} \right) \left(\frac{V_0}{c_0} \right)^2 \right]^{\frac{k}{k-1}} \tag{6–31}$$

where c_0 is the velocity of pressure propagation in the undisturbed flow. Let $M_0 = V_0/c_0$. Then Equation (6–31) becomes

$$\frac{p_s}{p_0} = \left[1 + \left(\frac{k-1}{2} \right) M_0^2 \right]^{\frac{k}{k-1}} \tag{6–32}$$

If the term $\left(\dfrac{k-1}{2} \right) M_0^2$ is less than unity, then the right side of Equation (6–32) can be expanded in terms of a convergent power series. The result of this expansion and the substitution $\rho_0 = k p_0 / c_0^2$ can be arranged in the form

$$p_s = p_0 + \frac{1}{2} \rho_0 V_0^2 \left[1 + \frac{M_0^2}{4} + \left(\frac{2-k}{24} \right) M_0^4 \ldots \right] \tag{6–33}$$

The important feature is a comparison of Equations (6–30) and (6–33). Equation (6–33) shows that the stagnation pressure for compressible flow is higher than the stagnation pressure for incompressible flow. If M is very small, p_s for compressible flow may not be very much different from that for incompressible flow; the difference depends upon the Mach number in the undisturbed stream. Figure 6–5 shows graphically the variation of stagnation pressure with velocity for incompressible and compressible flow. The results shown in Fig. 6–5 are based on standard air in the undisturbed stream. Note the difference in pressures at high velocities.

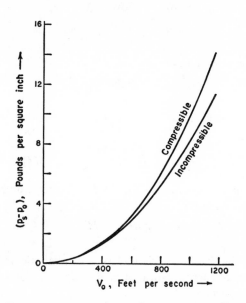

Fig. 6–5. Pressure at stagnation point p versus undisturbed velocity, $k = 1.4$.

6–7. Relation between area and velocity

An important and useful relation is that between area and velocity for a fluid flowing through a tube of varying cross section. Let A represent the cross-sectional area of the tube at any section, ρ the density, and V the average velocity. The equation of continuity states that

$$\rho V A = \text{constant} \qquad (6\text{–}34)$$

The relation between area and velocity can be obtained by combining the equation of continuity, the equation of state, and the energy equation. The case for a gas following the adiabatic process will be taken. Differentiating Equation (6–34) gives

$$\frac{dA}{A} = -\frac{dV}{V} - \frac{d\rho}{\rho} \qquad (6\text{–}35)$$

The energy Equation (6–19) can be expressed in the form

$$\frac{V^2}{2} + \left(\frac{k}{k-1}\right)\frac{p}{\rho} = \text{constant} \tag{6-36}$$

Differentiating Equation (6–36) gives

$$V\,dV + \left(\frac{k}{k-1}\right)\left[\frac{dp}{\rho} - \frac{p\,d\rho}{\rho^2}\right] = 0 \tag{6-37}$$

Equations (6–26), (6–13), and (6–27) show that

$$c = \sqrt{\frac{kp}{\rho}} = \sqrt{\frac{E}{\rho}} = \sqrt{\frac{dp}{d\rho}} \tag{6-38}$$

Then Equation (6–37) becomes

$$V\,dV + \frac{c^2\,d\rho}{\rho} = 0 \tag{6-39}$$

Combining Equations (6–35) and (6–39) gives a convenient form for compressible flow:

$$\frac{dA}{A} = -\frac{dV}{V}\left[1 - \frac{V^2}{c^2}\right] = -\frac{dV}{V}[1 - M^2] \tag{6-40}$$

where the ratio V/c is frequently called the *local* Mach number M. Differentiating Equation (6–29) gives

$$\frac{dT}{T} = 2\left[\frac{dV}{V} - \frac{dM}{M}\right] \tag{6-41}$$

It is assumed that the process is frictionless and adiabatic and that there is no change in potential energy. Noting Equation (6–20), the energy equation in differential form becomes

$$c_p\,dT + V\,dV = 0$$

$$\left(\frac{kR}{k-1}\right)dT + V\,dV = 0$$

$$\frac{dV}{V} = -\frac{1}{M^2(k-1)}\frac{dT}{T} \tag{6-42}$$

Eliminating dT/T from Equations (6–41) and (6–42) gives an expression for dV/V; putting this value in Equation (6–40) gives the differential equation

$$\frac{dA}{A} = -\frac{(1-M^2)\,dM}{M\left[\left(\dfrac{k-1}{2}\right)M^2 + 1\right]} \tag{6-43}$$

Equation (6–43) will be integrated between a section of minimum area or throat area A_c where $M = 1$ and any variable area where the Mach number is M. The final result is

$$\frac{A}{A_c} = \frac{1}{M}\left[\frac{2 + (k-1)M^2}{2 + (k-1)}\right]^{\frac{k+1}{2(k-1)}} \tag{6-44}$$

Figure 6–6 shows a plot of area ratio versus the Mach number for $k = 1.4$.

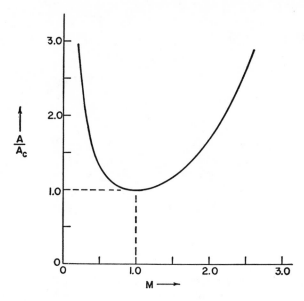

Fig. 6–6. Area ratio versus Mach number, $k = 1.4$.

Equation (6–40) and Fig. 6–6 show several important features. Flow in which the fluid velocity is *greater* than the velocity of sound differs from flow in which the fluid velocity is *less* than the velocity of sound. When M is less than 1, dA/dV is negative; this sign indicates that the velocity *increases* in a converging channel and *decreases* in a diverging channel. The flow can be assumed incompressible if M is very small in comparison with unity. When M is greater than 1, dA/dV is positive; this sign indicates that the velocity *increases* in a diverging channel and *decreases* in a converging channel.

6–8. Flow between converging-diverging streamlines with no friction

When a fluid flows through a channel or around a body, some mechanical energy is degraded into unavailable energy; frequently we say there is an "energy loss." This loss may be caused by the internal friction or viscosity of the fluid, by turbulent motion, eddying wakes, or by impact. No energy is actually destroyed. Some energy, however, is transformed into a form that is not available for maintaining the flow; from the point of view of the flow it is "lost." This section will discuss a case in which there is no loss or friction.

The word "nozzle" refers to a channel in which the fluid velocity is increased and the pressure is decreased. In a "diffuser" the fluid velocity is decreased and the pressure is increased.

Several fundamental features of practical importance can be conven-

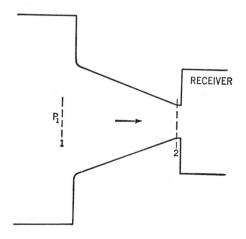

Fig 6-7. Converging tube.

iently illustrated by specific reference to the flow of a gas through a nozzle. Picture first the adiabatic frictionless flow in a short converging tube as represented in Fig. 6–7. The case is taken in which the velocity V_1 at entrance is small in comparison with the velocity V_2 at the throat or section of minimum area. p_1 is the pressure at the entrance, and p_2 is the pressure at the throat. Then Equation (6–21) becomes

$$V_2 = \sqrt{\left(\frac{2k}{k-1}\right)\frac{p_1}{\rho_1}\left[1 - \left(\frac{p_2}{p_1}\right)^{\frac{k-1}{k}}\right]}$$ (6–45)

Let W represent the mass rate of gas passing through the tube per unit time, and A_2 the area at the throat. Then $W = A_2 V_2 \rho_2$. Since $p_1/\rho_1^k = p_2/\rho_2^k$, W can be expressed as

$$W = A_2\sqrt{\frac{2k\rho_1 p_1}{k-1}\left[\left(\frac{p_2}{p_1}\right)^{2/k} - \left(\frac{p_2}{p_1}\right)^{\frac{k+1}{k}}\right]}$$ (6–46)

The variation of W with the ratio p_2/p_1, as given in Equation (6–46), is illustrated by the curved line (partly dotted and partly solid) in Fig 6–8.

Figure 6–8 shows that W reaches a maximum value for a certain pressure ratio p_c/p_1. p_c will be called a critical pressure. The critical pressure can be determined by differentiating W with respect to p_2 and setting the result equal to zero. This operation gives

$$\frac{p_c}{p_1} = \left(\frac{2}{k+1}\right)^{\frac{k}{k-1}}$$ (6–47)

For air at normal conditions, p_c/p_1 is 0.528 or about 0.53.

The dotted curve in Fig. 6–8 is not actually attained for the flow in the *converging* tube. If the pressure p_2 is decreased from the value p_1 to p_c, the weight rate of discharge increases from zero to a maximum, as indicated by the solid curve in Fig. 6–8. Picture the conditions when the pressure p_2 equals p_c. From Equation (6–19)

$$\frac{V_c^2}{2} = \left(\frac{k}{k-1}\right)\left(\frac{p_1}{\rho_1} - \frac{p_c}{\rho_c}\right)$$

where V_c is the fluid velocity at the critical pressure and v_c is the specific volume at the critical pressure. Then

$$V_c = \sqrt{\frac{2k}{k-1}\left(\frac{p_1}{\rho_1} - \frac{p_c}{\rho_c}\right)}$$

$$(6\text{--}48)$$

Inasmuch as $p_1/\rho_1^k = p_2/\rho_2^k$ and $p_c = p_1\left(\frac{2}{k+1}\right)^{\frac{k}{k-1}}$, Equation (6–48) becomes

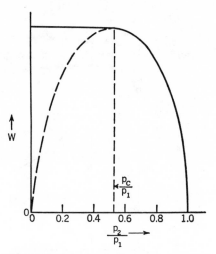

Fig. 6–8. Relation between mass rate of discharge and pressure ratio for converging tube.

$$V_c = \sqrt{\frac{kp_c}{\rho_c}} \qquad (6\text{--}49)$$

where ρ_c is the density at the critical pressure. Equation (6–49) shows that the *fluid velocity at the throat equals the velocity of sound at the critical pressure.* The Mach number at the throat is unity.

If the pressure in the receiver (following the throat) is reduced below p_c, the pressure in the receiver cannot be telegraphed back into the throat of the nozzle because the fluid in the throat is moving with the velocity of pressure propagation. For receiver pressures less than p_c, the pressure *in* the throat is always p_c, and the mass rate of discharge always equals the maximum value. On the other hand, if the pressure in the receiver is above p_c, then the fluid velocity at the throat is less than the velocity of sound; the receiver pressure can be telegraphed back into the throat.

The next step is to investigate the flow of gas in a converging-diverging nozzle. Attention will be directed to the pressure at different points along the nozzle. Assume that a certain nozzle shape is given, as shown in Fig. 6–9(a). Let A be the cross-sectional area at any point and x the distance along the axis of the nozzle. The mass rate of discharge W is the same for each cross section. Thus the ratio W/A becomes a known function of x. Figure 6–9(b) shows a plot of this function for one value of W; the W selected equals the maximum rate that can pass through the nozzle for given inlet and exit conditions.

The pressure at points along the nozzle can be determined by an adaptation of Equation (6–45). Let p_1 be the entrance pressure, p the pressure at

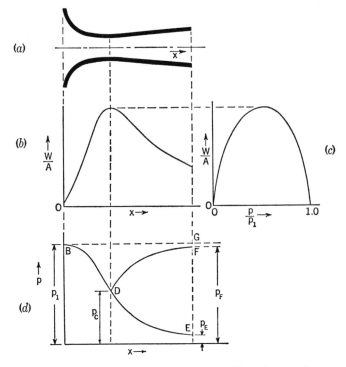

Fig. 6–9. Flow through a converging-diverging nozzle.

any point, and V the velocity at any point in the nozzle. With this notation, Equation (6–45) becomes

$$V = \sqrt{\left(\frac{2k}{k-1}\right)\frac{p_1}{\rho_1}\left[1 - \left(\frac{p}{p_1}\right)^{\frac{k}{k-1}}\right]} \qquad (6\text{--}50)$$

Since $W = AV\rho_1$ and $p_1/\rho_1^k = p_2/\rho_2^k$, Equation (6–50) gives

$$\frac{W}{A} = \frac{V}{v} = \sqrt{\left(\frac{2gk}{k-1}\right)\frac{p_1}{v_1}\left[\left(\frac{p}{p_1}\right)^{\frac{2}{k}} - \left(\frac{p}{p_1}\right)^{\frac{k+1}{k}}\right]} \qquad (6\text{--}51)$$

Figure 6–9(c) shows a plot of W/A against p/p_1 for the given W and nozzle shape. Figure 6–9(b) gives a certain value of W/A for each value of x. Figure 6–9(c) gives the corresponding value of p. Figure 6–9(d) shows the resulting pressure variation along the axis of the nozzle.

Point B in Fig. 6–9(d) represents the entrance pressure p_1. The fluid velocity increases in the converging tube as the pressure drops from p_1 to the critical pressure p_c (from point B to D). At the throat the fluid velocity equals the velocity of sound. Note that the process from B to D is unique. For the diverging portion *one* value of x gives one value of W/A and *two* values of the pressure p. The gas can expand adiabatically or it can be compressed adiabatically in the diverging portion. Flow beyond point D in Fig. 6–9 depends upon the nozzle exit conditions; there are two and only two possibilities for frictionless flow:

(1) If the gas expands adiabatically in the diverging portion, the process follows the curve D to E in Fig. 6–9(d). There is only one exit pressure p_E (for a given nozzle shape) which will make this expansion possible. Figure 6–6 shows that during this expansion process the fluid velocity *increases above* the velocity of sound. The channel is diverging, and the local Mach number at each point is greater than unity.

(2) If the gas is compressed adiabatically without friction losses, the pressure in the diverging portion follows the curve D to F in Fig. 6–9(d). There is only one exit pressure p_F which will make this compression possible. Beyond the throat the fluid velocity *decreases below* the velocity of sound; the local Mach number at each point beyond the throat is less than unity. The fluid velocity and the pressure at the exit are the same as those in the section of the same area in the converging portion of the nozzle.

Example. Air at 100 pounds per square inch absolute and 100 degrees Fahrenheit in a large tank enters a converging-diverging nozzle. The exit flow from the nozzle discharges into the atmosphere. The nozzle exit diameter is 2 inches. Let the subscript 1 represent conditions at the nozzle inlet, the subscript 2 represent conditions at the throat, and the subscript 3 the conditions at the nozzle exit. Assume a frictionless adiabatic process, neglect inlet velocity, and assume maximum rate of flow.

The ratio p_3/p_1 equals $14.7/100 = 0.147$. For critical flow, with a Mach number of one at the throat, the pressure ratio $p_2/p_1 = 0.528$. Thus there is subsonic flow in the converging portion, and supersonic flow in the diverging portion of the nozzle. The pressure at the throat is 52.8 pounds per square inch absolute. Using Equations (6-12) and (6-47) the temperature T_2 can be calculated by the relation

$$T_2 = T_1 \left(\frac{p_2}{p_1}\right)^{\frac{k-1}{k}} = T_1 \left(\frac{2}{k+1}\right) = 466 \text{ degrees Rankine}$$

The velocity at the throat equals the acoustic velocity, namely,

$$V_2 = c_2 = \sqrt{1.4(1716)466} = 1060 \text{ feet per second}$$

The density at the throat can be calculated by the equation of state:

$$\rho_2 = \frac{p_2}{RT_2} = \frac{52.8(144)}{1716(466)} = 0.0095 \text{ slug per cubic foot}$$

The temperature at the exit section of the nozzle can be calculated by Equation (6-12):

$$T_3 = T_1 \left(\frac{p_3}{p_1}\right)^{\frac{k-1}{k}} = 560(0.147)^{\left(\frac{0.4}{1.4}\right)} = 320 \text{ degrees Rankine}$$

The density at the exit section is

$$\rho_3 = \frac{14.7(144)}{1716(320)} = 0.00386 \text{ slug per cubic foot}$$

The velocity V_3 at the exit can be calculated by Equation (6-50), noting that $p_1/\rho_1 = RT_1$.

$$V_3 = \sqrt{\frac{2(1.4)}{0.4}(1716)560[1 - (0.147)^{\left(\frac{0.4}{1.4}\right)}} = 1690 \text{ feet per second}$$

The mass rate of flow can be calculated for the exit conditions:

$$\text{mass rate} = A_3\rho_3 V_3 = \frac{\pi}{4}\left(\frac{2}{12}\right)^2 0.00386(1690) = 0.1425 \text{ slug per second}$$

Knowing the steady flow rate, the velocity at the throat, and the throat density, we can calculate the throat area:

$$A_2 = \frac{\text{mass rate}}{\rho_2 V_2} = \frac{0.1425(144)}{0.0095(1060)} = 2.04 \text{ square inches}$$

6–9. Flow between converging-diverging streamlines with energy loss

Refer to Fig. 6–9(d). The foregoing article discussed two cases, one with the exit pressure p_E and the other with the exit pressure p_F. In actual cases the pressure may lie somewhere between these pressures. This article will cover this more general case of any exit pressure between p_E and p_F.

Let A_c be the cross-sectional area at the throat, A_2 the cross-sectional area of the flow at the exit, and p_2 the actual exit pressure somewhere between p_E and p_F. Assume that the flow in the converging part is without energy loss but that there may be a loss in the diverging part.

From the energy Equation (6–19) we can arrange

$$\frac{V_2^2}{2g} = \left(\frac{k}{k-1}\right)\left(\frac{p_1}{\rho_1} - \frac{p_2}{\rho_2}\right)$$

$$M_2^2 = \left(\frac{2}{k-1}\right)\left(\frac{p_1\rho_2}{p_2\rho_1} - 1\right)$$

(6–52)

where M_2 is the Mach number at the nozzle exit.

For the converging portion

$$\frac{p_1}{\rho_1^k} = \frac{p_c}{\rho_c^k}$$

$$V_c = \sqrt{\frac{kp_c}{\rho_c}} = \sqrt{\frac{2kp_1}{(k+1)\rho_1}}$$

and

$$\frac{p_c}{p_1} = \left(\frac{2}{k+1}\right)^{\frac{k}{k-1}}$$

From the equation of continuity we get

$$\rho_c A_c V_c = A_c \rho_c \sqrt{\frac{2kp_1}{(k+1)\rho_1}} = A_2 V_2 \rho_2$$

$$\frac{\rho_2}{\rho_1} = \left(\frac{A_c}{A_2}\right)^2 \left(\frac{p_1}{p_2}\right)\left(\frac{2}{k+1}\right)^{\frac{k+1}{k-1}}\frac{1}{M_2^2}$$

(6–53)

Substituting Equation (6–53) in Equation (6–52) gives the final result

$$M_2^4\left(\frac{k-1}{2}\right) + M_2^2 - \left(\frac{2}{k+1}\right)^{\frac{k+1}{k-1}}\left(\frac{p_1}{p_2}\right)^2\left(\frac{A_c}{A_2}\right)^2 = 0 \qquad (6\text{–}54)$$

A plot of M_2 as a function of the parameter $p_1 A_c/p_2 A_2$ for $k = 1.4$ is shown in Fig. 6–10. Note that the exit flow can be subsonic or supersonic depending

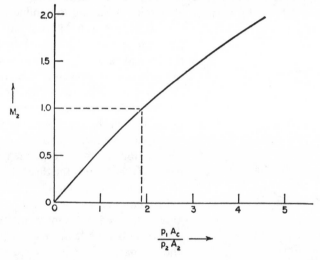

Fig. 6–10. Exit Mach number as a function of pressure ratio and area ratio, $k = 1.4$.

upon the combination of pressure ratio and area ratio. For a Mach number $M_2 = 1$, Equation (6–54) gives

$$\frac{p_1 A_c}{p_2 A_2} = \left(\frac{k+1}{2}\right)^{\frac{k}{k-1}} \qquad (6\text{–}55)$$

As an illustration, consider the case for $k = 1.4$. Then reference to Equation (6–55) and Fig. 6–10 gives the following results:

If $p_2/p_1 < 0.53\ (A_c/A_2)$, the exit flow is supersonic.
If $p_2/p_1 = 0.53\ (A_c/A_2)$, the exit flow is sonic.
If $p_2/p_1 > 0.53\ (A_c/A_2)$, the exit flow is subsonic.

The critical pressure ratio equal to 0.53 (A_c/A_2) distinguishes between two different types of exit flows. In each one of these types, various actions may occur. One case with the normal compression shock will be discussed in the next article. In other cases the fluid may break away from the walls; the jet may not fill the exit area completely. So-called oblique shocks (see Section 6–12) may result. Note that the area A_2 is the fluid flow area and not necessarily the exit area of the channel.

6–10. Normal compression shock

There are various compressibility effects which are associated with the "normal or plane *compression shock*." Some discussion will be given of the more prominent features of this action. One convenient method of introduction is the study of the flow between converging-diverging streamlines.

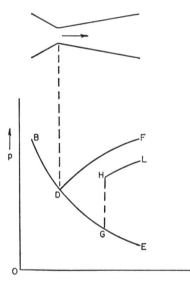

Fig. 6–11. Flow through a converging-diverging passage.

Refer to Fig. 6–11. Assume that the terminal pressure p_L is between p_F and the critical value for sonic exit flow; thus the exit flow is subsonic. Experimental evidence indicates the following action: just beyond the throat the pressure drops below that in the throat and the velocity increases above the velocity of sound. This action might be explained on the basis that the fluid at the throat has a high velocity and considerable inertia; the fluid tends to keep moving downstream at a high speed. At some point G, however, there is a sudden increase in pressure and a sharp decrease in velocity. The action from G to H is called the "normal or plane compression shock"; there is a sharp pressure rise normal to the flow velocity. Before the shock the flow is supersonic; after the shock the flow is subsonic. The process from H to L is a frictionless adiabatic compression. Considerable energy may be lost in the impact or shock.

In the foregoing discussion, specific reference to a nozzle was made in order to provide a definite illustration. The flow, however, could be that between any surface of converging-diverging streamlines (like that near an airplane wing or a compressor blade).

A compression shock takes place in such a short distance that it is accurate to regard the action as taking place between parallel streamlines. Thus imagine a pipe or tube of constant diameter; subscript 1 refers to conditions before the shock and subscript 2 to conditions after the shock. From the continuity equation

$$\rho_1 V_1 = \rho_2 V_2 \qquad (6\text{–}56)$$

A momentum study gives

$$p_1 - p_2 = \rho_2 V_2^2 - \rho_1 V_1^2 \qquad (6\text{–}57)$$

A convenient method for studying the shock is to rearrange the relations in terms of a Mach number. A functional relation can then be developed which expedites the solution of many difficult problems. Using the relation $p = \rho RT$ and the expression for the acoustic velocity, the momentum Equation (6–57) becomes

$$p_1 - p_2 = \frac{p_2 V_2^2}{RT_2} - \frac{p_1 V_1^2}{RT_1}$$

$$p_1 - p_2 = k[p_2 M_2^2 - p_1 M_1^2] \qquad (6\text{–}58)$$

$$\frac{p_1}{p_2} = \frac{1 + kM_2^2}{1 + kM_1^2}$$

Equation (6–58) is the equation of a so-called Rayleigh line.

Taking an energy balance next, Equation (6–19) can be put in the form

$$\frac{V_2^2}{2}\left(\frac{k-1}{k}\right) + \frac{p_2}{\rho_2} = \frac{V_1^2}{2}\left(\frac{k-1}{k}\right) + \frac{p_1}{\rho_1}$$

$$\frac{p}{\rho}\left[1 + \frac{(k-1)V^2\rho}{2kp}\right] = \text{constant} \qquad (6\text{–}59)$$

For a gas following the relation $p = \rho RT$

$$T\left[1 + \frac{(k-1)V^2\rho}{2kp}\right] = \text{constant}$$

$$\frac{T_1}{T_2} = \frac{2 + (k-1)M_2^2}{2 + (k-1)M_1^2} \qquad (6\text{–}60)$$

From the equation of continuity ρV is constant, the relation for the Mach number, and Equation (6–60), Equation (6–59) becomes

$$pV\left[1 + \frac{(k-1)V^2\rho}{2kp}\right] = \text{constant}$$

$$pM\sqrt{RT}\left[1 + \frac{(k-1)V^2\rho}{2kp}\right] = \text{constant} \qquad (6\text{–}61)$$

$$p_1 M_1\left[1 + \left(\frac{k-1}{2}\right)M_1^2\right]^{1/2} = p_2 M_2\left[1 + \left(\frac{k-1}{2}\right)M_2^2\right]^{1/2}$$

Equation (6–61) is the equation for a Fanno line.

Eliminating the pressure ratio between Equations (6–58) and (6–61) gives the following relation between the initial Mach number, the final Mach number, and the ratio of specific heats

$$\frac{1 + kM_1^2}{M_1\left(M_1^2 + \dfrac{2}{k-1}\right)^{1/2}} = \frac{1 + kM_2^2}{M_2\left(M_2^2 + \dfrac{2}{k-1}\right)^{1/2}} \qquad (6\text{–}62)$$

In eliminating the pressure ratio between Equations (6–58) and (6–61), one obvious solution is that for the case of $M_1 = M_2$. If the root $(M_2 - M_1)$ is factored out, the result is

$$M_2^2 = \frac{M_1^2(k - 1) + 2}{2kM_1^2 - k + 1} \qquad (6\text{--}63)$$

Figure 6–12 shows a plot of M_2 as a function of M_1 for $k = 1.4$, as calculated by Equation (6–63). If M_2 and M_1 are known, the pressure ratio can be calculated by Equation (6–58) and the temperature ratio by Equation (6–60).

6–11. Entropy study of the normal compression shock

The foregoing article is limited to the compression shock. Inspection of Equation (6–62) and Fig. 6–12 provokes this question: Is it possible to have

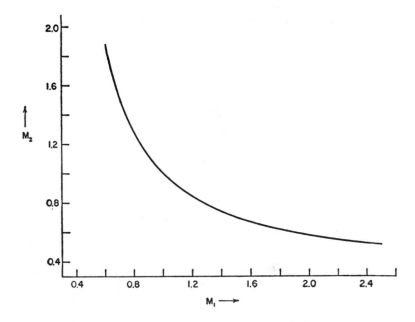

Fig. 6–12. M_2 versus M_1 for a compression shock, $k = 1.4$.

an "expansion shock," that is, can the flow go suddenly from subsonic to supersonic with a large pressure change? A convenient method of approach is to make a calculation of the so-called "entropy" change across the shock. A discussion of entropy can be found in the usual elementary book on thermodynamics. If there is any entropy increase in going from state 1 to state 2, there is an increase in unavailable energy; there has been some degradation of energy. If there is an entropy decrease, there is a decrease in unavailable energy. Some energy has been made available. The entropy change across the shock will be organized in terms of the initial Mach number.

Since entropy is a function of point or position, we can follow any possible path in going from state 1 to state 2, where 1 refers to conditions before the shock and 2 refers to conditions after the shock. A nonflow frictionless process will be selected. Let Q represent heat added, u internal energy, and s entropy. The nonflow equation in differential form is

$$dQ = du + p\, dv \tag{6-64}$$

Using the definition of entropy change, we get the relation

$$s_2 - s_1 = \int_1^2 \frac{dQ}{T} = c_v \int_1^2 \frac{dT}{T} + \int_1^2 \frac{p\, dv}{T} \tag{6-65}$$

Differentiation of the equation of state for the gas, and use of Equation (6–66), gives

$$(s_2 - s_1)\frac{1}{R} = \frac{1}{k-1} \log_e \frac{T_2}{T_1} + \log \frac{\rho_1}{\rho_2} \tag{6-66}$$

Each term in Equation (6–66) is a dimensionless ratio.

From the equation of state we can form the relation

$$\frac{T_2}{T_1} = \frac{p_2 \rho_1}{p_1 \rho_2} \tag{6-67}$$

By using the relations in the foregoing article, we can arrange the terms on the right-hand side of Equation (6–66) as a function of k and the initial Mach

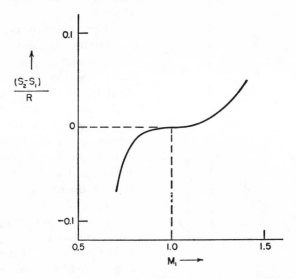

Fig. 6–13. Entropy change as a function of M_1 across a shock, for $k = 1.4$.

number M_1. The final result is

$$(s_2 - s_1)\frac{1}{R} = \frac{1}{k-1}\left\{\log_e\left(\frac{2}{(k+1)M_1^2} + \frac{k-1}{k+1}\right)\right.$$

$$\left. + \log_e\left(\frac{2kM_1^2}{k+1} - \frac{k-1}{k+1}\right)\right\} + \log_e\left(\frac{2}{(k+1)M_1^2} + \frac{k-1}{k+1}\right) \quad (6\text{-}68)$$

Figure 6–13 shows a plot of Equation (6–68) for $k = 1.4$. At $M = 1.0$ the entropy change is zero; note that at this point there is an inflection in the curve. At $M = 1.0$ infinitesimal pressure changes—either in expansion or in compression—are physically possible. This case is illustrated by the common action in speaking and similar acoustic cases. If the initial Mach number is greater than 1, Fig. 6–13 shows an entropy increase across the compression shock; this, too, is a physically possible process. If the initial Mach number, however, is less than 1, Fig. 6–13 shows an entropy decrease; this is physically impossible. Thus the finite compression shock is the only one physically realizable.

6–12. Normal and oblique shocks

In a normal compression shock the sharp pressure rise is across a line normal to or at right angles to the flow velocity. As illustrated by the "oblique" shock in Fig. 6–14(a), frequently a sharp change or shock line is noticed which is oblique to the approach velocity q_1. The velocity after the shock is q_2.

The velocity and pressure changes are normal to the shock line. The resultant velocity q_1 can be broken up into a component V_1 normal to the shock line and a component U parallel to the shock line. One might imagine a normal shock and observation of it from a vehicle moving parallel to the line of the shock. Then the fluid appears to the moving observer as if it were oblique to the shock. As illustrated in Fig. 6–14(b), there is no change in U in passing across the shock line, but there is a change from V_1 to V_2. Thus the change from q_1 to q_2 is caused by the change in normal components. The change in normal

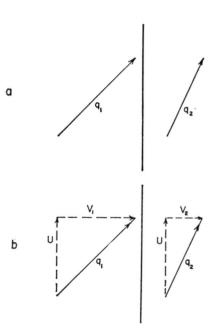

Fig. 6–14. Oblique shock.

velocities can be investigated by the relations in Sec. 6-10. Note that q_2 makes a smaller angle with the shock line than q_1. Note also that the flow after an oblique shock can be supersonic.

6–13. Particle action at subsonic and supersonic velocities

In the general case, across the compression shock there is a finite and relatively large pressure change in a very short distance. There is one limiting case of an infinitesimal pressure change in a very short distance; this action might be called an "infinitestimal disturbance." In the usual speaking and other acoustic cases there are many examples of actions in which there is a very small pressure change in a short distance; analyzing these actions by means of infinitesimal disturbances gives results which agree closely with experimental data. The rest of this article will be devoted to infinitesimal disturbances.

As illustrated in Fig. 6–15, imagine an infinitesimal particle or a point disturbance moving through a fluid with a constant velocity V, smaller than the velocity of sound c; this is called *subsonic* motion. A pressure or sound wave with the velocity c is produced when the particle is at A. This pressure or sound wave has a spherical front with a center at A. After a time interval t this wave front has traveled a distance ct. During this time interval the particle has moved a distance Vt, from A to B. The pressure wave reaches B before the particle; the pressure wave has been "telegraphed" ahead. The upstream fluid particles have some opportunity for adjustment to the motion before the particle, initially at A, reaches their positions. The intermediate points between A and B are sources of other spherical pressure waves. Since V is less than c, these other waves will always be contained within the sphere of radius ct.

Imagine some craft moving steadily through an expanse of fluid at a subsonic velocity. A pressure signal travels ahead at sound velocity minus craft velocity with respect to the craft, whereas a pressure signal travels backward at a speed equal to the sum of craft velocity and sound velocity with respect to the craft. The distribution of pressure signals is not symmetrical. Every point in the immediate region around the body, however, is reached by a signal.

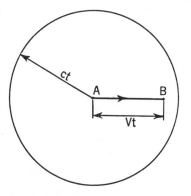

Fig. 6–15. Wave front produced by a particle moving at subsonic velocity.

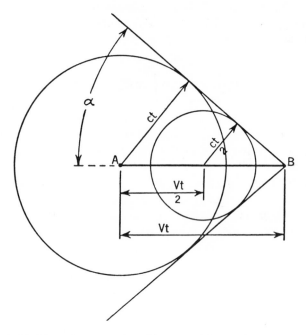

Fig. 6–16. Wave fronts produced by a particle moving at a supersonic velocity.

As illustrated by Fig. 6–16, imagine a particle or point disturbance moving with a velocity V that is *greater* than the velocity of sound; this is called *supersonic motion*. The particle reaches point B before the pressure or sound wave; the pressure wave is not moving fast enough to be telegraphed ahead of the particle. Intermediate points between A and B are sources of other spherical waves. The entire system of spherical pressure waves results in a conical sound front with the vertex at B. The half angle α at the cone vertex is frequently called the Mach angle.

$$\alpha = \sin^{-1}\frac{ct}{Vt} = \sin^{-1}\frac{1}{M} \qquad (6\text{–}69)$$

where M is the Mach number.

For the supersonic motion illustrated in Fig. 6–16 all action is restricted to the interior of a cone that includes all the spherical pressure wave fronts. Following Kármán's suggestion, the outside of this cone is called the *zone of silence*. The cone that separates the zone of action from the zone of silence is the Mach cone. The Mach cone or Mach wave can be regarded as a very weak oblique shock.

When the particle A moves with the velocity of sound, the motion is sonic and the angle α in Fig. 6–16 becomes 90 degrees. The Mach cone becomes a flat surface.

6–14. Optical studies of compressibility effects

Various optical systems have proved very helpful in studying compressible flow. In these systems light from a suitable source is passed through the region of gas in motion. Small differences in density are made visible by means of differences in intensity of the light which has passed through the gas. The deflection or refraction of a light beam passing through a body of gas depends upon the density of the gas. A mass of gas can act as a weak lens. Sudden changes in density result in definite bands of shadow or intense illumination.

Figure 6–17 illustrates one arrangement of optical apparatus. Light from a source at A passes through lens B and emerges as a parallel beam. The compressible fluid passes through the region where the light beams are parallel. For example, in this region the fluid may be confined by transparent walls, as indicated by the dotted lines C and D. The lens E is used to bring the light rays to a focus at point F. The light rays form an image on the screen G. Changes in fluid density in the flow region change the index of refraction of the flowing gas; this results in lines or bands of shadow or intense illumination on the image screen.

In the "schlieren" system there is a knife-edge at the focal point F. The knife-edge acts as an optical filter. Rays can be bent either away from or toward the knife-edge. Some of the rays interrupted by the knife-edge in the case of no gas flow may pass over the knife-edge when gas in flowing. This optical filtering effect helps to give good contrast on the image screen. In the "shadow" method there is no knife-edge. Mirrors can be used in place of lenses.

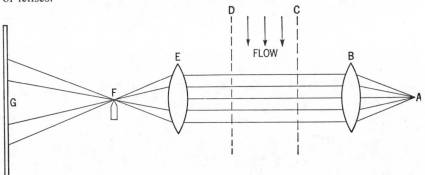

Fig. 6–17. Diagrammatic arrangement of apparatus for schlieren method.

Figures 6–18 to 6–24 show shadow pictures of various bodies taken at the Ballistic Research Laboratory, Ordinance Research and Development Center, Aberdeen Proving Ground. The apparatus was so arranged that the illuminating spark occured while the body or projectile was between a spark and a photographic plate. The body leaves an ordinary shadow; the shock waves and wake also give a shadow because of refraction effects. Each figure shows the head shock wave, the tail wave, and the eddies in the wake.

In Figs. 6–18, 6–19, 6–20, and 6–21, directly forward of the stagnation point of the body, there is short yet definite part of the bow or nose wave which is normal to the projectile motion; this short portion is a normal compression shock. The nose wave curves backward from the normal compression shock; in this curved portion are a series of oblique shocks.

In Figs. 6–23 and 6–24, at some distance from the forward point of the projectile, the nose wave is straight. This straight portion is the Mach line, across which the pressure change is infinitesimal. The velocity of the projectile can be calculated with fair accuracy from the angle of the nose wave some distance from the projectile; V can be calculated by Equation (6–69) if α and c are known.

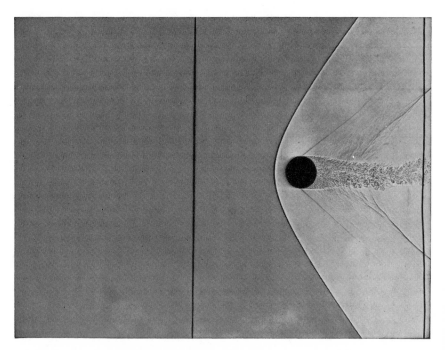

Fig. 6–18. Shadow photograph of sphere ($\frac{9}{16}$ inch diameter) at a Mach number of 1.30. (Courtesy of Ballistic Research Laboratories, Aberdeen, Md.)

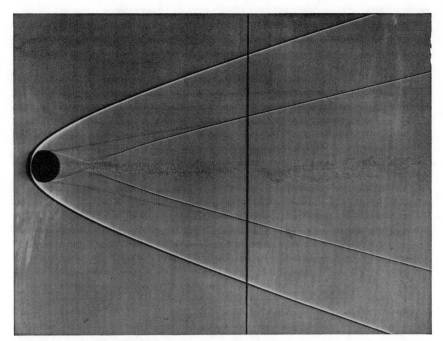

Fig. 6–19. Shadow photograph of sphere ($\frac{9}{16}$ inch diameter) at a Mach number of 3.96. (Courtesy of Ballistic Research Laboratories, Aberdeen, Md.)

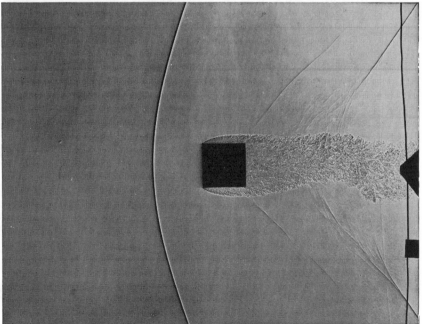

Fig. 6–20. Shadow photograph of right cylinder (20 mm. diameter) at a Mach number of 1.15. (Courtesy of Ballistic Research Laboratories, Aberdeen, Md.)

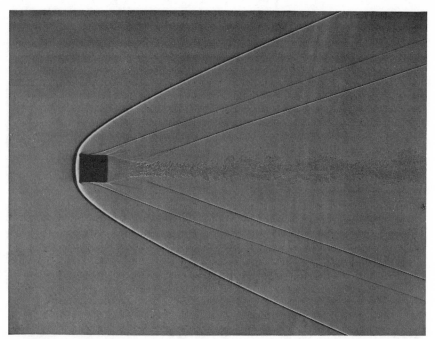

Fig. 6-21. Shadow photograph of a right cylinder (0.50 in. diameter) at a Mach number of 3.55. (Courtesy of Ballistic Research Laboratories, Aberdeen, Md.)

Fig. 6-22. Shadow photograph of a cone-cylinder (20 mm. diameter) at a Mach number of 1.54. (Courtesy of Ballistic Research Laboratories, Aberdeen, Md.)

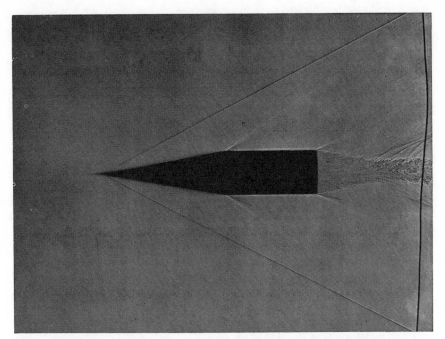

Fig. 6–23. Shadow photograph of a cone-cylinder at a Mach number of 2.32. (Courtesy of Ballistic Research Laboratories, Aberdeen, Md.)

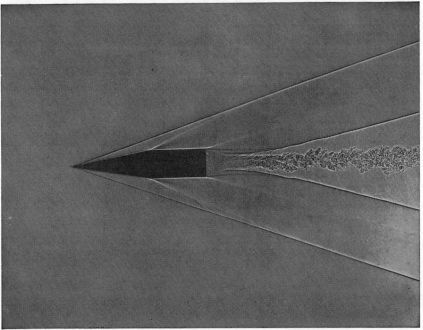

Fig. 6–24. Shadow photograph of a cone-cylinder (0.50 inch diameter) at a Mach number of 3.62. (Couurtesy of Ballistic Research Laboratories, Aberdeen, Md.)

6–15. Velocity and temperature measurements

Various applications require a measurement of velocity and temperature in a stream of fluid. Different factors need to be noted, depending upon the type of flow. Imagine a simple pitot tube or a temperature probe, as indicated in Fig. 6–25. The velocity is V_0 and the density is ρ_0 some distance ahead of the pitot tube. The pressure p_0 in the upstream undisturbed fluid is called the "static pressure." The fluid is brought to rest or zero velocity at the stagnation point S where the "total" or "stagnation pressure" is p_S.

Picture first incompressible flow at a very low Mach number. Then the dynamic equation gives the familiar relation

$$p_S - p_0 = \tfrac{1}{2}\rho_0 V_0^2$$
$$V_0 = \sqrt{\frac{2(p_S - p_0)}{\rho_0}}$$

$$(6\text{–}70)$$

The velocity can be calculated by Equation (6–70).

Picture next the case of compressible flow, but subsonic and without shock. Then Equation (6–31) shows that V_0 is calculated by the expression

$$V_0 = \sqrt{\left(\frac{2k}{k-1}\right)\frac{p_0}{\rho_0}\left[\left(\frac{p_S}{p_0}\right)^{\frac{k-1}{k}} - 1\right]}$$

$$(6\text{–}71)$$

Note that ρ_0 in Equation (6–71) requires a static temperature and a static pressure for calculation.

Let T_0 represent the "static temperature" in the undisturbed stream approaching the probe. This static temperature could be measured by an instrument moving along with the stream. At the stagnation point, where the velocity is zero, the "stagnation temperature" is T_S. Assuming an adiabatic process the energy equation between the undisturbed flow and the stagnation point gives the result

$$\frac{V_0^2}{2} = h_S - h_0 = c_p(T_S - T_0)$$
$$T_S - T_0 = \frac{V_0^2}{2c_p}$$

$$(6\text{–}72)$$

where h is the enthalpy and c_p the specific heat at constant pressure.

Fig. 6–25. Simple pitot tube or temperature probe in a stream.

$T_S - T_0$ is called the "ideal adiabatic temperature rise" between the undisturbed stream and the stagnation point S. The temperature probe itself may indicate a temperature T_A somewhere between the static temperature T_0 and the stagnation temperature T_S. The "recovery factor," R.F., is defined as the ratio of the actual temperature rise divided by the ideal adiabatic temperature rise, or

$$\text{R.F.} = \frac{T_A - T_0}{T_S - T_0} \tag{6–73}$$

The recovery factor can be found from a calibration, as for instance, from a calibration in a wind tunnel.

Picture next the flow around a pitot tube mounted in a supersonic stream, as indicated in Fig. 6–26. Upstream the supersonic velocity is V_0 and the Mach number is M_0. There is a short region forward of the stagnation point in which there is a normal compression shock; across this shock there is a pressure rise from p_0 to p_1 while the Mach number changes from M_0 to the subsonic value M_1. The pressure ratio is given by Equation (6–58) as

$$\frac{p_1}{p_0} = \frac{1 + kM_0^2}{1 + kM_1^2} \tag{6–74}$$

The subsonic flow slows down to zero velocity at the stagnation point S. For this flow we can use the shockless compressible-flow relation given by Equation (6–32), namely

$$\frac{p_S}{p_1} = \left[1 + \left(\frac{k-1}{2} \right) M_1^2 \right]^{\frac{k}{k-1}} \tag{6–75}$$

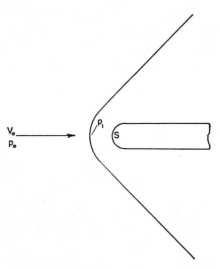

Fig. 6–26. Pitot tube in a supersonic stream.

Using Equations (6–74) and (6–75), the overall pressure ratio then becomes

$$\frac{p_S}{p_0} = \frac{p_S p_1}{p_1 p_0} \qquad (6\text{–}76)$$

We can use Equation (6–63) to express M_1 in terms of M_0. Then the pressure ratio p_S/p_0 can be expressed in terms of k and the upstream Mach number. The final result is

$$\frac{p_S}{p_0} = M_0^2 \left[\frac{k+1}{2}\right]^{\frac{k}{k-1}} \left[\frac{2kM_0^2 - k + 1}{M_0^2(k+1)}\right]^{1-\frac{k}{k-1}} \qquad (6\text{–}77)$$

If p_S, p_0, and k are known, M_0 can be calculated by means of Equation (6–77). Note that Equation (6–77) is for approaching supersonic flow.

6–16. Effect of compressibility on lift and drag forces

As fluid flows around a body, the fluid exerts a total or resultant force on the body. This total force is frequently broken into two components. The "lift" force L is the component of the total force at right angles to the undisturbed, approaching stream. The "resistance" or "drag" force D is the component of the total force parallel to the undisturbed stream.

As illustrated in Fig. 6–27, let V_0 represent the undisturbed velocity and ρ_0 the undisturbed density. Then the lift coefficient C_L is defined as the dimensionless ratio

$$C_L = \frac{L}{\frac{1}{2}\rho_0 V_0^2 A} \qquad (6\text{–}78)$$

where A is some characteristic area. As illustrated in Fig. 6–27, the geometric chord of length b is an arbitrary line usually established by the designer in laying out the section. The area in Equation (6–78) is commonly taken as the chord area for a certain length or span of the section. The drag coefficient C_D is defined as the dimensionless ratio

$$C_D = \frac{D}{\frac{1}{2}\rho_0 V_0^2 A} \qquad (6\text{–}79)$$

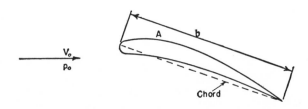

Fig. 6–27. Notation for body in a stream.

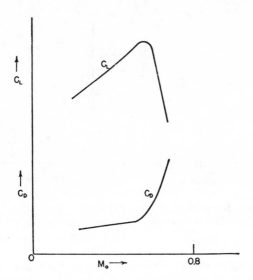

Fig. 6–28. Lift coefficient and drag coefficient as a function of Mach number for a vane section.

Let M_0 represent the approach Mach number, that is $M_0 = V_0/c_0$, where c_0 is the acoustic velocity of the approaching stream. Figure 6–28 illustrates a typical trend. The lift coefficient first increases as M_0 increases from zero. At a certain value of the Mach number the lift coefficient reaches a maximum value. Beyond this Mach number the lift coefficient drops sharply. The drag coefficient first increases gradually as M_0 increases from zero. At a certain M_0 (that for maximum C_L) the drag coefficient starts to increase rapidly. This simultaneous drop in lift and increase in drag apparently is caused by a compression-shock wave extending from the surface to a certain distance from the surface. A "critical" Mach number M_0' is defined as the Mach number M_0 of the undisturbed flow for which the local velocity at some point on the surface reaches the local velocity of sound.

As illustrated in Fig 6–27, let p_0 represent the static pressure in the undisturbed stream and p_A the pressure at some point, as A, on the body. Then the "pressure coefficient" C_p is defined as the dimensionless ratio

$$C_p = \frac{p_A - p_0}{\frac{1}{2}\rho_0 V_0^2} \tag{6–80}$$

Figure 6–29 illustrates the pressure distribution along the low-pressure side of a vane or blade at a low Mach number M_0. The pressure coefficient is plotted for each point along the chord. There is no sharp pressure change along the surface. Figure 6–30 illustrates the pressure distribution for the same blade at a higher Mach number M_0. Starting at the leading edge the pressure first drops; correspondingly, the velocity increases. At point B

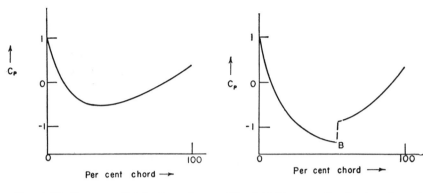

Fig. 6–29. Pressure distribution along the low-pressure side of a blade or vane at a low Mach number M_0.

Fig. 6–30. Pressure distribution along the low-pressure side of a blade or vane at high subsonic Mach number M_0.

there is a compression shock, a pressure rise, and a reduction in velocity. Thus pressure-distribution measurements and force measurements illustrate certain characteristic features.

6–17. Analogy between open-channel flow and flow of compressible fluids

There are features associated with the propagation of a pressure wave through a compressible fluid that are somewhat similar to those involved in the travel of a slight disturbance at the free surface of a liquid. Some of these analogies will be discussed briefly in the following paragraphs.

Imagine liquid at rest, at a depth y, as illustrated in Fig. 6–31(a). Next, picture a surface disturbance of small amplitude traveling with a wave velocity c_w as indicated in Fig. 6–31(b). The next step is to calculate this wave velocity.

The equation of continuity for flow through a channel having a simple, rectangular cross section is

$$Vy = \text{constant} \qquad (6\text{–}81)$$

The energy equation, with no friction, is

$$y + \frac{V^2}{2g} = \text{constant} \qquad (6\text{–}82)$$

Differentiating Equations (6–81) and (6–82) gives

$$V \, dy + y \, dV = 0 \qquad (6\text{–}83)$$

$$dy + \frac{V \, dV}{g} = 0 \qquad (6\text{–}84)$$

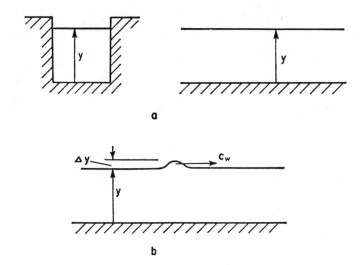

Fig. 6–31. Notation for surface wave action.

Applying Equations (6–83) and (6–84) to the case of a surface wave disturbance as shown in Fig. 6–31(b) gives

$$c_w = \sqrt{gy} \qquad (6\text{–}85)$$

For example, if a stone were dropped into still water, waves would move with velocity c_w. Equation (6–85), however, should not be applied to ocean waves or waves in very deep channels.

For "critical" open-channel flow, the liquid velocity V equals the velocity of propagation of a small surface wave. If a wave were started, it could not progress upstream but would remain stationary because the two velocities are equal. In "subcritical" flow the liquid velocity V is *less* than c_w. In subcritical flow small surface waves would travel upstream. In "supercritical" flow V is *greater* than c_w; small surface waves would be swept downstream.

The flow of a compressible fluid is supersonic if the fluid velocity V is greater than the velocity of sound c. On the other hand, the compressible flow is subsonic if V is less than c.

The continuity equation for steady flow in an open channel has the form

$$Vy = \text{constant} \qquad (6\text{–}86)$$

This relation is analogous to the continuity equation for the steady flow of a compressible fluid through a channel or pipe of constant cross section

$$V\rho = \text{constant} \qquad (6\text{–}87)$$

if the density ρ of the compressible fluid is taken as analogous to the depth y of the liquid flow.

Consider next the energy equation, in differential form without friction, for each flow. Assuming a horizontal channel bottom, the total head for open-channel flow equals $y + (V^2/2g)$. Differentiating this relation gives

$$g\,dy + V\,dV = 0 \qquad (6\text{-}88)$$

Using Equation (6-84) with Equation (6-88) gives an energy relation for open-channel flow, whereas Equation (6-89) gives an energy equation for compressible flow. These relations are

$$c_w^2 \frac{dy}{y} + V\,dV = 0 \qquad (6\text{-}89)$$

and

$$c^2 \frac{d\rho}{\rho} + V\,dV = 0 \qquad (6\text{-}90)$$

The two foregoing equations are similar. Thus both the continuity equation and the energy equation for compressible flow are analogous to those for open-channel flow if depth y and density ρ are regarded as corresponding quantities.

The "hydraulic jump" in open-channel flow is similar to the compression shock in compressible flow. Before the hydraulic jump the liquid moves with a high velocity at a low depth; behind the hydraulic jump the liquid moves with a low velocity at a high depth. Before the compression shock the gas moves with a high velocity at a low density (low pressure); behind the shock the fluid moves with a low velocity at a high density (high pressure).

6–18. Types of flow around a body

Picture the flow around a body. The flow around the body is called "subsonic" if the Mach number is less than one at every point in the field of flow. The flow around the body is "supersonic" if the Mach number is greater than one at every point in the field of flow. Intermediate between these two types is "transonic" flow. In transonic flow around a body there is a region of supersonic flow and a region of subsonic flow.

Figure 6–32 illustrates subsonic flow. Figure 6–33 illustrates super-

Fig. 6–32. Subsonic flow around a body.

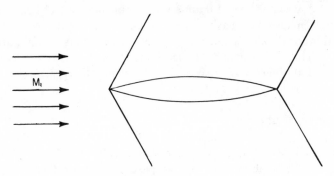

Fig. 6–33. Supersonic flow around a body.

sonic flow around a body with sharp leading and trailing deges. There is an oblique shock attached at the leading edge; the flow is supersonic behind the oblique shock. Figure 6–34 illustrates a case of transonic flow.

In Fig. 6–34 the upstream Mach number M_1 is less than unity; the upstream flow is subsonic. Because of the shape of the body, however, the velocity of flow over the body is at first increased. Thus, when the fluid passes over the body, there is a supersonic flow and then a compression shock. The flow is subsonic behind the compression shock. Such transonic flow may exist over a compressor blade. Along part of the blade the pressure may be low and the velocity high.

In Fig. 6–26 the upstream flow is supersonic. Because the leading edge has a finite radius of curvature or is somewhat blunt instead of mathematically sharp, there is a head or bow wave detached from the body. Directly upstream from the stagnation point on the body there is a short length of the head wave which is a normal compression shock. There is subsonic flow between this normal compression shock and the stagnation point on the body.

In some cases the term "supersonic" may be used in the restricted sense for a Mach number range up to about 5. The term "hypersonic" flow may be used for higher Mach numbers, as for Mach numbers ranging from about 5 to values higher than 15. The exact distinction between supersonic and

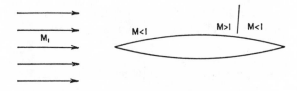

Fig. 6–34. Transonic flow around a body.

hypersonic flow cannot be set up in a very simple manner. The exact distinction might be made on the basis of the character or type of the partial differential equation for the flow regime. For supersonic flow the partial differential equation might be linear and of the elliptic type, whereas for hypersonic flow the partial differential equation is of a hyperbolic type.

6–19. Flow of vapors

If we analyze the flow of a gas which follows the simple relation $p = \rho RT$, we usually obtain relatively simple equations for the direct solution of problems. Various relations for the flow of this ideal gas have been worked out in the preceding paragraphs. The following gives some discussion of the method of attack for the flow of vapors.

It is to be emphasized that the basic equations, as energy, continuity, and dynamic or momentum, apply to the flow of *any* fluid. If we have a problem involving the flow of a vapor. we simply use the basic equations together with data from suitable vapor tables and charts.

The energy equation has the form

$$q + W = h_2 - h_1 + \frac{V_2^2 - V_1^2}{2} + (z_2 - z_1)g \qquad (6\text{–}91)$$

As an example, imagine adiabatic flow through a horizontal, converging-diverging channel, with no work W. The energy equation takes the special form

$$\frac{V_2^2 - V_1^2}{2} = h_1 - h_2 \qquad (6\text{–}92)$$

Say we are given the velocity at section 1, and the pressures and temperatures at sections 1 and 2. The velocity at section 2 becomes

$$V_2 = \sqrt{V_1^2 + 2(h_1 - h_2)} \qquad (6\text{–}93)$$

For a vapor we can consult suitable vapor charts or tables, determine the enthalpy difference, and solve for the velocity V_2.

As another example, picture an orifice or valve in a constant diameter, horizontal pipe. Section 1 is upstream from the orifice, and section 2 is downstream from the orifice. If the process is adiabatic, $z_2 - z_1 = 0$, W is zero, and $V_2 = V_1$, then the energy relation becomes

$$h_1 = h_2 \qquad (6\text{–}94)$$

The enthalpy at section 1 equals the enthalpy at section 2. If we know all the properties at one section, but do not know one property at the other section, the one unknown property can be found by using Equation (6–94).

Certain terms are useful in dealing with vapor flow problems. A brief reference to these terms will be given here.

A "reversible" process is a frictionless process in which the system is always infinitesimally close to a state of equilibrium. The term "entropy" is a state property of a fluid; other properties of a fluid are pressure, temperature, specific volume, and enthalpy. Let dQ represent heat added during a reversible heating (or cooling) process. Let T represent absolute temperature. Then the entropy change $S_2 - S_1$ is defined as

$$S_2 - S_1 = \int_1^2 \frac{dQ}{T} \tag{6-95}$$

In a frictionless, reversible, adiabatic process there is no change in entropy; the process is isentropic.

In studying the flow of vapors, one very useful chart is the Mollier diagram, a plot of enthalpy as a function of entropy. Charts with numerical values are usually included with vapor tables. Figure 6–35 illustrates this chart. The saturated-vapor line (shown solid) separates the wet vapor region from the superheated vapor region. The dotted line is a line of constant pressure. Other lines, as lines of constant temperature, are normally included. A frictionless adiabatic, or an isentropic, process follows a vertical line on the enthalpy-entropy diagram. A constant enthalpy process follows a horizontal line.

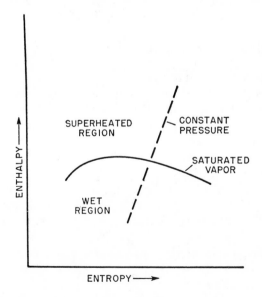

Fig. 6–35. Diagrammatic plot of enthalpy versus entropy for a vapor.

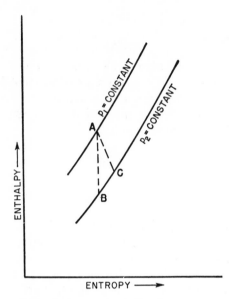

Fig. 6-36. Comparison of two processes on the enthalpy-entropy diagram.

As an example, picture the flow of a vapor in a nozzle from section 1, where the pressure is p_1, to section 2, where the pressure is p_2. In Fig. 6-36 there is a line $p_1 = $ constant, and line p_2 is constant. Say the condition at section 1 is represented by point A, at a certain temperature and with a certain enthalpy. Assume expansion through the nozzle follows the process A to B; this is a frictionless adiabatic process. The process, as an alternate, could go from pressure p_1 to p_2 following an adiabatic flow, but with friction; the final state is indicated by point C. The process without friction has no entropy change. With friction there is an entropy increase.

COMPRESSIBLE FLOW IN CHANNELS WITH FRICTION AND HEAT TRANSFER

6-20. General relations for pipe flow

The relations for incompressible pipe flow are relatively simple and straightforward. In many cases an incompressible-flow study alone is sufficient. In other cases, however, the compressibility effects are important and must be taken into account. As an illustration, when a compressible fluid flows steadily in a long pipe, for subsonic flow the pressure generally decreases because of frictional resistance. The density along the pipe decreases, and the velocity along the pipe increases. In order to analyze an action of this sort, we must put the basic relations in differential form and integrate.

Figure 6-37(a) illustrates the notation for a horizontal pipe of diameter D. The flow is in the direction of positive *l*. The pressure at any point is *p*, and the velocity at any point is *V*. The infinitesimal element has a length *dl* and density ρ. The pressure on the left face of the element is *p*, and the pressure on the right face is $p + dp$. Figure 6-37 (b) shows a free body diagram. The mass of the element is

$$\frac{\pi D^2}{4} \rho \, dl$$

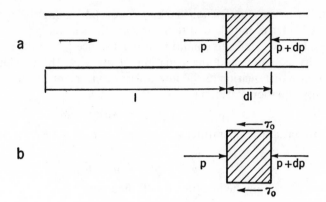

Fig. 6–37. Notation for pipe flow.

For steady flow the acceleration is a function of length only. Then the acceleration becomes

$$\frac{dV}{dt} = \frac{dV}{dl} \qquad \frac{dl}{dt} = V\frac{dl}{dt}$$

The shear stress τ_0 acts on the outer surface of the cylindrical mass of diameter D. The net force acting on the mass in the positive l direction equals mass time acceleration, or

$$[p - (p + dp)]\frac{\pi D^2}{4} - \tau_0 \pi D \, dl = \left(\frac{\pi D^2}{4}\rho \, dl\right)V\frac{dV}{dl} \qquad (6\text{–}96)$$

Using Equation (5–68), Equation (6–96) reduces to

$$dp + \rho V \, dV + f\rho\frac{V^2 \, dl}{2D} = 0 \qquad (6\text{–}97)$$

Equation (6–97) is the basic differential equation; each term has the units of pressure.

6–21. Isothermal flow in pipes with friction

This article discusses the isothermal, or constant-temperature, case of pipe flow. Equation (6–97) can be arranged in the form

$$\frac{2}{\rho V^2}\,dp + \frac{2\,dV}{V} + \frac{f\,dl}{D} = 0 \qquad (6\text{–}98)$$

Several terms in Equation (6–98) require investigation before integrating. The friction factor f is a function of the Reynolds number $\rho VD/\mu$. For many cases it is common to regard the dynamic viscosity μ as a function of temperature only; thus for an isothermal process the dynamic viscosity is constant. For a constant-area channel the continuity equation shows that ρV

is constant. Therefore, the friction factor f is constant along the pipe. Let the subscript 1 refer to initial conditions or conditions at the pipe entrance, and let the subscript 2 refer to condition at the distance l from the entrance. The continuity equation and the equation of state can be combined to put the first term of Equation (6–98) into a form convenient for integration. The continuity equation states that

$$A_1 \rho_1 V_1 = A_2 \rho_2 V_2 = A\rho V$$

For gases at constant temperature, $p_1/\rho_1 = p/\rho = p_2/\rho_2 = RT$. Thus

$$\frac{\rho_1}{\rho} = \frac{V}{V_1} = \frac{p_1}{p} \qquad \frac{1}{\rho V^2} = \frac{p}{\rho_1 V_1^2 p_1}$$

Equation (6–98) can now be integrated with the limits

$$\frac{2}{\rho_1 V_1^2 p_1} \int_2^1 p \, dp + 2 \int_2^1 \frac{dV}{V} + \frac{f}{V} \int_l^0 dl = 0$$

$$p_1^2 - p_2^2 = p_1 \rho_1 V_1^2 \left[2 \log_e \frac{V_2}{V_1} + \frac{fl}{D} \right]$$

(6–99)

Let M_1 represent the initial Mach number, which is calculated by the relation

$$M_1 = \frac{V_1}{c_1} = \frac{V_1}{\sqrt{kRT_1}}$$

(6–100)

It is convenient to rearrange Equation (6–99) such that the dimensionless ratio fl/D is a function of the other dimensionless ratios k, M_1, and a pressure ratio. A regrouping of terms gives

$$\frac{fl}{D} = \frac{1}{kM_1^2} \left[1 - \left(\frac{p_2}{p_1} \right)^2 \right] - 2 \log_e \frac{p_1}{p_2}$$

(6–101)

Using the equations of state and continuity, Equation (6–98) can be arranged in the form

$$\frac{dp}{dl} = \frac{\dfrac{pf}{2D}}{1 - \dfrac{p}{\rho V^2}}$$

(6–102)

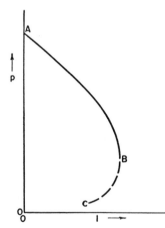

Fig. 6–38. Diagrammatic plot of pressure versus length for flow of gas in a pipe.

Figure 6–38 illustrates a plot of pressure p versus length l. From Equation (6–102), if the ratio $p/\rho V^2$ is greater than unity, then the slope of the curve, or dp/dl, is negative. A negative slope exists in the region between points A and B in Fig. 6–37; the pressure drops with an increase in length. When $p/\rho V^2$ is less than unity, the slope of the curve, dp/dl, is positive; this case exists in the region between points C and B. At point B the slope of the curve is vertical, and

$p/\rho V^2 = 1$. Thus there is a limiting or critical condition, as represented by point B.

Let p_0 be the limiting or critical pressure, V_0 the limiting velocity, and ρ_0 the critical density. From the equation of state, the equation of continuity, and the relation $p_0 = \rho_0 V_0^2$ there results

$$\frac{p_0}{p_1} = \frac{V_1}{V_0} \qquad \frac{p_0}{p_1} = M_1\sqrt{k} \tag{6-103}$$

Figure 6–39 shows a plot of pressure ratio p_2/p_1 versus fl/D for several initial Mach numbers. Point B illustrates the limiting condition for $M_1 = 0.10$. The dotted line represents limiting conditions for different initial Mach numbers.

The pressure drop for the flow of an incompressible fluid in a horizontal pipe can be written as

$$p_1 - p_2 = \rho_1 \frac{flV_1^2}{2D}$$
$$\frac{p_2}{p_1} = 1 - \frac{k}{2}M_1^2\left(\frac{fl}{D}\right) \tag{6-104}$$

Equation (6–104) was employed to plot the straight dot-dash lines in Fig. 6–39. A comparison of a dot-dash line and a solid line for the same initial Mach number brings out the difference between compressible and in-

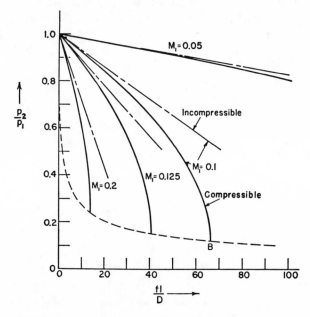

Fig. 6–39. Variations of pressure ratio with fl/D and Mach number for isothermal flow, for gases with $k = 1.4$.

compressible flow. The difference is not very great at low Mach numbers and short lengths. The difference becomes appreciable, however, at the high Mach numbers and long lengths.

Note that isothermal flow involves both heat transfer and friction. For all subsonic Mach numbers, pipe friction causes the temperature to decrease. As the fluid passes down the pipe, the pressure drops because of friction and the temperature tends to drop. For Mach numbers up to $1/\sqrt{k}$, the addition of heat to the fluid causes the temperature to increase. It is possible for the two effects to cancel each other and to obtain a constant temperature. At $M = 1/\sqrt{k}$ the addition or subtraction of heat does not affect the temperature; therefore the temperature must decrease because of friction. For $1 > M > 1/\sqrt{k}$, the extraction of heat causes the temperature to rise, but heat cannot be extracted by forced convection at a rate that is high enough to cancel the effect of friction. Therefore the temperature cannot remain constant.

6–22. Flow of gases in insulated pipes

This article will be devoted to the steady flow of a gas in a horizontal pipe so insulated that no heat passes through the pipe walls. If the flow were frictionless, then the process would follow the simple frictionless adiabatic equation $p/\rho^k = $ constant. With friction, however, the process does not follow this simple relation.

With the notation shown in Fig. 6–40, the energy Equation (6–19) shows that

$$\frac{V_2^2}{2}\left(\frac{k-1}{k}\right) + \frac{p_2}{\rho_2} = \frac{V_1^2}{2}\left(\frac{k-1}{k}\right) + \frac{p_1}{\rho_1}$$

$$\frac{p}{\rho}\left[1 + \frac{(k-1)V^2\rho}{2kp}\right] = \text{constant} \qquad (6\text{–}105)$$

Equation (6–105) is the equation for a gas flowing in a pipe with friction. If W is the constant mass rate of flow and A is the constant pipe area, then $W = \rho A V$, and

$$\frac{p}{\rho} + \left(\frac{k-1}{k}\right)\frac{W^2}{2A^2\rho^2} = \text{constant} \qquad (6\text{–}106)$$

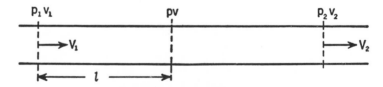

Fig. 6–40. Notation for pipe flow.

Note that the factor $[(k - 1)/k](W^2/2A^2)$ is a constant for a particular problem of steady flow.

Differentiation of the equation of continuity, $W = AV\rho = $ constant, gives

$$d\rho = -\rho \frac{dV}{V} \tag{6-107}$$

For convenience in intermediate steps, let Equation (6–106) be written in the form

$$\frac{p}{\rho} + \frac{b}{\rho^2} = G \tag{6-108}$$

where b and G are constants.

Equations (6–107) and (6–108) can be employed to arrange the basic Equation (6–97) in integrable form. The final result is

$$\int_0^l \frac{f\,dl}{D} = 2\int_{p_1}^{p_2} \frac{dp}{\sqrt{p^2 + 4bG}} - \frac{A^2}{GW^2}\int_p^{p_2} (p + \sqrt{p + 4bG})\,dp \tag{6-109}$$

Several terms in Equation (6–109) require investigation before an integration can be made. D, G, and b are constants; f is a function of the Reynolds number $\rho VD/\mu$. For a constant diameter pipe, ρV is constant, and thus the Reynolds number is primarily a function of dynamic viscosity. The change in viscosity and f is usually small, praticularly at high Reynolds numbers. f will be regarded as a constant for integration purposes. In a practical problem, the initial and final Reynolds numbers can be computed, and an average value of f used. The integration of Equation (6–109) can be expressed in the form

$$\frac{fl}{D} = \frac{k-1}{kB}\left[1 - \left(\frac{p_2}{p_1}\right)^2 - \frac{p_2}{p_1}\sqrt{\left(\frac{p_2}{p_1}\right)^2 + B} + \sqrt{1 + B}\right]$$

$$+ \frac{k+1}{k}\log_e \frac{\dfrac{p_2}{p_1} + \sqrt{\left(\dfrac{p_2}{p_1}\right)^2 + B}}{1 + \sqrt{1 + B}} \tag{6-110}$$

where $B = (k - 1)M_1^2[2 + (k - 1)M_1^2]$.

Before discussing the integrated relation further, it is helpful to investigate the slope of a curve of pressure versus length. Differentiation of Equation (6–105) and use of Equation (6–109) gives the final relation

$$\frac{dp}{dl} = -\frac{f\rho V^2}{2D}\left[\frac{1 + (k-1)M^2}{1 - M^2}\right] \tag{6-111}$$

Equation (6–111) illustrates several fundamental features of pipe flow. If M is low and can be neglected in comparison with unity, then Equation (6–111) reduces to the familiar expression for incompressible flow. If M is less than unity, Equation (6–111) shows a negative value for dp/dl, as illustrated by the variation between A and B in Fig. 6–38; this means that the pressure drops in the direction of flow. When M is greater than unity, Equation (6-111) shows a positive value for dp/dl, as illustrated by the dotted curve between C and B in Fig. 6-38. A critical or limiting condition is reached when the Mach number is unity; at this condition the fluid velocity equals the velocity of sound. The pressure at B is the lowest pressure that can be reached in the pipe for subsonic flow. This limiting condition occurs only at the end of the pipe.

At the limiting pressure p_0 the limiting Mach number M_0 is unity. From Equation (6-105)

$$\frac{p_0}{\rho_0}\left[1 + \frac{k-1}{2}\right] = \frac{p_1}{\rho_1}\left[1 + \left(\frac{k-1}{2}\right)M_1^2\right] \qquad (6\text{-}112)$$

From the equation of continuity

$$\frac{W}{A} = \rho V = \rho_1 V_1 = \rho_0 V_0 \qquad (6\text{-}113)$$

The acoustic velocity at the initial state is $c_1 = \sqrt{kp_1/\rho_1}$, and the acoustic velocity at the limiting condition is $c_0 = V_0 = \sqrt{kp_0/\rho_0}$. Combining these two relations with Equation (6-112) gives

$$\frac{\rho_1 p_0}{\rho_0 p_1} = \frac{c_0^2}{c_1^2} = \left(\frac{k-1}{k+1}\right)\left[\frac{2}{k-1} + M_1^2\right]$$

$$\frac{c_0}{V} = \left\{\left(\frac{k-1}{k+1}\right)\left[1 + \frac{2}{M_1^2(k-1)}\right]\right\}^{\frac{1}{2}} \qquad (6\text{-}114)$$

Using the relations expressed in Equation (6-113) results in

$$\frac{\rho_1}{\rho_0} = \left\{\left(\frac{k-1}{k+1}\right)\left[1 + \frac{(k-1)M_1^2}{2}\right]\right\}^{\frac{1}{2}} \qquad (6\text{-}115)$$

Substituting the foregoing relation in Equation (6-112) gives

$$\frac{p_0}{p_1} = M_1^2\left\{\left(\frac{k-1}{k+1}\right)\left[1 + \frac{2}{(k-1)M_1^2}\right]\right\}^{\frac{1}{2}} \qquad (6\text{-}116)$$

Equations (6-115) and (6-116) provide means for calculating the limiting pressure and specific volume.

Figure 6-41 shows a plot of pressure ratio against fl/D for different initial Mach numbers. The solid curves were plotted using Equation (6-110). The dotted curve represents limiting conditions. The dot-dash lines are for incompressible flow.

6-23. Supersonic diffusers

In Sec. 6-8 a treatment is presented of the flow in a converging-diverging nozzle. In such a nozzle the gas starts at a low velocity, is accelerated to the sonic velocity in the converging part, and is accelerated to supersonic velocities in the diverging part. With proper inlet and exit pressures and with

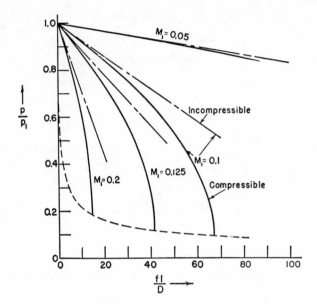

Fig. 6–41. Variation of pressure ratio with fl/D and Mach number for adiabatic flow, for gases with $k = 1.4$.

a proper shape, the process can be smooth and continuous with no shock. The continuous process is stable. The energy losses in such a nozzle can be made relatively small.

Let us picture the reverse process, that of a supersonic diffuser. Supersonic flow enters a converging-diverging passage. One might wonder about the possibility of obtaining a smooth, continuous deceleration through the velocity of sound to subsonic velocities, and a corresponding compression of the gas. No experimental data are available to show that the continuous deceleration is possible. Instead, experimental results show a process with a normal shock. Tests indicate that a normal shock in the throat of the diffuser is stable, whereas a normal shock in the converging part is unstable. Thus it is concluded that smooth continuous deceleration from supersonic to subsonic flow is unattainable.

A simplified study of diffuser efficiency will be presented. This study is helpful in gaining a general picture of the transformation taking place in the diffuser. As illustrated in Fig. 6–42(a) the static pressure is p_1 at the inlet where the flow is supersonic, the static pressure at any variable position is p, and the exit static pressure is p_2 where the flow is subsonic or at a very low velocity. Let us first take the process from 1 to 2 as a frictionless adiabatic. Then the following relations hold:

$$p_1 v_1{}^k = p v^k = p_2 v_2{}^k \tag{6–117}$$

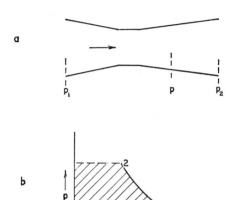

Fig. 6-42. Supersonic diffuser with a frictionless process.

where v, in this case, is taken as $1/\rho$. Let us form the integral of $v\,dp$ between the limits 1 and 2; this integral gives the shaded area in Fig. 6-42(b). Using Equation (6-117) and the relation $pv = R_0 T$, this integral becomes

$$\int_1^2 v\,dp = \left(\frac{k}{k-1}\right)p_1 v_1\left[\left(\frac{p_2}{p_1}\right)^{\frac{k-1}{k}} - 1\right] = c_p(T_2 - T_1) \qquad (6\text{-}118)$$

c_p is the specific heat at constant pressure. The energy equation shows that

$$c_p(T_2 - T_1) = \frac{V_1^2 - V_2^2}{2} \qquad (6\text{-}119)$$

Thus, for a frictionless adiabatic process,

$$\int_1^2 v\,dp = \frac{V_1^2 - V_2^2}{2} \qquad (6\text{-}120)$$

If compression follows a frictionless adiabatic process from p_1 to p_2, all of the available kinetic-energy change is converted into a useful increase in pressure flow-work.

With friction and impact, however, not all the available kinetic energy is transformed into useful increase in flow-work; some of the available kinetic energy is degraded into unavailable energy.

Picture a diffuser in which the actual inlet pressure is p_1, the actual outlet pressure is p_2, and the actual inlet temperature is T_1. If the compression process were without energy loss, the temperature at p_2 would be T_2. With losses, however, the actual temperature at p_2 is T_2'; T_2' is higher than T_2. Let V_2' represent the velocity at the diffuser exit for the actual flow with the actual temperature T_2'. The energy equation shows that

$$c_p(T_2' - T_1) = \frac{V_1^2 - (V_2')^2}{2}$$

The term $c_p(T_2 - T_1)$ is the enthalpy change for a frictionless adiabatic process. The term $c_p(T_2' - T_1)$ is the enthalpy change for the actual process. A diffuser efficiency E is defined as the ratio

$$E = \frac{c_p(T_2 - T_1)}{c_p(T_2' - T_1)} \qquad (6\text{–}121)$$

An alternate form for the diffuser efficiency is

$$E = \frac{\int_1^2 v\, dp}{\dfrac{V_1^2 - (V_2')^2}{2}} \qquad (6\text{–}122)$$

E might be called a "simple" or "practical" efficiency term.

Figure 6–43 illustrates the notation for one type of supersonic diffuser. The inlet Mach number M_1 is above one and the corresponding pressure is p_1. Behind the normal shock the Mach number is M_0, less than one, and the corresponding pressure is p_0. Behind the shock the subsonic flow slows down in the diverging passage to p_2. Assume that the exit velocity is negligible in comparison with the inlet velocity. Then the diffuser efficiency becomes

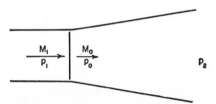

Fig. 6–43. Notation for a supersonic diffuser.

$$E = \frac{\left(\dfrac{2}{k-1}\right)\left[\left(\dfrac{p_2}{p_1}\right)^{\frac{k-1}{k}} - 1\right]}{M_1^2} \qquad (6\text{–}123)$$

Section 6–10 gives the following relations for conditions across the shock:

$$\frac{p_0}{p_1} = \frac{1 + kM_1^2}{1 + kM_0^2} \qquad (6\text{–}124)$$

$$M_0^2 = \frac{M_1^2 + \dfrac{2}{k-1}}{\left(\dfrac{2k}{k-1}\right)M_1^2 - 1} \qquad (6\text{–}125)$$

The next step is to evaluate the loss in the subsonic section. Let us assume an average and constant value of E_0 for the efficiency of the subsonic diffuser. Expressed in terms of differentials, we can write

$$E_0 = \frac{v\, dp}{c_p\, dT}$$

$$E_0 c_p\, dT = v\, dp = \frac{R_0 T}{p}\, dp$$

Integrating between limits gives

$$\frac{p_2}{p_0} = \left(\frac{T_2'}{T_0}\right)^{\frac{E_0 k}{k-1}} \tag{6-126}$$

Neglecting the exit velocity, the energy equation for the subsonic diffuser section gives

$$c_p T_0 + \frac{V_0^2}{2} = c_p T_2'$$

$$\frac{T_2'}{T_0} = 1 + M_0^2\left(\frac{k-1}{2}\right) \tag{6-127}$$

Combining Equations (6–126) and (6–127) gives the pressure ratio

$$\frac{p_2}{p_0} = \left[1 + M_0^2\left(\frac{k-1}{2}\right)\right]^{\frac{E_0 k}{k-1}} \tag{6-128}$$

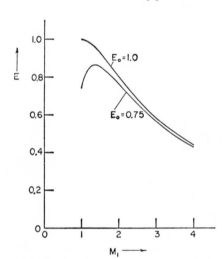

The pressure ratio across the entire diffuser is given by the relation

$$\frac{p_2}{p_1} = \left(\frac{p_2}{p_0}\right)\left(\frac{p_0}{p_1}\right) \tag{6-129}$$

For a given inlet Mach number, we can calculate the overall pressure ratio by means of Equations (6–124), (6–128), and (6–129). Substitution of this overall pressure ratio in Equations (6–123) gives the diffuser efficiency.

Figure 6–44 shows a plot of E versus M_1 as calculated by the foregoing relations for several values of E_0. The curve for $E_0 = 1.0$ represents a frictionless subsonic diffuser. A value of $E_0 = 0.75$ probably represents a practical value for the subsonic diffuser.

Fig. 6–44. Efficiency E versus Mach number for supersonic diffusers, $k = 1.4$.

A combination of a supersonic nozzle and a supersonic diffuser, as illustrated in Fig. 6–45, presents a problem in starting. If the flow in a supersonic nozzle is without shock, the energy loss across the nozzle is relatively small. But to start this flow requires certain features. Picture first the fluid at rest in the channel. As the downstream pressure is lowered, sonic flow is reached in the nozzle throat. As the downstream pressure is lowered further, a normal compression shock forms behind the nozzle throat as illustrated in Fig. 6–45. Behind this shock the flow is subsonic. In the converging part of the diffuser the flow is then accelerated.

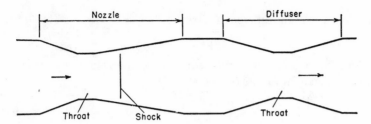

Fig. 6–45. Combination of supersonic nozzle and supersonic diffuser.

Across the shock there is an appreciable energy loss. Imagine the case in which the diffuser throat has the same area as the nozzle throat. After the shock, the total pressure or total available energy is not the same as that before the shock. This means that the acceleration in the nozzle cannot be exactly reversed; the flow in the nozzle throat cannot be exactly re-established in the diffuser throat. Picture the case in which the normal shock is in the diverging part of the nozzle and the flow at the diffuser throat is sonic. In this case the shock is blocked in between the two throats. More mass rate of flow through the diffuser throat is necessary in order to move the shock further downstream. This can be done by making the diffuser-throat area larger than the nozzle-throat area.

6–24. General case of compressible flow in a variable-area channel with heat transfer and friction

In the foregoing sections, various specific cases for one-dimensional compressible flow have been worked out and certain characteristics have been demonstrated. The most general case is that of compressible flow in a variable-area channel with heat transfer and friction. This general case is very complicated; for some conditions it is very difficult and tedious to solve. An approach to this general case, however, can be outlined by a study of the basic equations in differential form.

As illustrated in Fig. 6–46, picture the flow in some channel in which the area varies. At a certain section the velocity is V, the area is A, the density is ρ, the absolute static pressure is p, and the absolute static temperature is T.

The equation of continuity ($\rho V A =$ con-

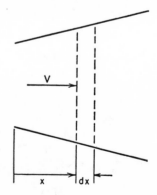

Fig. 6–46. Notation for general case.

stant) can be put in the differential form

$$d(\rho VA) = 0 \tag{6-130}$$

We will assume a gas following the equation of state $p/\rho = RT$. This can be put in the differential form as

$$d\left(\frac{p}{R\rho T}\right) = 0 \tag{6-131}$$

We will neglect changes in potential energy. Using Equation (6–16) the energy equation in differential form becomes

$$dQ = c_p dT + VdV \tag{6-132}$$

where dQ is the heat added to the fluid and c_p is the specific heat at constant pressure.

Across an infinitesimal element there is a velocity change dV and a pressure change dp; let dF represent the work done against friction or the frictional head loss per unit mass. From the dynamic or momentum equation, the net force acting on an infinitesimal element equals mass rate of flow times the velocity change, or

$$-dp - \rho dF = \rho V \, dV \tag{6-133}$$

Equations (6–130), (6–131), (6–132), and (6–133) are the tools we use in tackling the general case.

It is convenient in manipulating the equations to use as a variable the square of the Mach number M. Let $N = M^2$, or

$$N = M^2 = \frac{V^2}{kRT} \tag{6-134}$$

where k is the ratio of specific heats. It is also convenient to express the relations in logarithmic differential form. For example, the equation of continuity ($\rho VA =$ constant) can be expressed in the form

$$\log \rho + \log V + \log A = \log \text{constant}$$

Differentiating each term gives the logarithmic differential form

$$\frac{d\rho}{\rho} + \frac{dV}{V} + \frac{dA}{A} = 0$$

Let us put Equation (6–131) in logarithmic form and differentiate it, divide Equation (6–132) by $c_p T$ and make use of Equation (6–7), and then divide Equation (6–134) by p. Consequently, Equations (6–130), (6–131), (6–132), and (6–133) take the form

$$\frac{d\rho}{\rho} + \frac{dV}{V} + \frac{dA}{A} = 0 \tag{6-135}$$

$$-\frac{dp}{p} + \frac{d\rho}{\rho} + \frac{dT}{T} = 0 \tag{6-136}$$

$$\frac{dQ}{c_p T} = \frac{dT}{T} + \frac{V^2 \, dV(k-1)}{VkRT} \tag{6-137}$$

$$-\frac{\rho \, dF}{p} - \frac{dp}{p} = \frac{kV^2}{kRT}\frac{dV}{V} \tag{6-138}$$

A logarithmic differentiation of Equation (6–134) gives

$$\frac{dV}{V} = \frac{1}{2}\left(\frac{dN}{N} + \frac{dT}{T}\right) \tag{6-139}$$

The factor dp/p can be eliminated by solving Equations (6–135) and (6–136) simultaneously. Equation (6–139) can be used to eliminate dV/V from Equations (6–135) to (6–138). The result is

$$\frac{dA}{A} + \frac{1}{2}\frac{dN}{N} - \frac{1}{2}\frac{dT}{T} + \frac{dp}{p} = 0 \tag{6-140}$$

$$\frac{dQ}{c_p T} = \left[1 + N\left(\frac{k-1}{2}\right)\right]\frac{dT}{T} + \left(\frac{k-1}{2}\right)\frac{N \, dN}{N} \tag{6-141}$$

$$\frac{dF}{RT} = -\frac{dp}{p} - \frac{kN \, dN}{2N} - \frac{kN}{2}\frac{dT}{T} \tag{6-142}$$

By solving Equations (6–140), (6–141), and (6–142) simultaneously we can find a value for dN/N, a value for dp/p, and a value for dT/T. One systematic method of solving these equations is to use determinants. The results are listed in Table 6–1. Each coefficient in the table represents the partial derivative of the variable in the left-hand vertical column with respect to the variable in the top horizontal row. For example,

$$\frac{dN}{N} = \frac{\dfrac{dQ}{c_p T}(1+kN) + \dfrac{dF}{RT}[2+(k-1)N] - \dfrac{dA}{A}[2+(k-1)N]}{1-N} \tag{6-143}$$

TABLE 6-1

COEFFICIENTS FOR BASIC EQUATIONS

	$\dfrac{dQ}{c_p T}$	$\dfrac{dA}{A}$	$\dfrac{dF}{RT}$
$\dfrac{dN}{N}$	$\dfrac{1+kN}{1-N}$	$-\dfrac{[2+(k-1)N]}{1-N}$	$\dfrac{2+(k-1)N}{1-N}$
$\dfrac{dp}{p}$	$\dfrac{-kN}{1-N}$	$\dfrac{kN}{1-N}$	$-\dfrac{(k-1)N+1}{1-N}$
$\dfrac{dT}{T}$	$\dfrac{1-kN}{1-N}$	$\dfrac{(k-1)N}{1-N}$	$-\dfrac{(k-1)N}{1-N}$
$\dfrac{d\rho}{\rho}$	$-\dfrac{1}{1-N}$	$\dfrac{N}{1-N}$	$-\dfrac{1}{1-N}$
$\dfrac{dV}{V}$	$\dfrac{1}{1-N}$	$-\dfrac{1}{1-N}$	$\dfrac{1}{1-N}$

The first three horizontal rows in Table 6–1 can be used to derive other relations. For example, Equation (6–136) gives

$$\frac{d\rho}{\rho} = \frac{dp}{p} - \frac{dT}{T}$$

Knowing the coefficients for dp/p and dT/T, we can find the coefficients for $d\rho/\rho$ directly from the Table. Using Equation (6–139), the same technique can be used for dV/V.

Each flow variable is a function of a single variable parameter, such as the distance x along the axis of the tube or channel; hence the term "one-dimensional." x is regarded as positive in the direction of flow. The equations represented in Table 6–1 can be written in a different form involving differentiation with respect to x. For example, the first relation for dN/N can be written as

$$\frac{dN}{dx} = \frac{\frac{dQ}{dx}[1 + kN]\frac{N}{c_pT} + \frac{dF}{dx}[2 + (k-1)N]\frac{N}{RT} - \frac{dA}{dx}[2 + (k-1)N]\frac{N}{A}}{1 - N}$$

(6–144)

Equations can be written for dp/dx, dT/dx, and the other parameters. Except at $N = 1$, a solution exists for these equations as soon as the functions Q, F, and A are specified in terms of x. Solutions may be obtained formally or by numerical or graphical methods. It may be a tedious job to get some solutions, but solutions exist.

In practical applications it is frequently useful to determine the trend of flow variables with respect to heat addition, friction, or area variation without bothering to get quantitative information from integrated forms. For example, inspection of Equation (6–144) shows that in subsonic flow the effect of positive dQ/dx, positive dF/dx, and negative dA/A is to increase N. For supersonic flow, the effect is to decrease N.

As another example, picture compressible flow in a constant-diameter pipe with no friction. The first three rows in Table 6–1 give

$$\frac{dN}{dx} = \frac{N}{c_pT}\left(\frac{1 + kN}{1 - N}\right)\frac{dQ}{dx}$$

(6–145)

$$\frac{dp}{dx} = -\frac{kN}{1 - N}\frac{p}{c_pT}\frac{dQ}{dx}$$

(6–146)

$$\frac{dT}{dx} = \frac{1 - kN}{1 - N}\frac{1}{c_p}\frac{dQ}{dx}$$

(6–147)

Equations (6–145) and (6–146) show that in a subsonic flow, as heat is added along the pipe, the Mach number increases and the pressure drops. Refer to Equation (6–147) for subsonic flow. When kN is less than one, addition of heat along the pipe results in a temperature increase. When kN is greater than one, for subsonic flow, an addition of heat causes a temperature drop.

As another example, consider adiabatic frictionless flow. The fifth row in Table 6–1 gives the relation

$$\frac{dV}{V} = -\left(\frac{1}{1 - M^2}\right)\frac{dA}{A}$$

$$\frac{dA}{A} = -\frac{dV}{V}(1 - M^2)$$

which is identical to Equation (6–40).

The term dF was defined as the work done against friction or the frictional head loss per unit mass. For incompressible and compressible flow in pipes we express dF in the form

$$dF = \frac{fV^2\,dx}{2D} \tag{6-148}$$

where D is the internal pipe diameter, dx is the distance along the pipe axis, and f is the usual pipe-friction factor. The pipe-friction factor is a function of the Reynolds number. Substituting Equation (6–148) in Equation (6–133) gives the same form as that in Equation (6–97).

In some cases it is not too difficult to get an integrated form. As an example, consider compressible frictionless flow in a pipe with heat transfer. Say it is desired to get the relation between pressures and the Mach number. From the first and second rows in Table 6–1 we get the relations

$$\frac{dN}{N} = \left(\frac{1 + kN}{1 - kN}\right)\frac{dQ}{c_p T}$$

$$\frac{dp}{p} = \left(-\frac{kN}{1 - N}\right)\frac{dQ}{c_p T}$$

Eliminating the heat transfer between these two relations gives

$$\frac{dp}{p} = -\frac{k\,dN}{1 + kN}$$

$$\frac{p_2}{p_1} = \frac{1 + kM_1^2}{1 + kM_2^2} \tag{6-149}$$

PROBLEMS

6-1. How long does it take a pressure wave to travel 1 mile through water at ordinary conditions?

6-2. What pressure must be added to water at atmospheric pressure to reduce its volume 0.5 per cent?

6-3. Take the bulk modulus E for benzene at 32 degrees Fahrenheit as 170,000 pounds per square inch. What is the specific gravity if the pressure is increased from atmospheric by an amount equal to 2000 pounds per square inch?

6-4. In the atmosphere at 15,000 feet the temperature is 5.5 degrees Fahrenheit and at 30,000 feet the temperature is −40.8 degrees Fahrenheit. What is the acoustic velocity at each altitude?

6-5. A prototype rocket is to travel at 3000 feet per second through air at −40 degrees Fahrenheit. A model is to be tested in a tunnel with air at 90 degrees Fahrenheit. What should be the corresponding air speed in the tunnel for dynamically similar conditions?

6-6. A rocket moves through standard air at 2800 feet per second, and an airplane travels at 650 miles per hour. What is the Mach number for each?

6-7. In an undisturbed stream of carbon dioxide the pressure is 15.0 pounds per square inch absolute, the density is 0.0032 slug per cubic foot, and the velocity is 500 feet per second. If a body were held stationary in the stream, what would be the stagnation-point pressure on the basis of compressible flow? What would be the stagnation pressure if the fluid were assumed incompressible?

6-8. A body moves through standard air at 600 miles per hour. What are the gage pressure and the temperature at the forward stagnation point?

6-9. At point A in the undisturbed portion of an airstream which flows past a body, ρ is 0.002378 slug per cubic foot, the pressure is 14.7 pounds per square inch absolute, the velocity is 450 feet per second, and c is 1120 feet per second. The pressure at a point B on the body is 8.0 pounds per square inch absolute. Calculate the Mach number for each point for frictionless adiabatic flow.

6-10. A photograph of a bullet shows a Mach angle of 28 degrees. Estimate the speed of the bullet for standard air.

6-11. Air flows through a converging tube having a throat area of 0.50 square inch. The entrance pressure is 140 pounds per square inch absolute, the entrance temperature is 120 degrees Fahrenheit, and the nozzle discharges into the atmosphere at 14.7 pounds per square inch absolute. If the entrance velocity is negligible, what is the weight discharge per unit time?

6-12. Air enters a converging-diverging nozzle at 160 pounds per square inch absolute and a temperature of 120 degrees Fahrenheit. For a frictionless process, what is the Mach number at the cross section where the pressure is 50 pounds per square inch absolute? Neglect entrance velocity.

6-13. Starting with Equation (6-32), derive Equation (6-33).

6-14. When the receiver pressure is equal to or less than the critical pressure, the following relation is sometimes given for the flow of air through an orifice or a short converging tube:

$$W = 0.53 \frac{Ap_1}{\sqrt{T_1}}$$

W represents the weight rate of flow per unit time, A the throat area, p_1 the entrance pressure, and T_1 the absolute temperature at entrance. This relation is commonly called Fliegner's equation. Derive this special relation, starting with Equation (6-45).

6-15. Air at 10 pounds per square inch absolute and 80 degrees Fahrenheit enters a passage at 1600 feet per second; the cross-sectional area is 0.20 square foot. At section 2, farther downstream, the pressure is 18 pounds per square inch absolute. Assuming a frictionless adiabatic process, calculate the Mach number at section 2.

6-16. A pitot tube is installed on a vehicle moving through air at 14.0 pounds per square inch absolute and a temperature of 40 degrees Fahrenheit. The stagnation-point pressure on the nose of the pitot is 7.4 pounds per square inch gage. What is the speed of the vehicle? Assume a frictionless adiabatic process.

6-17. Air enters a converging-diverging nozzle with a pressure of 100 pounds per square inch absolute and a temperature of 90 degrees Fahrenheit. In the diverging part of the nozzle at a certain section just before a shock, the pressure is 20 pounds per square inch absolute, and the acoustic velocity is 914 feet per second. What is the pressure just after the shock?

6-18. Methane flows through a converging nozzle. At inlet the pressure is 30 pounds per square inch absolute. Discharge is into the atmosphere. What is the Mach number at the throat?

6-19. What is the acoustic velocity for mercury? Take the bulk modulus as 3,800,000 pounds per square inch.

6-20. A certain crude oil has a bulk modulus of 218,000 pounds per square inch, and a specific gravity of 0.885. What is the acoustic velocity for this oil?

6-21. Air flows through a thermally insulated channel. At section 1 the linear velocity is 450 feet per second, the pressure is 120 pounds per square inch absolute, and the temperature is 110 degrees Fahrenheit. At section No. 2 the temperature is 60 degrees Fahrenheit. What is the Mach number at section No. 1, and what is the Mach number at section No. 2?

6-22. Air at a static temperature of 60 degrees Fahrenheit and 15 pounds per square inch absolute approaches a thermometer element with a velocity of 1000 feet per second. Assume a frictionless adiabatic process. What are the stagnation temperature and stagnation pressure?

6-23. Carbon dioxide flows through an insulated 6-inch diameter pipe at 78 pounds per second. At inlet the pressure is 120 pounds per square inch absolute and the temperature is 110 degrees Fahrenheit. What is the Mach number at inlet and the Mach number at a point along the pipe where the temperature is 60 degrees Fahrenheit?

6-24. Air at 14.7 pounds per square inch absolute occupies a volume of 1.25 cubic feet. For a frictionless, nonflow, adiabatic process, what would be the pressure if the volume were reduced to 0.50 cubic foot?

6-25. What is the critical pressure ratio for helium?

6-26. For a frictionless, adiabatic, nonflow process, derive the functional relation between temperature and specific volume.

6-27. Air is at 14.7 pounds per square inch absolute and 70 degrees Fahrenheit. For a frictionless, nonflow, adiabatic process, what would be the temperature if the pressure were increased to 52.6 pounds per square inch absolute?

6-28. A certain gas ($R = 3100$, $k = 1.31$) is to be used in a gas turbine. The gas expands from 45 pounds per square inch gage and 85 degrees Fahrenheit to 10 pounds per square inch gage. Neglecting kinetic energy changes, what is the maximum work that can be obtained from this turbine?

6-29. Atmospheric air at standard conditions is drawn through a converging nozzle into a tank where the pressure is 4 pounds per square inch absolute; the flow rate is 2 pounds per minute. What will be the flow rate if the pressure in the tank is reduced to 2 pounds per square inch absolute?

6-30. In a certain flow the Mach number is greater than one. Under what conditions can a pressure wave travel upstream?

6-21. Air at 16.0 pounds per square inch absolute and 100 degrees Fahrenheit enters a channel which converges and then diverges. The throat area is 20 square inches and the exit area is 35 square inches. Neglect inlet velocity and assume a frictionless adiabatic process. What is the maximum weight rate of flow through this channel?

6-32. Picture a flow in which elastic and surface tension forces are involved. Derive a number proportional to the ratio elastic force divided by surface tension force. Let l represent length, ρ density, c acoustic velocity, and σ surface tension.

6-33. Picture a flow in which three types of forces act: elastic, viscous, and inertial. Derive a number proportional to the ratio of pressure force divided by elastic force. Derive a number proportional to the ratio of viscous force divided by elastic force.

6-34. Air enters a venturi meter at 50 pounds per square inch gage and 80 degrees Fahrenheit. Inlet diameter is 8.0 inches and the throat diameter is 4.8 inches. The throat pressure is 37 pounds per square inch gage. For frictionless flow, what is the weight rate of flow?

6-35. Air at 60 degrees Fahrenheit and 20 pounds per square inch absolute approaches a compressor blade with a velocity of 200 feet per second. Assume a frictionless adiabatic process. The pressure at point A on the blade is 15 pounds per square inch absolute. What is the velocity at point A?

6-36. Air at 60 degrees Fahrenheit and 20 pounds per square inch absolute approaches a compressor blade with a velocity of 700 feet per second. Assume an adiabatic process. At point A on the blade the air temperature is 85 degrees Fahrenheit. What is the velocity at point A?

6-37. At a certain nozzle section the velocity is 900 feet per second at 60 pounds per square inch absolute and 100 degrees Fahrenheit. Assume a frictionless adiabatic process. What must be the temperature at exit for an exit Mach number of 2.0? The fluid is air.

6-38. At a certain nozzle section the velocity is 800 feet per second, the pressure is 60 pounds per square inch absolute and the temperature is 100 degrees Fahrenheit. What must be the pressure at exit if the Mach number at exit is to be 3.0? Assume a frictionless adiabatic process. The fluid is air.

6-39. Air enters a 6-inch diameter horizontal pipe at 200 pounds per square inch absolute and 59 degrees Fahrenheit with a velocity of 190 feet per second.

The friction coefficient f is 0.016. Assume isothermal flow. What is the Mach number and the distance from entrance at the section where the pressure is 75 pounds per square inch absolute?

6-40. Air enters a horizontal pipe at 600 pounds per square inch absolute, a temperature of 200 degrees Fahrenheit, and a velocity of 300 feet per second. What is the limiting pressure for isothermal flow? What is the limiting pressure for adiabatic flow?

6-41. Air enters a 4-inch diameter horizontal pipe at 500 pounds per square inch absolute, at 200 degrees Fahrenheit, and with a velocity of 400 feet per second. Assume isothermal flow. The pipe is the maximum possible length. What is the heat transfer rate to the air, and what is the total friction force?

6-42. Air enters a 6-inch diameter horizontal pipe at 80 pounds per square inch absolute, at 59 degrees Fahrenheit, and with a velocity of 150 feet per second. Assume isothermal flow. The pipe length is 1400 feet. At the pipe outlet the pressure is 40 pounds per square inch absolute. What is the wall shear stress at the pipe inlet? What is the total friction force?

6-43. A diffuser consists of a short piece of circular pipe joined to a conical diverging section. Area of pipe section is one square foot. Air at 10 pounds per square inch absolute and 59 degrees Fahrenheit flows through the straight section with a velocity of 2240 feet per second. A normal compression shock forms at the junction between the straight pipe and the diverging section. The subsonic diffuser efficiency E_0 is 60 per cent. Assume that the velocity at exit from the diverging channel is negligible. What is the temperature at the end of the diffuser? For the same overall pressure ratio as the actual process, what would be the exit temperature for a reversible adiabatic process?

6-44. Air enters a converging nozzle with a static pressure p_1 and a negligible velocity. Assume a frictionless isothermal process. The static pressure p_2 at the throat is such that the highest possible velocity at the throat is obtained. What is the numerical value of the pressure ratio p_2/p_1?

CHAPTER SEVEN

Flow Measurement

I have seen so much of the danger arising from presenting results or rules involving variable coefficients in the form of algebraic formulas which the hurried or careless worker may use far beyond the limit of the experimental determination that I present the results mainly in the form of plotted curves which cannot be thus misused and which clearly show the margin of uncertainty and the limitations of the data.
—JOHN R. FREEMAN.[1]

Various problems arise concerning the use of instruments for fluid measurement and control. The present chapter discusses the more common devices which are not covered in other parts of this book.

Force, length, and time are quantities which can be measured very accurately. It is good technique to refer all derived measurements, such as pressure, velocity, and rate of discharge, as directly as possible to the primary standards of force, length, and time. For example, in order to determine the rate of discharge from a pump or fluid machine, good accuracy can be obtained by weighing the discharge during a measured time interval. In some cases, however, economy and convenience may dictate the use of a secondary instrument whose accuracy depends solely on a calibration or assumed coefficients.

7–1. Pitot tubes

Various types of pitot tubes are frequently used for measuring velocity. As shown in Fig. 7-1, V_0 is the velocity some distance ahead of the pitot tube. The pressure p_0 in the upstream undisturbed fluid is called the *static pressure*. The streaming fluid is brought to a state of zero velocity at the nose or stagnation point of the instrument. The fluid at rest in the pitot tube can be connected to some manometer or pressure gage, to give a measurement of the *total or stagnation pressure* p_s. Applying the dynamic equation for incompressible frictionless flow between point O and the nose gives

---

[1] *Experiments upon the Flow of Water in Pipes and Pipe Fittings.* A.S.M.E., New York, 1941.

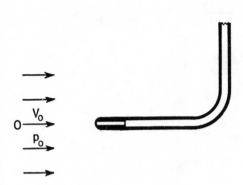 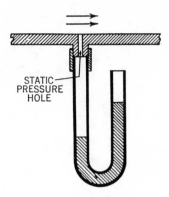

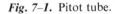

| *Fig. 7–1.* Pitot tube. | *Fig. 7–2.* Illustration of static-pressure measurement by means of a manometer. |

$$p_0 + \rho \frac{V_0^2}{2} = p_s \tag{7-1}$$

or

$$V_0 = \sqrt{\frac{2(p_s - p_0)}{\rho}} \tag{7-2}$$

Note that the density in Equation (7-1) is the density of the fluid flowing. The pitot may be connected to a manometer containing a fluid which is either the same as that or different from that of the flowing stream. If the difference in pressure $p_s - p_0$ can be measured, the velocity V_0 can be computed by Equation (7-2).

In general, it is not difficult to measure the total pressure p_s accurately. On the other hand, it is sometimes difficult to obtain an accurate measurement of the static pressure or the difference $p_s - p_0$ in a stream. If the pitot tube is inserted in a pipe, p_0 may be measured by means of a static-pressure opening in the pipe wall, provided the static-pressure distribution across the pipe is constant.

Figure 7-2 illustrates how the static pressure at a wall can be measured. Care must be taken to have the opening at right angles to the wall, and to avoid burrs. Any type of pressure-measuring instrument could be employed. The fluid flows past the static-pressure hole, opening, or *tap*, but the fluid remains at rest in the hole itself if its dimensions are small enough.

Static-pressure holes might be included in a velocity-measuring instrument, as in the combined pitot tube shown in Fig. 7-3. A differential gage across the static-pressure and total-pressure connections will give the dynamic pressure $\frac{1}{2}\rho V_0^2 = p_s - p_0$ directly.

One type of pitot tube, a *direction-finding tube*, is convenient and useful for measurements of velocity in both direction and magnitude. This instrument can be explained by referring to the two-dimensional flow around a

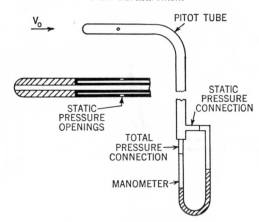

Fig. 7–3. Combined pitot tube.

circular cylinder whose axis is perpendicular to the flow some distance ahead of the cylinder. Let p be the pressure at any point on the surface of the cylinder. Experiments show that the distribution of the pressure difference $(p - p_0)$ is somewhat as represented by the radial ordinates in Fig. 7-4. At the stagnation point A, the pressure difference

$$p - p_0 = \tfrac{1}{2}\rho V_0^2.$$

The pressure difference decreases for successive points from A to B (or C). At points B and C, $p - p_0 = 0$; if there were an opening in the surface of the cylinder at the *critical angle*, the pressure transmitted to a gage would be truly static. Experiments show that this critical angle is $39\tfrac{1}{4}$ degrees for an average range of turbulent flow.

One possible construction of the direction-finding pitot is shown in Fig. 7-5; the pitot consists of a cylindrical tube, with two holes and compartments; the axis of the tube is perpendicular to the stream. In the position shown in Fig. 7-5(*a*), the fluid is flowing in the plane of the diagram; a pressure read-

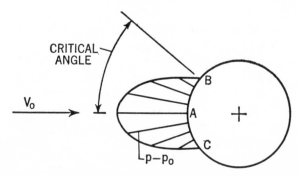

Fig. 7–4. Pressure distribution for two-dimensional flow around a cylinder.

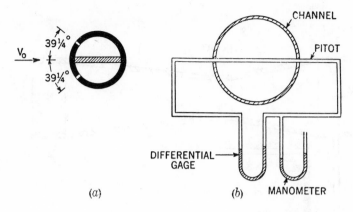

Fig. 7–5. Cylindrical or direction-finding pitot tube.

ing from either compartment will give p_0. If the tube is rotated about its axis so that one opening is in line with V_0, this opening will give p_s. The velocity can then be calculated by Equation (7-2).

Each opening in the tube can be connected to one side of a differential gage, as shown in Fig. 7-5(a). In Fig. 7-5(b) the axis of the cylindrical tube is in the plane of the diagram. The axis of the channel and the direction of the flowing fluid is perpendicular to the plane of the diagram. The pitot tube in a stream of unknown direction can be rotated about its axis until the pressure at each hole is p_0, that is, until the differential pressure is zero. In this position the bisector of the angle between the holes gives the flow direction. Thus this pitot provides means for measuring the velocity vector in both magnitude and direction.

7–2. Venturi meter

If a direct weighing or a volumetric measurement of rate of discharge is not possible or convenient, then some secondary method might be used. A venturi meter, nozzle, orifice, or weir may be used for measuring or controlling the rate of flow.

The venturi meter, as indicated in Fig. 7-6, consists of a venturi tube and a suitable differential pressure gage. The venturi tube has a converging portion, a throat, and a diverging portion. The function of the converging portion is to increase the velocity of the fluid and lower its static pressure. A pressure difference between inlet and throat is thus developed, which pressure difference is correlated with the rate of discharge. The diverging cone or diffuser serves to change the area of the stream back to the entrance area, and to convert velocity head into pressure head; it is desirable to have the diffuser loss as low as possible.

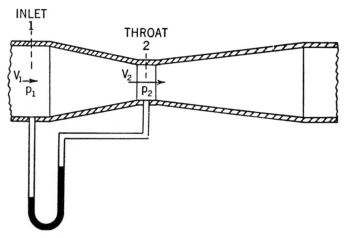

Fig. 7–6. Venturi meter.

Different venturi-tube proportions and different arrangements of pressure gages are found in practice. The diameter of the throat is usually between one-half and one-fourth of the inlet diameter.

Using the notation shown in Fig. 7-6, assume incompressible flow and no friction losses. Then the dynamic equation becomes

$$\frac{p_1}{\gamma} + \frac{V_1^2}{2g} = \frac{p_2}{\gamma} + \frac{V_2^2}{2g} \tag{7-3}$$

Use of the continuity equation $Q = A_1 V_1 = A_2 V_2$ with Equation (7-3) gives

$$\frac{p_1}{\gamma} - \frac{p_2}{\gamma} = \frac{V_2^2}{2g}\left[1 - \left(\frac{A_2}{A_1}\right)^2\right]$$

$$\text{ideal } Q = A_2 V_2 = \frac{A_2}{\sqrt{1 - \left(\frac{A_2}{A_1}\right)^2}} \sqrt{2g\left(\frac{p_1 - p_2}{\gamma}\right)} \tag{7-4}$$

Note that the specific weight γ in Equations (7-3) and (7-4) is the specific weight of the fluid flowing through the meter. The venturi may be connected to a manometer containing a fluid which is different from that of the flowing stream. A measurement of the pressure difference $p_1 - p_2$ offers a method for determining the rate of discharge Q. Equation (7-4) is an ideal relation which assumes uniform velocity distributions at inlet and throat, and no friction. Differences between the actual and ideal flows can be taken into account by an experimentally determined discharge coefficient.

The actual discharge Q is commonly expressed as $Q = CA_2 V_2$, where the discharge coefficient C is dimensionless. Dimensional analysis and dynamic similarity show that C is a function of Reynolds number. Some experimental values of C are shown in Fig. 7-7. D_2 is the internal diameter at the venturi throat.

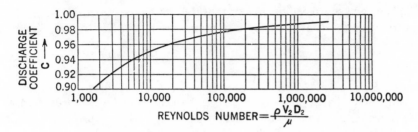

Fig. 7-7. Discharge coefficients *C* for venturi meters.[2] (Ratio of inlet to throat diameter from 2 to 3.)

Figure 7-7 is intended only to give some idea of the order of magnitude of the discharge coefficient. For accurate test work it is advisable to calibrate the venturi meter in its actual service location. It is good technique to calibrate the meter by weighing the discharge during a measured time interval.

7-3. Flow nozzles

Nozzles are employed to form jets and streams for a variety of purposes, as well as for fluid metering. Usually the term "flow nozzle" refers to a nozzle placed in or at the end of a pipe for purposes of metering. As illustrated in Fig. 7-8, the flow nozzle may be considered as a venturi tube that has been

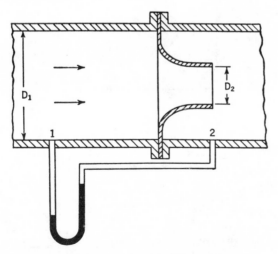

Fig. 7-8. Flow nozzle with differential gage.

[2] Data adapted from *Fluid Meters, Their Theory and Application.* A.S.M.E., New York, 1959.

simplified and shortened by omitting the diffuser on the outlet side. The final equation for the venturi meter can be applied to the flow nozzle, to give

$$\text{actual } Q = \frac{CA_2}{\sqrt{1 - \left(\frac{A_2}{A_1}\right)^2}} \sqrt{2g\left(\frac{p_1 - p_2}{\gamma}\right)} \qquad (7\text{-}5)$$

The rate of discharge can be correlated with the measured pressure difference.

Some of the terms in Equation (7-5) are frequently grouped into a single term; this term is called the *flow coefficient* K, and is defined as

$$K = \frac{C}{\sqrt{1 - \left(\frac{A_2}{A_1}\right)^2}} \qquad (7\text{-}6)$$

The dimensionless term $1/\sqrt{1 - (A_2/A_1)^2}$ is called the *velocity-of-approach factor*. Then Equation (7-5) can be written as

$$\text{actual } Q = KA_2\sqrt{2g\left(\frac{p_1 - p_2}{\gamma}\right)} \qquad (7\text{-}7)$$

Dimensional analysis and dynamic similarity show that K and C is a function of Reynolds number and other factors. Some sample values are given in Fig. 7-9.

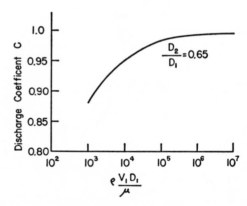

Fig. 7–9. Discharge coefficients for A.S.M.E. long-radius high-ratio flow nozzles as a function of Reynolds number.[3]

7–4. Jet flow

Some features of jet flow will be reviewed before discussing the practically important case of orifice meters. Figure 7-10 shows an orifice with

[3] *Power Test Codes*, PTC 19.5.4-1959. A.S.M.E., New York, 1959.

a free discharge. Applying the dynamic
equation between sections 1 and 2 for
incompressible frictionless flow gives

$$\frac{p_1}{\gamma} + \frac{V_1^2}{2g} + h = \frac{p_2}{\gamma} + \frac{V_2^2}{2g} \qquad (7\text{-}8)$$

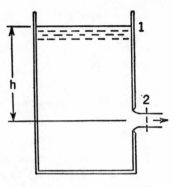

If the free surface 1 and the jet at section
2 each were open to the atmosphere, then
p_1 would equal p_2. If the area A_1 were
large in comparison with A_2, the continuity
equation shows that V_1 would be small
in comparison with V_2. Assume that V_1
can be neglected. Then Equation (7-8)
becomes

Fig. 7–10. Orifice with free discharge.

$$\text{ideal } V_2 = \sqrt{2gh} \qquad (7\text{-}9)$$

V_2 in Equation (7-9) is an "ideal" velocity because of the conditions taken.
Equation (7-9) is identical with the relation for the velocity attained by a
freely falling body.

Because of friction the actual jet velocity is less than the ideal. The ratio
of the actual velocity to the ideal velocity is called the *coefficient of velocity*
C_V. The actual $V_2 = C_V\sqrt{2gh}$. The streamlines converge as they approach
the orifice, and continue to converge beyond the orifice. At a certain distance
from the plane of the orifice, the jet has a minimum area where all the streamlines are parallel. This effect is shown in the photograph in Fig. 7-11. This

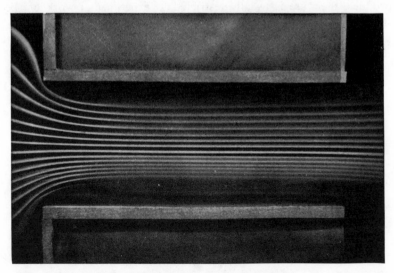

Fig. 7–11. Example of the converging of streamlines in a jet. Flow is
from left to right.

minimum section of the jet is termed the *vena contracta*. The ratio of the minimum jet area to the orifice or opening area is called the *coefficient of contraction* C_c. If A is the orifice area, an ideal rate of discharge Q can be expressed as

$$\text{ideal } Q = A\sqrt{2gh}$$

Friction and contraction being taken into account, the actual discharge is found to be

$$\text{actual } Q = C_V C_c A\sqrt{2gh} = CA\sqrt{2gh} \tag{7-10}$$

where C is the discharge coefficient. For sharp-edged circular orifices C_V ranges from about 0.95 to about 0.99, and C_c ranges from about 0.61 to 0.72. The discharge coefficient C is probably the one of most practical use, and the one that can be found most readily experimentally.

7–5. Orifice meter

The thin-plate or sharp-edged orifice is frequently employed for metering. A common arrangement consists of a thin flat plate which is clamped between the flanges at a joint in a pipe line: the flat plate has a circular hole concentric with the pipe. Pressure connections, or "taps" for attaching separate pressure gages or a differential pressure gage, are made at holes in the pipe walls on both sides of the orifice plate. In some cases the pressure connections are integral with the orifice plate. The inlet edge of the orifice is sometimes rounded. Because round-edged orifices are difficult to specify and reproduce, their use is usually restricted to experimental work or special instruments.

Using the notation shown in Fig. 7-12, the equation for the venturi tube can be applied to the orifice meter, to give

$$\text{actual } Q = \frac{A_2 C}{\sqrt{1 - \left(\dfrac{A_2}{A_1}\right)^2}} \sqrt{2g\left(\frac{p_1 - p_2}{\gamma}\right)} \tag{7-11}$$

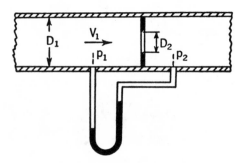

Fig. 7–12. Orifice meter with differential manometer.

where A_2 is the orifice area, A_1 the pipe area, and C the discharge coefficient.

Some sample values are shown in Fig. 7-13. The term *vena contracta taps* signifies that the center of the downstream pressure tap is placed at the minimum pressure position, which is assumed to be at the *vena contracta*. The center of the inlet pressure tap is located between one-half and two pipe diameters from the upstream side of the orifice plate; usually a distance of

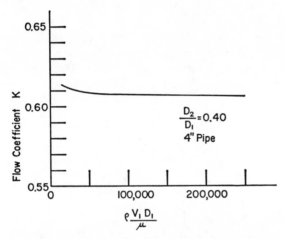

Fig. 7–13. Flow coefficient K for orifice meters with *vena contracta* pressure taps.[4]

one pipe diameter is employed. Figures 7-9 and 7-13 show a general tendency found in the behavior of various flow meters, a tendency for the coefficient to approach a constant value as the Reynolds number is increased. The behavior is similar to the tendency for the pipe friction factor to approach a constant value at high Reynolds numbers.

7–6. Weir

Weirs are used for the measurement and control of the flow of water in open channels. With the notation shown in Fig. 7-14, the height H of the undisturbed water level above the crest of the weir is correlated with the volumetric discharge per unit time. It is necessary to ventilate the weir to allow air to pass freely under the jet. Since H in general is small, its measurement has to be carried out accurately. One method is to have a micrometer screw with a sharp conical point, or hook, entirely submerged in the water. This point is screwed up until it touches the surface, and the height H is observed. Another method involves a height-measuring device in which

[4] Data from *Fluid Meters, Their Theory and Application.* A.S.M.E., New York, 1959.

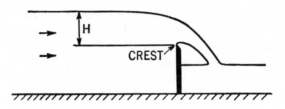

Fig. 7–14. Weir.

the contact between the water and the movable measuring point completes a simple electric circuit and lights a small lamp.

With the notation shown in Fig. 7-15, the ideal velocity of a stream filament or layer issuing from a weir notch at a distance z below the undisturbed surface is

$$\text{ideal } V = \sqrt{2gz} \tag{7-12}$$

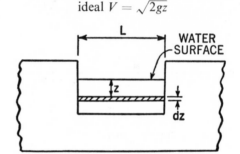

Fig. 7–15. Front view of flow over a rectangular weir.

If this layer thickness is dz and its width L, the ideal volume rate of flow for an element is

$$\text{ideal } dQ = LV\,dz = L\sqrt{2gz\,dz} \tag{7-13}$$

Integrating Equation (7-13) between the limits $z = 0$ and $z = H$ gives

$$\text{ideal } Q = L\sqrt{2g} \int_{0}^{H} z^{1/2}\,dz = \tfrac{2}{3}L\sqrt{2gH^3} \tag{7-14}$$

The actual rate of flow is less than the ideal, so that

$$\text{actual } Q = \tfrac{2}{3}CL\sqrt{2gH^3} \tag{7-15}$$

where C is a discharge coefficient. Applying a similar method of analysis to other types of weirs gives the general relations shown in Fig. 7-16. F is a proportionality factor which, in general, is different for each weir.

The best plan is to calibrate the weir in place, under the same conditions for which it is to be used. If this procedure is not possible, weir coefficients might be estimated after consulting reports of original experimental investigations.

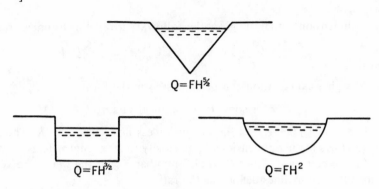

$$Q = FH^{5/2}$$

$$Q = FH^{3/2} \qquad\qquad Q = FH^2$$

Fig. 7–16. General form of relations for different types of weirs. *F* is a proportionality factor.

A brief discussion of the triangular weir will be given; such a study helps to illustrate a method of organizing data. Figure 7-17 shows the notation. The actual discharge dQ through the infinitesimal area $x\,dz$ is

$$dQ = xC\sqrt{2gz}\,dz$$

By similar triangles, $x/L = (H-z)/H$. Then the total discharge becomes

$$Q = C\sqrt{2g}\,\frac{L}{H}\int_0^H (Hz^{1/2} - z^{3/2})\,dz$$

Integrating, and noting that $L = 2H \tan \theta/2$, the final result is

$$Q = \frac{8}{15}C\sqrt{2g}\tan\frac{\theta}{2}H^{5/2}$$

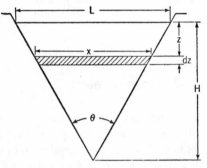

Fig. 7–17. Notation for triangular weir.

For a particular weir, the foregoing relation may be expressed as

$$Q = Kg^{\frac{1}{2}}H^{\frac{5}{2}}$$

where K is a dimensionless flow coefficient.

There is a question about determining the value of the discharge coefficient C or the flow coefficient K. For a particular weir, the flow problem involves the following variables: the discharge Q, the head H, the gravitational acceleration g, the density ρ, the dynamic viscosity μ, and the surface tension σ of the interface between the fluid passing over the weir and the surrounding air. There are six variables and three primary units. Thus the physical equation can be expressed in terms of three dimensionless ratios. A number of solutions to this problem are possible. The following discussion gives one solution.

A dimensional analysis gives the following three dimensionless ratios:

$$\pi_1 = \frac{Q}{H^{5/2}g^{1/2}}, \qquad \pi_2 = \frac{Q\rho}{H\mu}, \qquad \pi_3 = \frac{gH^2\rho}{\sigma}$$

Thus the physical equation can be expressed in the form:

$$\pi_1 \text{ is some function of } \pi_2 \text{ and } \pi_3$$

The ratio $Q/H^{5/2}g^{1/2}$ is the dimensionless flow coefficient K. The quotient Q/H has the net dimensions of a velocity times a length. Thus the term $Q\rho/H\mu$ can be regarded as a Reynolds number N_R for the weir. The Weber number W is sometimes defined as the ratio

$$W = \frac{V^2\rho a}{\sigma}$$

where V is a characteristic velocity and a is a characteristic length. The product gH^2 has the dimensions of a velocity squared times a length. Thus the ratio $gH^2\rho/\sigma$ can be regarded as a Weber number for the weir. Hence from dimensionless analysis we can say that the flow coefficient (or the discharge coefficient C) is some function of a Reynolds number and a Weber number, or

$$K = \phi(\mathrm{Re}, W)$$

7–7. Anemometers

There are some mechanical devices which consist primarily of a rotating element whose angular speed of rotation is correlated with the linear velocity of flow. These devices require calibration. When used for air flow, such devices are commonly termed *anemometers*, whereas when used for water flow they are called *current meters*. The rotating element may consist of a series of cups or a series of vanes (similar to a windmill), mounted on a shaft held in bearings. The rotating shaft may be connected to some sort of revolution counter, and an observation made of the number of revolutions in a measured time interval.

Various forms of hot-wire anemometers have been developed for the measurement of velocity and velocity fluctuations. Such instruments have been used primarily in air streams. In the hot-wire anemometer there is a wire (as platinum) of small diameter which is heated by passing an electric current through it. When exposed to moving air the heated wire cools, and consequently its electric resistance changes. The velocity of flow is correlated with certain electrical measurements. Two types of circuits have been devised: a constant-voltage circuit and a constant-resistance circuit.

The constant-voltage arrangement is illustrated diagrammatically in Fig. 7-18. R_1, R_2, and R_3 are electric resistances in the Wheatstone bridge circuit. The hot wire forms the remaining resistance in the bridge. *A* repre-

sents a battery and B a variable electric
resistance. The voltage across the bridge
(across points CD) is kept constant after
the circuit is so adjusted that the galvanometer indicates zero current when the
heated wire is in stationary air. As the air
flows, the hot wire cools, the resistance
changes, and the galvanometer deflects.
The galvanometer deflection is correlated
with the air velocity by calibration. This
method is particularly adapted for low
air velocities.

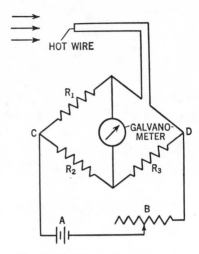

Fig. 7-18. Schematic arrangement
of constant-voltage hot-wire anemometer.

Figure 7-19 shows a schematic
arrangement of the constant-resistance
hot-wire anemometer. As air flows, the
temperature of the wire is kept constant
by varying the resistance in the battery
across the bridge. The bridge voltage is
varied so that the galvanometer reading
remains zero. The voltmeter reading is correlated with the air velocity by
calibration.

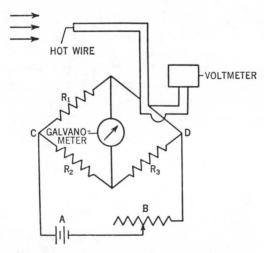

Fig. 7-19. Schematic arrangement of constant-resistance hot-wire
anemometer.

7-8. Some volumetric meters

There are certain flow meters which are designated as *volumetric meters*.
In this type of instrument the fluid passes through the meter in successive

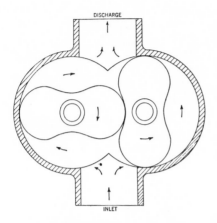

Fig. 7–20. Two-lobe rotary meter.

and more or less completely isolated volumes. The rotary meter illustrated in Fig. 7-20 is one example. The two lobes, impellers, or rotors are mounted on parallel shafts and rotate in opposite directions. A pair of timing gears, located at one end of the shafts, maintains the proper relation between the lobes throughout rotation. Fluid enters the space between a lobe and the case, and moves from inlet to discharge. An indication of the volumetric displacement (such as cubic feet) can be obtained by means of a revolution counter connected to the shaft of one lobe.

Such a rotary unit can be employed as a meter by itself, or it can be employed to calibrate such other flow meters as orifices and flow nozzles. In general, it is not difficult to calibrate a flow meter handling a liquid; an accurate calibration can be made by weighing the discharge during a time interval. The calibration of flow meters handling air or gas, however, presents a much more difficult problem.

The *nutating-disk meter* or *wobble-plate meter* is another example of a volumetric meter. Such a device is frequently used to meter the water supply for domestic use. The movable element in Fig. 7-21 is a circular disk attached to a central ball. A pole or shaft *A* is fastened to the ball; this shaft is held in an inclined position by a cam or roller. The disk is mounted in a chamber which has a spherical side wall, a conical top surface, and a conical bottom surface. The disk is prevented from rotating about its own axis by a radial

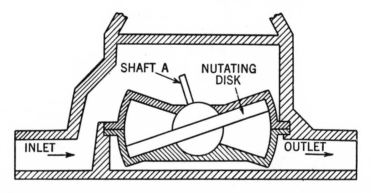

Fig. 7–21. Diagrammatic sketch of the measuring chamber in a nutating-disk meter.

slot which engages a radial partition in the chamber. Liquid enters through an opening in the spherical wall on one side of the partition and leaves on the other side of the partition. As liquid flows through the measuring chamber (alternately above and then below the disk), the disk wobbles or executes a nutating motion (nodding in a circular path without revolving about its own axis). The shaft A generates a cone with the apex down. The top of shaft A actuates a revolution counter (not shown in Fig. 7-21) through a crank and a set of gears. The dial indication is correlated by calibration with the volume that has passed through the meter.

7–9. Salt-velocity method of measuring discharge

The salt-velocity method of measuring the rate of water flow in a conduit is based on the fact that salt in solution increases the electric conductivity of water. Salt, or a salt solution, is introduced into the conduit at some convenient point. One or more pairs of electrodes are arranged at one or more sections, and the flow of the solution is recorded by electric instruments. The time required for the salt solution to travel a measured distance is determined. The volume rate of flow is calculated from these data and the dimensions of the conduit.

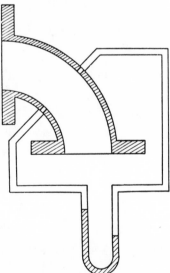

7–10. Pipe elbow as a flow meter

A diagrammatic sketch of an elbow meter is shown in Fig. 7-22. The dynamic action of the flowing fluid causes a difference between the pressures at the inside and outside curves of the elbow. Pressure connections from the inside and outside curves of the elbow are led to a differential pressure gage. The differential pressure is correlated with the average velocity of flow in the connecting pipe or the rate of discharge. For accurate results an elbow meter should be calibrated in the actual service location.

7–11. Rotameter

Fig. 7–22. Diagrammatic sketch of elbow meter.

The *rotameter* is a flow meter which derives its name from the fact that a rotating free float is the indicating element. As shown in Fig. 7-23, a transparent tapered tube is set in a vertical position with the large end at the top.

The fluid flows vertically upward through the tube. Inside the tapered tube is a freely suspended "float" of plumb-bob shape. When no fluid is passing, the float rests on a stop at the bottom end. As flow commences, the float rises toward the larger end of the tube, to make a passage for the fluid. The float rises only a short distance if the rate of flow is small; the float rise is greater for a higher rate of flow. There is a corresponding float position for each flow. Sometimes this type of instrument is classified as an *area type* flow meter, because of the varying annular space between the float and tube. Slantwise slots are cut in the head of the float, to cause the float to rotate and to maintain a position in the center of the tube.

FLOW MEASUREMENT

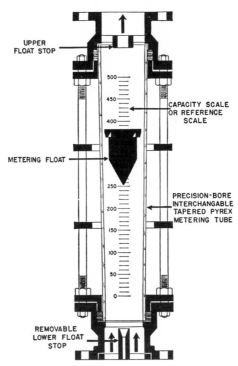

Fig. 7–23. Rotameter. (Courtesy Fisher & Porter Co.)

7–12. Capillary-tube viscometer

A viscometer or viscosimeter is an instrument for measuring viscosity. The operation of a viscometer depends upon the existence of laminar flow under certain controlled conditions. An ideal viscometer would be one in which the flow involved is completely determined by the fluid viscosity. In actual viscometers this ideal is not completely realized, and it is necessary to apply certain correction factors or to calibrate the instruments with fluids of known viscosity.

Probably the best scientific method for determining dynamic viscosity is the *capillary-tube* or *transpiration* method. In this method a measurement is made of the time required for a certain amount of fluid to flow through a capillary (or small-bore) tube of known length and diameter under a measured, constant pressure difference. The Hagen-Poiseuille law can be applied, if the flow is laminar, to calculate the viscosity μ.

$$\mu = \frac{\pi \Delta p D^4}{128 Q l} \tag{7-16}$$

where D is the internal diameter of the tube, Δp is the pressure difference in the length l, and Q is the volume rate of flow.

If the pressure is measured at the ends of a tube, certain corrections for changes in velocity distribution and inlet loss must be made. The corrections depend upon the particular apparatus used. Since viscosity depends upon temperature, it is necessary to control and specify the temperature in all viscosity measurements.

7–13. Other types of viscometers

Figure 7-24 shows the essential features of the Couette or rotational type of viscometer. The fluid to be tested is placed in the annular space between two concentric circular cylinders. One cylinder is rotated with respect to the other. Measurements of torque and velocity gradient can be correlated with viscosity, as by a calibration with liquids of known viscosity. Effects of the ends of the cylinder must be considered or reduced if the readings of this device are to be used for accurate determinations of viscosity. A particular instrument of this type may be useful for relative measurements, as in comparing the action of different fluids.

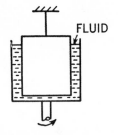

Fig. 7–24. Rotational type of viscometer. Fluid is placed in the annular space between two concentric circular cylinders.

Commercial variations of the rotational type of viscometer are the MacMichael and the Stormer. In the MacMichael instrument the outer cylinder is rotated at a constant speed, and the inner cylinder is supported by a torsion wire. A measurement of the angular twist of the wire is used to obtain a reading proportional to the viscous forces acting. In the Stormer instrument the outer cylinder is stationary, and a constant torque is applied to the inner rotating cylinder by means of a weight and pulley arrangement. A measurement is made of the time required for a definite number of revolutions of the inner cylinder. Sometimes a horizontal flat disk, a circular open bottom cup, or a fork is immersed in the fluid instead of an inner circular cylinder. Rotational viscometers are sometimes used for checking the manufacture of suspensions, paints, and food products.

In the falling-body type of viscometer, a body (a sphere, for example) is allowed to fall through a mass of fluid in a cylindrical tube. The time taken for the body to fall a certain distance (in which the velocity is constant) is measured, and these measurements are correlated with viscosity. This type of instrument can be calibrated with fluids whose viscosities are known. Fluids with unknown viscosities can then be tested within the range of calibration.

The Saybolt viscometer, as illustrated diagrammatically in Fig. 7-25, is widely used for comparative industrial purposes in this country, particularly for petroleum products and lubricants. Two instruments have been standardized by the American Society for Testing Materials;[5] these instruments are of the same general design but of different dimensions. One is the Saybolt Universal, and the other is the Saybolt Furol (contraction for "fuel and motor oils").

The liquid to be tested is placed in the central cylinder, which has a short, small-bore tube and cork at its lower end. Surrounding the central cylinder is a bath of liquid for maintaining the temperature of the test liquid.

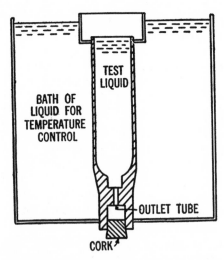

Fig. 7–25. Diagrammatic sketch of a Saybolt viscometer.

[5] "Standard Method of Test for Viscosity by Means of the Saybolt Viscometer," in *A.S.T.M. Standards*, 1939, Part III, p. 216.

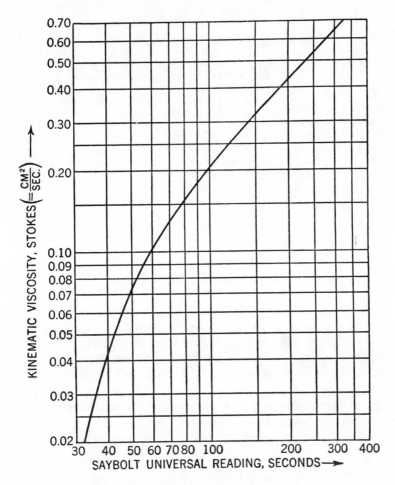

Fig. 7–26. Empirical correlation between Saybolt Universal Reading and kinematic viscosity at 100° Fahrenheit.

After thermal equilibrium is established, the cork is pulled, and the time required for 60 milliliters of the fluid to flow out is measured. This time, in seconds, is called the Saybolt reading. This Saybolt reading can be correlated empirically with kinematic viscosity. One correlation at 100 degrees Fahrenheit, based on experimental results, is shown in Fig. 7-26.[6]

The Society of Automotive Engineers has adopted certain S.A.E. numbers for some ranges of Saybolt readings. Table 7-1 shows S.A.E. numbers which are commonly used for designating crankcase oils, as far as viscous properties only are concerned.

[6] *A.S.T.M. Standards*, 1939, Part III, p. 215.

TABLE 7-1

CRANKCASE OIL CLASSIFICATION[7]

S.A.E. viscosity number	Range of Saybolt Universal readings, seconds			
	At 130° Fahrenheit		At 210° Fahrenheit	
	Minimum	Maximum	Minimum	Maximum
10	90	Less than 120		
20	120	Less than 185		
30	185	Less than 255		
40	255	. . . .	. . .	Less than 80
50	. . .	. . . .	80	Less than 105
60	. . .	. . . .	105	Less than 125
70	. . .	. . . .	125	Less than 150

7-14. Meters for measuring mass rate of flow

Various meters, as the pitot tube, orifice, flow-nozzle, and venturi, are in common use; meters of this sort have certain advantages in measuring linear velocities and volume rates of flow. In various applications, however, it is highly desirable to obtain a direct measurement of mass-rate of flow. In some cases, for example, it would be desirable to have available a meter which would give a measurement of mass-rate independent of such factors as pressure, temperature, viscosity, and density. One attractive possibility is to use a rotating element and measure the torque.

Equation (3-43) shows that the torque T can be expressed in the form

$$T = \text{(mass rate)} \, (R_2 V_{2U} - R_1 V_{1U}) \qquad (7\text{-}17)$$

Imagine a rotating member with vanes, blades, or passageways. The angular speed of rotation is ω. If the inlet relative flow is radial, and if the exit relative flow is radial, then

$$V_{2U} = U_2 = R_2\omega, \qquad V_{1U} = U_1 = R_1\omega$$

For these conditions, Equation (7-17) becomes

$$\text{(mass rate)} = \frac{T}{(R_2^2 - R_1^2)\omega} \qquad (7\text{-}18)$$

Thus we can measure the mass rate directly by measuring the torque, the two radii, and the angular speed.

---

[7] Data from 1947 *S.A.E. Handbook*. Society of Automotive Engineers, New York, p. 492.

7-15. Compressible flow through fluid meters

Let W represent the weight rate of flow (as pounds per second). Since $W = A_2 V_2 \gamma = Q\gamma$ then Equation (7-7) can be used to give

$$W = KA_2\gamma\sqrt{2g\left(\frac{p_1 - p_2}{\gamma}\right)} \qquad (7\text{-}19)$$

It is common practice in fluid metering work to generalize Equation (7-19) by multiplying the right side by an *expansion* factor, and by specifying the specific weight γ as that at the inlet. Let Y represent the expansion factor. Thus, for both incompressible and compressible flow, the fundamental relation can be written as

$$W = KA_2\gamma_1 Y\sqrt{2g\left(\frac{p_1 - p_2}{\gamma_1}\right)} \qquad (7\text{-}20)$$

For incompressible fluids $Y = 1$. If the fluid is frictionless, then the discharge coefficient C in the flow coefficient K is unity. For flow with friction, the discharge coefficient should be determined from suitable experiments. Tables and charts giving values of expansion factors can be found in the reference literature. It is customary to use theoretically determined values of Y for flow nozzles and venturi tubes, and to use experimentally determined values of Y for orifice meters. The minimum jet area and the pressure at various points in an orifice meter are not the same as those in a venturi tube of the same diameter operating with the same pressure drop.

Consider the determination of Y for a gas flowing through a venturi meter or a flow nozzle. An adiabatic process is assumed; this assumption is reasonable for many practical applications. Then Equation (6-21) applies, and can be written as

$$\frac{V_2^2 - V_1^2}{2} = \left(\frac{k}{k-1}\right)\frac{gp_1}{\gamma_1}\left[1 - \left(\frac{p_2}{p_1}\right)^{\frac{k-1}{k}}\right] \qquad (7\text{-}21)$$

A combination of the continuity equation $A_2 V_2/v_2 = A_1 V_1/v_1$ and the adiabatic relation $p_1 v_1^k = p_2 v_2^k$ gives

$$V_1 = V_2\frac{A_2}{A_1}\left(\frac{p_2}{p_1}\right)^{1/k} \qquad (7\text{-}22)$$

V_1 can be eliminated from Equation (7-21) by using Equation (7-22). Thus

$$V_2 = \sqrt{\frac{2g\left(\dfrac{k}{k-1}\right)\dfrac{p_1}{\gamma_1}\left[1 - \left(\dfrac{p_2}{p_1}\right)^{\frac{k-1}{k}}\right]}{1 - \left(\dfrac{A_2}{A_1}\right)^2\left(\dfrac{p_2}{p_1}\right)^{2/k}}} \qquad (7\text{-}23)$$

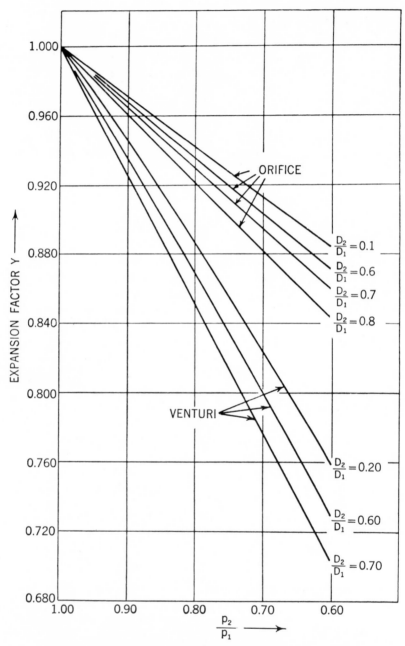

Fig. 7–27. Expansion ratios plotted against pressure ratio for $k = 1.4$. Orifice values are for flange, radius, and *vena contracta* taps. (Data from *Fluid Meters, Their Theory and Application*. A.S.M.E., 1959.)

For compressible flow, then,

$$W = CA_2\gamma_2 \sqrt{\frac{\left(\frac{2gk}{k-1}\right)\frac{p_1}{\gamma_1}\left[1 - \left(\frac{p_2}{p_1}\right)^{\frac{k-1}{k}}\right]}{1 - \left(\frac{A_2}{A_1}\right)^2\left(\frac{p_2}{p_1}\right)^{2/k}}} \tag{7-24}$$

Equating the W in Equation (7-20) to the W in Equation (7-24) and solving for Y gives

$$Y = \left(\left(\frac{p_2}{p_1}\right)^{2/k}\left(\frac{k}{k-1}\right)\left[\frac{1 - \left(\frac{p_2}{p_1}\right)^{\frac{k-1}{k}}}{1 - \left(\frac{p_2}{p_1}\right)}\right]\left[\frac{1 - \left(\frac{A_2}{A_1}\right)^2}{1 - \left(\frac{A_2}{A_1}\right)^2\left(\frac{p_2}{p_1}\right)^{2/k}}\right]\right)^{\frac{1}{2}} \tag{7-25}$$

At first thought it might appear that the introduction of the factor Y provides a complication. Closer inspection, however, reveals the fact that Y is a function of three dimensionless ratios: the pressure ratio p_2/p_1, the area ratio A_2/A_1, and the ratio of the specific heats k. If values of Y are once determined, they will be convenient for further calculations. Some values of the expansion factor are plotted in Fig. 7-27. The curves marked "venturi" are also applied to flow nozzles. D_2 is the throat or orifice diameter, and D_1 is the inlet or pipe diameter.

PROBLEMS

7-1. Air of standard density flows through a pipe. A differential manometer connected to a combined pitot tube shows a head difference of 0.180 inch of water. What is the velocity? Assume frictionless flow.

7-2. Carbon dioxide flows through a smooth pipe 8 inches in diameter. At a certain cross section the gas pressure is 14.7 pounds per square inch absolute and the temperature is 60 degrees Fahrenheit. At this section in the center of the pipe is a pitot tube connected to a U-tube manometer containing an oil with a specific gravity of 0.90. The manometer deflection is 0.06 inch. Assume that the flow is incompressible. What is the volume rate of flow through the pipe?

7-3. How can the discharge coefficient for a venturi meter be greater than unity even if there are frictional losses?

7-4. Benzene flows through a horizontal venturi meter having an inlet of 8-inch diameter and a throat of 3.5-inch diameter. The differential pressure is measured with a manometer having benzene on top of mercury. The mercury level in the throat leg is 4.0 inches above the mercury level in the inlet leg. If $C = 0.99$, what is the volume rate of discharge?

7-5. An A.S.M.E. long-radius flow nozzle, with a throat diameter of 2.75 inches, is installed in a water line of 6-inch diameter. A differential gage across the nozzle shows a pressure difference of 2.80 pounds per square inch. Calculate the rate of discharge if the discharge coefficient is 0.99.

7-6. The actual velocity in the *vena contracta* of a jet of liquid issuing from an orifice of 3-inch diameter is measured as 28.0 feet per second. If the head is 15 feet, and the actual discharge is 0.90 cubic foot per second, determine C_v and C_c.

7-7. Turpentine at 50 degrees Fahrenheit (specific gravity = 0.87) flows through an orifice meter with *vena contracta* taps. The pipe diameter is 4.0 inches, and the orifice diameter is 1.6 inches. A differential gage across the meter indicates a pressure difference equivalent to a head of 8.0 inches of water. What is the rate of discharge?

7-8. Standard air flows through an orifice meter. The pipe diameter is 6.0 inches and the orifice diameter is 1.80 inches. Assume incompressible flow. The discharge coefficient is 0.98. The differential gage across the orifice gives a reading of 1.40 inches of water. What is the volume rate of flow?

7-9. Benzene flows through an orifice meter. The horizontal pipe diameter is 10 inches and the orifice diameter is 5.5 inches. Take the flow coefficient as 0.64. The U-tube differential manometer across the orifice is filled with benzene and mercury. For a flow of 2.0 cubic feet per second, what is the height difference between the mercury levels in the gage?

7-10. A liquid (specific gravity = 0.96) flows through a tube of 0.25 inch diameter at a rate of 0.22 cubic inch per second. The measured pressure drop in a length of 22.0 inches is 0.031 pound per square inch. What is the dynamic viscosity of the liquid?

7-11. The Saybolt Universal reading for an oil at 100 degrees Fahrenheit, having a specific gravity 0.92, is 106 seconds. What is the dynamic viscosity of this oil?

7-12. Standard air flows through a 2.4-inch diameter orifice in a 6-inch diameter pipe. The differential pressure across the orifice is equivalent to 3 inches of water. For incompressible, frictionless flow, what is the stagnation pressure in the pipe?

7-13. Water flows through a vertical venturi meter having an inlet diameter of 4 feet and a throat diameter of 1.5 feet. The throat is 4.5 feet above the inlet. At inlet is a U-tube manometer having mercury and water. One leg of this manometer is open to the atmosphere; the level between mercury and water is 6 inches below that of the level between mercury and air. At the throat is a U-tube manometer containing mercury and water with one leg open to the atmosphere; the level between mercury and water is 1 inch below the level between mercury and air. Assuming no friction, what is the volume rate of flow?

7-14. Standard air flows through a 4-inch diameter orifice at the end of an 8-inch diameter pipe. The discharge is into the atmosphere. Assuming frictionless, incompressible flow at 11.9 cubic feet per second, what is the net axial force of the orifice plate on the fluid?

7-15. Water flows with a velocity of 15 feet per second through a 10-inch diameter pipe. In this pipe is a venturi meter with a throat diameter of 6 inches. Neglect friction. What is the magnitude of the axial force of the converging portion of the venturi on the fluid? Venturi inlet pressure is 50 pounds per square inch gage.

7-16. A pitot tube is placed in a stream of air at 14.0 pounds per square inch absolute and 240 degrees Fahrenheit. The U-tube manometer connected to the pitot indicates a differential pressure equivalent to 2.0 inches of water. What is the velocity of the air stream?

7-17. Standard air approaches a pitot tube with a velocity of 132 feet per second. A water U-tube manometer is used to measure the differential pressure. What is the height difference?

7-18. Standard air flows through a 12-inch diameter pipe. An orifice of diameter 4.8 inches is in this pipe. Assume incompressible, frictionless flow at a rate of 14.5 cubic feet per second. Across the orifice is a U-tube manometer with water. What is the height difference in the manometer?

7-19. Standard air flows through a horizontal 12-inch diameter pipe. A venturi meter in this pipe has a throat diameter of 4 inches. A U-tube manometer connected to the inlet and throat shows a pressure difference equal to 3 inches of water. Assuming frictionless, incompressible flow, what is the volume rate?

7-20. In Fig. 7-28 the tube extends from the wall of the container into the body of the fluid; such a re-entrant tube is sometimes called a Borda mouthpiece. Let A be the area of the opening, A_c the area of the jet as it leaves the opening, and let the contraction coefficient C_c be the ratio A_c/A. Assume incompressible, frictionless flow. Show that the contraction coefficient equals 0.5.

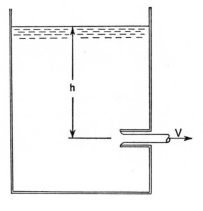

Fig. 7–28. Borda mouthpiece.

7-21. Air enters a venturi meter at 64.7 pounds per square inch absolute and a temperature of 70 degrees Fahrenheit. The pipe diameter is 6.0 inches and the throat diameter is 3.6 inches. The pressure at the throat is 19.4 pounds per square inch less than that at inlet. Discharge coefficient is 0.95. What is the weight rate of flow?

7-22. Air approaches an orifice with a gage pressure of 60 pounds per square inch and a temperature of 100 degrees Fahrenheit. The pipe diameter is 8.0 inches and the orifice diameter is 5.6 inches. The pressure at the *vena contracta* tap is 22.4 pounds per square inch less than in the inlet pipe. Discharge coefficient is 0.97. What is the weight rate of flow?

CHAPTER EIGHT

Fluid
Machinery

"... For in each one of these fields there is, with all our ignorance, a
core of definite, established, noncontroversial knowledge already taken,
insofar as it goes, as a dependable guide to correct thinking and correct
acting.
It is these continuously growing cores of knowledge, coupled with the
attitude of world loyalty, i.e., the combination of science and religion,
that provides today the sole basis for rational living; and in spite of
man's frailties this attitude and these cores are slowly guiding us
forward, so that we have actually in the United States attained within
a hundred years, and primarily because of science and its applications, a
higher standard of living for the common man than has existed in any
time or place in history.*

—R. A. MILLIKAN.[1]

8–1. Terminology

In a pump, fan, or compressor mechanical work is transformed into
fluid energy. A turbine is essentially a machine for transforming fluid energy
into mechanical work at some rotating shaft.

The word "pump" is frequently used for a machine handling a liquid.
The words "fan" and "compressor" are used frequently for machines handling a gas. In a fan the density change is so small that the gas is regarded as
incompressible. In a compressor the density change is appreciable and taken
into account. The term "fan" is usually used when the machine develops a
relatively low pressure difference, of the order of a few inches of water to less
than one pound per square inch. The compressor is frequently used when the
pressure is conveniently measured in pounds per square inch instead of inches
of water.

Sometimes various qualifying terms are applied to a machine. For
example, a fan may be called an "exhauster" or a "blower." In an exhauster,

---

[1] *The Autobiography of Robert A. Millikan.* Prentice-Hall, Inc., Englewood Cliffs,
N. J., 1950.

254

piping may be connected only to the inlet to the fan, and the fan discharges directly into the atmosphere. In a blower, piping may be connected only to the outlet of the fan, and the fan inlet is open directly to the atmosphere.

8–2. Types of machines

Fluid machines may be divided into two main types: the positive-displacement or static type; and the dynamic or kinetic type.

In the positive-displacement or static type the characteristic action is a volumetric change or a displacement of fluid. Static pressure is developed by a displacement action rather than by a velocity or kinetic-energy change. The reciprocating pump, the gear pump, and the rotary pump are examples of positive-displacement machines.

In the dynamic or kinetic type there is a dynamic or kinetic action between some mechanical element and a fluid; there is a velocity change and a corresponding force acting. In the dynamic type there is a conversion of kinetic energy into static pressure, or a conversion from static pressure to kinetic energy. Examples of the dynamic type are the centrifugal pump, the axial-flow compressor, and the fluid coupling; these are sometimes called "turbo-machines." A turbo-machine is characterized by a dynamic action between a fluid and one or more rotating elements. For example, a centrifugal blower might be called a turbo-blower, and a hydraulic turbine runner a turbo-runner.

8–3. Reciprocating pumps and fluid motors

In the familiar reciprocating pump a piston moves back and forth in a cylinder. Figure 8–1 shows a diagrammatic sketch of one type. As the piston moves toward the left, the discharge or outlet valve is closed, the inlet check valve is open, and the fluid enters the cylinder. As the piston moves toward the right, the inlet valve is closed, the discharge check valve is open, and fluid flows from the cylinder through the discharge pipe. If leakage past the piston or slip is neglected, then the pump discharges a volume of fluid equal to the volume displaced by the piston; thus we say this is a positive-displacement machine. The discharge pressure of a reciprocating pump is governed by the discharge piping, or load. For a given or constant volumetric rate of flow, the discharge pressure may be low or high, depending on the pressure necessary to force the flow through the system connected to the pump. The pressure that can be developed is limited by the strength of the pump and the power of the driving unit.

On some reciprocating pumps there is an air chamber on the discharge, to make the flow more steady and to make the pump operation quiet by

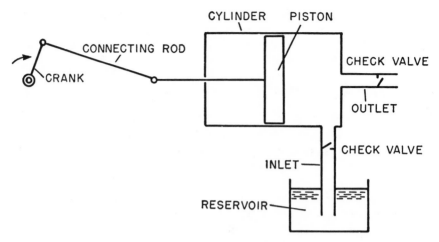

Fig. 8–1. Diagrammatic sketch of a reciprocating pump.

cushioning the discharge. The air in the chamber is compressed during discharge. When the piston reaches the end of the stroke, expansion of this air tends to keep the fluid in motion until the reverse stroke starts.

In general, reciprocating pumps are most efficient for relatively small rates of discharge or capacities, high pressures, and high suction lifts. They are built for practically every type of service and with a wide variety of materials to meet these services. Generally speaking, reciprocating pumps are not particularly well suited for handling very viscous or dirty liquids, because of the tendency toward clogging of suction and discharge valves. Reciprocating pumps are usually operated at slow speeds (40 to 200 crankshaft revolutions per minute) because of the reciprocating motion and the valves.

The action of a reciprocating pump can be reversed to give a reciprocating *fluid motor.* A combination of pump, fluid motor, and interconnecting piping or channels may be called a fluid power transmission. A displacement-type fluid power transmission has one positive-displacement pump in which mechanical work is converted into liquid flow against pressure, and a second positive-displacement pump acting in reverse, as a motor, in which the liquid flow is transformed into mechanical work. The displacement-type power transmission differs from the so-called *hydrodynamic* or dynamic transmission. The fluid coupling is an example of a hydrodynamic or dynamic transmission. In the fluid coupling the kinetic energy of the liquid, rather than pressure alone, is used to transmit power.

8–4. Some other positive-displacement machines

The action in a rotary pump is one of rotation and not reciprocation. The rotary pump is commonly classified as a positive-displacement machine.

The flow from a reciprocating pump is pulsating, whereas the flow from many types of rotary pumps is fairly steady. A rotary pump is *not* a centrifugal pump. The gear pump shown in Fig. 8-2 includes a pair of meshed gears in a casing. As the gears rotate, the fluid is trapped between the gear teeth and the case, and is carried around to the discharge. During each revolution of the gears a certain volume of fluid is transferred from suction to discharge, depending upon the size of the spaces between the gear teeth and the case. The pressure developed by a rotary pump is, as with a reciprocating pump, whatever is required to force the fluid through the discharge piping.

Rotary pumps are suitable for certain classes of service under low and medium head. The absence of valves is an advantage in the handling of thick, viscous liquids. Rotary pumps are not well suited for the handling of grit or abrasives because of the close clearances between gears and case. The

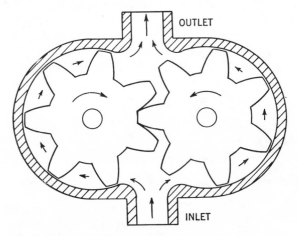

Fig. 8-2. Essential features of one type of rotary pump.

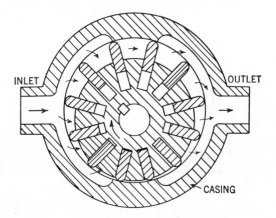

Fig. 8-3. Vane pump.

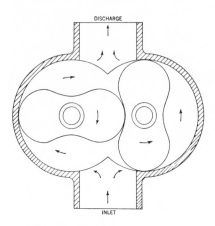

Fig. 8-4. Two-lobe rotary pump.

action of the device shown in Fig. 8-2 can be reversed, to result in a fluid "motor."

The vane pump, as illustrated in Fig. 8-3, is another type of positive-displacement machine. The rotating member, with its sliding vanes, is set off center in the casing. The entering fluid is trapped between the vanes (which ride on the inside of the case) and is carried to the outlet.

Different designs of lobe-type rotary pumps have been devised. Figure 8-4 shows a two-lobe type. The two lobes, mounted on parallel shafts, rotate in opposite directions. A pair of timing gears, located at one end of the shafts, maintains the proper relation between the lobes throughout rotation. Fluid is drawn into the space between the lobe and case and is pushed from inlet to outlet. Such units are used as blowers for moving air or gas, or in reverse as meters for measuring gas flows.

8-5. Performance characteristics of positive-displacement pumps and fluid motors

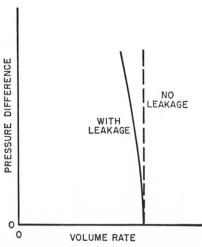

Fig. 8-5. Characteristics of a positive-displacement pump at constant speed.

Assume that a positive-displacement pump is operated at constant speed of the driver shaft. The dotted line in Fig. 8-5 shows the relation between pressure difference Δp across the pump versus volume rate of flow through the pump with no leakage or vaporization of liquid; with no leakage or vaporization the volume rate is constant. The pressure difference developed depends on the pressure necessary to force the flow through the system connected to the pump. The solid curve in Fig. 8-5 shows that leakage may reduce the net volume rate leaving the pump.

In describing performance criteria for positive-displacement machines, let us use the following notation:

Q = volume rate of flow from pump

D = volumetric displacement in pump, in volumetric units per revolution of driver

n = speed of rotation of driver, in revolutions per unit time

C_s = coefficient of slip, dimensionless

Δp = pressure differential across pump

μ = dynamic viscosity of liquid being pumped

Q_r = loss in delivery due to vaporization of liquid in inlet region

T = torque required to drive pump

C_D = viscous drag coefficient, dimensionless

C_f = dry-friction coefficient, dimensionless

T_c = frictional torque, which is independent of speed and pressure

Following the work of W.E. Wilson[2] the performance of a pump can be described in terms of two relatively simple equations, as

$$Q = Dn - \frac{C_s D \Delta p}{2\pi\mu} - Q_r \qquad (8\text{--}1)$$

$$T = \frac{D\Delta p}{2\pi} + C_D D \mu n + \frac{C_f D \Delta p}{2\pi} + T_c \qquad (8\text{--}2)$$

Similar equations can be written for a fluid motor, differing only in the signs of the terms after the first on the right-hand side of each equation. In all cases the systems of units used must be consistent.

With no leakage, no slip, and no vaporization, the theoretical or ideal volume rate of flow would be Dn. With no friction of any type, the theoretical or ideal torque would be $\Delta p D/2\pi$.

The useful or power output P_0 of the pump can be expressed as

$$P_0 = Q\Delta p \qquad (8\text{--}3)$$

The actual power input P_i to the pump can be expressed as

$$P_i = 2\pi T n \qquad (8\text{--}4)$$

The theoretical or ideal power P_t, with no slip, no leakage, and no vaporization, can be expressed as

$$P_t = \Delta p D n \qquad (8\text{--}5)$$

The power output is always less than the ideal power. The power input is always greater than the ideal.

Efficiency can be defined in a general fashion as the ratio between an ideal quantity and an actual quantity. This leads to the distinguishing of three different efficiencies: volumetric efficiency, torque efficiency, and the overall

---

[2] *Positive-Displacement Pumps and Fluid Motors*, by W. E. Wilson. Pitman, New York, 1950.

efficiency. The volumetric efficiency is the ratio between the actual delivery and the ideal delivery. The torque efficiency is the ratio between the ideal torque and the actual torque.

The overall efficiency E of the pump is the ratio of power output to power input. The over-all efficiency can be expressed in the form

$$E = \frac{1 - C_s \dfrac{\Delta p}{2\pi \mu n} - \dfrac{Q_r}{Dn}}{1 + C_D \dfrac{2\pi \mu n}{\Delta p} + C_f + \dfrac{2\pi}{D} \cdot \dfrac{T_c}{\Delta p}} \tag{8-6}$$

The various coefficients can be determined experimentally; values can be found in the reference literature. One type of plot that gives a compact form of data is the plot of overall efficiency E versus the dimensionless parameter $\mu n / \Delta p$. Figure 8–6 shows an example for a spur-gear pump.

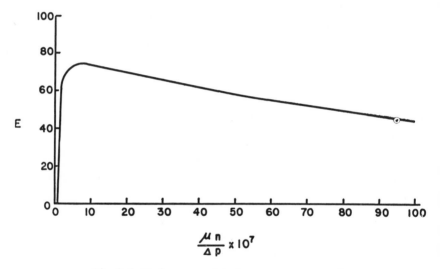

Fig. 8–6. Performance data for spur-gear pump.[3]

8–6. Dynamic machines

Dynamic machines are frequently classified according to the motion of the fluid through the rotating member.

The action in a centrifugal pump, fan, or compressor depends upon centrifugal action or the variation of pressure due to rotation; this fact explains the particular name given to such a machine.

[3] *Positive-Displacement Pumps and Fluid Motors*, by W. E. Wilson. Pitman, New York, 1950.

The essential parts of a centrifugal pump, fan, or compressor are a rotating member with vanes, or *impeller*, and a case surrounding it. The impeller may be driven by a high-speed electric motor, an internal-combustion engine, or a steam turbine. Belt drives and direct connections are used. As indicated in Fig. 8–6, fluid is led through the inlet pipe to the center or *eye* of the rotating impeller. The rotating impeller throws the fluid into the *volute*, where it is led through the discharge *diffuser* to the discharge piping. The fluid leaves the impeller with a high velocity. An important function of the pump passageways is to develop available pressure by an efficient conversion of kinetic energy.

Centrifugal pumps are sometimes designated as (*a*) volute-type pumps

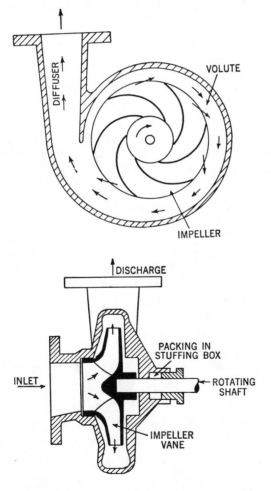

Fig. 8–7. Features of a volute-type centrifugal pump.

or (*b*) diffuser-type pumps. Figure 8–7 shows a volute type pump: the fluid is discharged directly from the impeller into the volute. In the diffuser type (Fig. 8–8), there is a diffuser, consisting of a series of fixed guide vanes, surrounding the impeller. The function of the diffuser is to guide the fluid and reduce its velocity; there is a reduction in kinetic energy and an increase in static pressure in the diffuser. The diffuser tends to make the static pressure distribution around the impeller uniform. Sometimes the diffuse-type pump is called a "turbine" pump, probably because its construction is similar to that of turbines having guide vanes. In practice, however, the term "turbine" pump[4] is also applied to a pump differing in construction from the diffuser pump shown in Fig. 8–8.

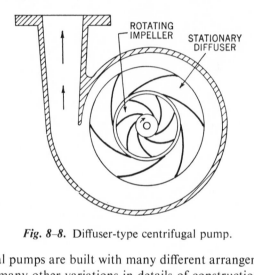

Fig. 8–8. Diffuser-type centrifugal pump.

Centrifugal pumps are built with many different arrangements of impellers, and with many other variations in details of construction. In the *single suction* pump, fluid enters the impeller eye from only side of the impeller. Figure 8–7 shows a single-suction pump. In the *double-suction* pump, fluid enters from both sides of the impeller. A pump may be *staged* with several impellers on one shaft, being then essentially several pumps in series. In a two stage pump, for example, two impellers can be mounted on the same shaft in one casing. The discharge from the first impeller enters the inlet of the second impeller. The same weight of fluid per unit time flows through each stage, but each stage increases the pressure.

In the "radial-flow" machine the fluid enters the impeller at the hub and flows radially to the periphery; Fig. 8–7 is an example. In the "mixed-flow" machine the fluid enters the impeller and flows in both an axial and radial direction, usually into a volute-type casing.

---

[4] A discussion of turbine pumps can be found in *Pumps*, by F. A. Kristal and F. A. Annett. McGraw-Hill, New York. 1940, p. 116.

In an "axial-flow" machine the fluid passes through the runner essentially without changing its distance from the axis of rotation. An axial-flow fan or pump may consist of a single runner in a cylindrical casing, or it may consist of a runner with one or two sets of fixed guide vanes. The runner may be driven by a high-speed electric motor, an internal combustion engine, or a gas turbine. Belt drives and direct connections are used. Figure 8–9 is a diagrammatic sketch of an axial-flow fan or pump.

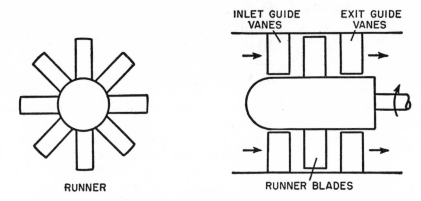

Fig. 8–9. Axial-flow fan or pump.

Dynamic machines differ from positive-displacement machines in many respects. For example, the discharge valve of a centrifugal pump can be closed completely without causing the pressure to rise above a certain value. If the discharge valve is closed, the rotating impeller simply churns the fluid, and the temperature of the fluid rises. If the discharge valve of a reciprocating pump were closed, the pump would be stopped or something would burst. During normal operation the flow from a single positive-displacement machine is nonsteady. The discharge from a centrifugal or axial-flow machine is relatively smooth and steady during normal operation. Centrifugal and axial-flow pumps are available that can handle various liquids, mashes, sewage, and liquids containing sand, gravel, and rocks of moderate size.

8-7. Performance terms for dynamic machines

The term "head," with various qualifications, is used in practice. Head of the fluid flowing is expressed in length units (such as feet) and equals energy per unit weight of fluid (such as foot-pounds per pound of fluid). The vertical distance from the surface of the supply source to the centerline of the pump shaft is called a *static* head. In Fig. 8–10, the distance from the pump shaft center line *A* to the level in the suction tank is variously called *negative static*

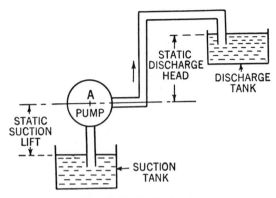

Fig. 8–10. Static heads.

inlet head, negative static suction head, or simply *static suction lift*. If the supply level is above the pump shaft center line, the static inlet head is positive.

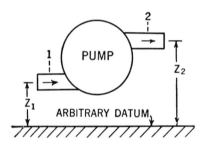

Fig. 8–11. Notation for pump.

The difference in *total* energy between the discharge and inlet openings of the pump or fan, which is the energy added to the stream, represents the energy added by the pump or fan. The term "total head," or "total dynamic head," is employed to designate the energy added by the pump or fan to unit weight of the flowing stream. Figure 8–11 shows the notation; the subscript 1 refers to inlet, and the subscript 2 refers to discharge. For steady incompressible flow, the total head H is

$$H = \frac{p_2 - p_1}{\gamma} + \frac{V_2^2 - V_1^2}{2g} + (z_2 - z_1) \qquad (8\text{--}7)$$

The average velocity at exit V_2 is different from the velocity at inlet V_1 if the pipe diameters are different. The inlet pressure p_1 may be higher or lower than atmospheric. Calculating H should not cause any difficulty if it is borne in mind that $p_2 - p_1$ is a pressure *difference*.

The power added to the fluid by the pump or fan is WH, where W is the weight flowing per unit time. For example, if the total head for a centrifugal pump handling an oil is 300 feet, the pump adds 300 foot-pounds of energy to each pound of oil flowing. Multiplying 300 by the number of pounds of oil flowing per second gives the power added to the oil in foot-pounds per second. The efficiency of a pump or fan is equal to the power increase furnished to the fluid by the pump or fan divided by the power input measured at the pump or fan shaft. Pump input horsepower is sometimes called *brake horsepower*.

The *capacity* or *discharge* of a pump or fan refers to the volume of fluid handled per unit time. *Normal* or *rated* capacity usually represents the capacity at maximum pump efficiency.

Example. Consider the steady flow of water through a pump with an 8-inch diameter inlet, a 6-inch diameter outlet, and a flow rate of 3.8 cubic feet per second. A U-tube manometer with mercury and water is arranged with one leg connected to the pump inlet and the other leg connected to the pump outlet; the pump outlet is above the inlet. The mercury level in the inlet leg (of the manometer) is 42 inches above the mercury level in the outlet leg. What power does the pump add to the water?

From the volume rate and the areas at inlet and outlet, the average linear velocities are: $V_1 = 10.9$ feet per second and $V_2 = 19.3$ feet per second. The manometer reading gives the combination

$$\frac{p_2 - p_1}{\gamma} + z_2 - z_1 = \frac{42}{12}(13.55 - 1) = 43.9 \text{ feet}$$

The head H, as given by Equation (8-7), is thus

$$H = 43.9 + \frac{1}{64.4}[(19.3)^2 - (10.9)^2] = 47.84 \text{ foot-pounds per pound}$$

Then the power becomes

$$\frac{47.84(3.8)62.4}{550} = 20.6 \text{ horsepower}$$

8-8. Basic equations for dynamic machines

In studying fluid dynamic machines we have available for use the equation of state, the equation of continuity, the energy equation, and the dynamic or momentum equation. A convenient form of the momentum equation is brought out in Article 3-14. Considering the steady flow action between runner and fluid, Equation (3-44) states that the rotor acts on the fluid with the torque T to increase the angular momentum of the fluid

$$T = \frac{m}{dt}(R_2 V_{2U} - R_1 V_{1U}) \tag{8-33}$$

where m/dt is mass rate of flow.

If the angular speed of a runner is given, then the values of U_1 and U_2 can be calculated for each radius. If the velocities V_1 and V_2 are given in direction and magnitude, then Equation (3-44) can be used to calculate the torque. For the notation taken the torque is positive for a pump, fan, or compressor; the torque is negative for a turbine runner.

Let P represent power or rate of doing work. Then

$$P = T\omega = \frac{\omega m}{dt}(R_2 V_{2U} - R_1 V_{1U}) \tag{8-8}$$

where ω is angular speed of the rotor in radians per unit time. Let W be the work or work transfer between runner and fluid; W could be expressed in such units as foot-pounds per pound of fluid. Thus

$$W = \frac{Pdt}{gm} = \frac{\omega}{g}(R_2 V_{2U} - R_1 V_{1U}) \tag{8-9}$$

Using the relations $U_2 = R_2\omega$ and $U_1 = R_1\omega$ there results

$$W = \frac{1}{g}[U_2 V_{2U} - U_1 V_{1U}] \tag{8-10}$$

8–9. Performance characteristics of dynamic machines

Figure 8–12 shows the performance characteristics of a centrifugal pump tested at constant shaft speed (the fluid is water). The efficiency rises to a peak value of 91.7 per cent and then drops. As illustrated in Fig. 8–13, a

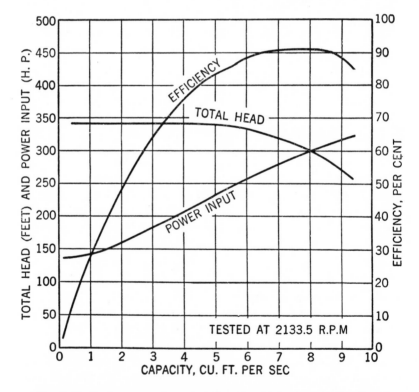

Fig. 8–12. Performance characteristics of a single-stage-suction Byron Jackson centrifugal pump. Test made at California Institute of Technology for the Metropolitan Water District of Southern California.

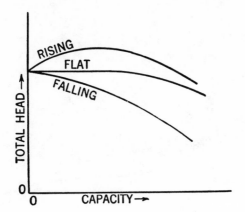

Fig. 8–13. Types of centrifugal-pump characteristics.

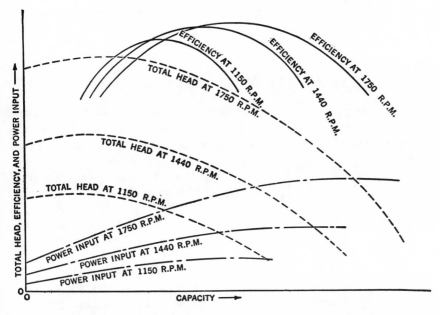

Fig. 8–14. Performance characteristics of a centrifugal pump at different constant speeds.

centrifugal pump may have a *rising, flat,* or *falling* characteristic for the relation between total head and capacity, depending upon the design. Figure 8–14 illustrates the general features of the pump characteristics obtained with different constant-speed tests. The normal capacity increases as the speed is increased. Figure 8–15 illustrates the characterstics of an axial-flow pump.

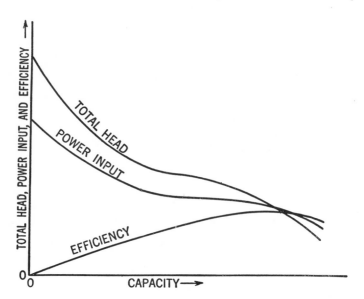

Fig. 8–15. Characteristics of an axial-flow pump at constant shaft speed.

8–10. Combination of dynamic machine and system

The pump or fan is usually only one part of a system. The entire arrangement should be analyzed as a unit in order to obtain the most economical combination of equipment. Equation (8–7) defines the total head developed by a pump or fan. The pump may be required to lift the fluid a certain vertical distance and to overcome system flow losses, such as those in the intake and discharge pipes.

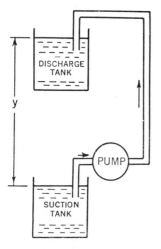

As an illustration, consider the system indicated in Fig, 8–16, with a certain static head y. The head-capacity characteristic for the pump at a certain speed is given in Fig. 8–17. The dotted horizontal line through point A represents the static head that must be overcome regardless of the flow. The head loss in pipe, fittings, and valves can be determined at various rates of flow. Addition of these system flow and the static head y at various capacities gives the service characteristic curve AC. The point of intersection B gives the head and capacity Q_1 at which the system will operate.

Fig. 8–16. Diagrammatic layout of a pumping system.

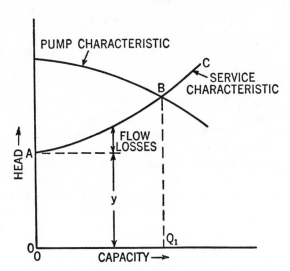

Fig. 8-17. Determination of the operating point of a pumping system.

The system cannot have a capacity greater than Q_1. If it is desired to reduce the capacity below Q_1, some throttling of the flow, as with a valve, would be necessary. Service characteristics for different pipe diameters can be determined, and the corresponding operating capacities, heads, power inputs, and efficiencies found. A study of the effects with different pipe sizes is useful in determining the most economical pipe size.

For a liquid pump with a static suction lift, as shown in Fig. 8-16, the pressure at the pump inlet must be below atmospheric in order to cause flow from the suction tank to the pump. Theoretically, the maximum suction lift is about 33.9 feet for water at 50 degrees Fahrenheit. Actually, however, this maximum suction lift is never realized. There are pressure losses due to the flow. If the temperature of the liquid is high, the liquid may boil or vaporize at the low inlet pressure. Trying to pump a vapor with a pump designed for liquids may cause difficulties. Practical limits of suction lifts for water at 60 degrees Fahrenheit are frequently set at about 15 feet for some centrifugal pumps and 22 feet for rotary and reciprocating pumps. The distances are less for higher temperatures.

8-11. Similarity relations

In order to obtain information regarding the flow phenomena in or around a structure or machine, called the original or *prototype*, it is often convenient, economical, and sound engineering to experiment with a copy or *model* of the prototype. The model may be geometrically smaller than, equal to, or larger than the prototype in size.

Certain laws of similarity must be observed in order to insure that the model-test data can be applied to the prototype. These laws, in turn, provide means for correlating and interpreting test data. For the flow around two bodies, there is this question: Under what conditions is the flow around one body mechanically similar to the flow around the other? Mechanical similarity implies not only geometric similarity but also similarity with respect to the forces acting, or dynamic similarity.

The flow around two bodies can be similar only if the body shapes are geometrically similar; this geometric similarity is a necessary condition, but it is not a sufficient condition. The streamline pattern for one body must also be similar to the streamline pattern for the other body. For corresponding points with respect to the bodies, the velocity direction in one flow must be the same as the velocity direction in the other flow. The velocity direction at any point in the field of flow is determined completely by the ratio of the forces acting on a particle at the point. Therefore, mechanical similarity is realized when the ratio of forces acting on a fluid particle in one flow is the same as the ratio of forces acting at a corresponding point in the other flow.

Consider a pair of geometrically similar pumps or fans of different sizes, but having similar flow patterns. Assume steady, incompressible flow. As illustrated in Fig. 8–18, V is the resultant velocity and U is the peripheral

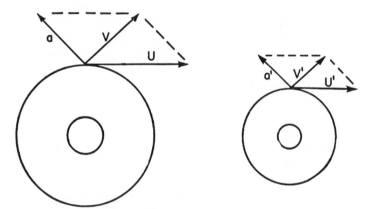

Fig. 8–18. Velocity diagrams for different runners.

velocity for one runner, whereas V' and U' are the corresponding velocities for the other runner. For similar flow patterns, and thus similar velocity triangles, the ratio V/U for one machine must equal the ratio V'/U' for the other machine.

Let Q represent volume rate of flow, N angular speed (as revolutions per unit time), and D diameter for one machine. The peripheral velocity U is proportional to the product ND. The runner exit area is proportional to D^2. Q is proportional to the radial component of V times the runner exit area. Thus Q is proportional to VD^2. Then the ratio V/U is directly proportional to Q/ND^3. Let Q', N', and D' refer to the other member of the pair. Thus for geometrically similar machines with similar flow patterns, the following relation holds

$$\frac{Q}{ND^3} = \frac{Q'}{N'(D')^3} = \text{constant} \qquad (8\text{–}11)$$

The same relation can be worked out for turbines, with the velocity directions reversed.

As shown by Equation (8–10) the work input to the runner is proportional to the product $V_{2U}U_2$. Let us consider a pair of geometrically similar fans or pumps with similar flow patterns. Further, let us assume that these machines have essentially the same efficiency. Then the fluid head H developed by one machine is proportional to ND; likewise U_2 is proportional to ND. Thus H is directly proportional to $(ND)^2$. For similar flows and the same efficiency, the proportionality factor for one machine equals that for the other machine. Thus

$$\frac{H}{(ND)^2} = \frac{H'}{(N'D')^2} = \text{constant} \qquad (8\text{–}12)$$

The foregoing relations are used in adapting the test results of a fan or pump. It may be desired to get a fan or pump characteristic at a constant runner speed. The actual measurements of head, capacity, and power may be made at runner speeds slightly different from the desired speed. Measurements at the different speeds can be converted to values at the desired constant speed by the foregoing relations.

8–12. Specific speed

Various factors or parameters are used in practice, in correlating data on performance and design. Among the factors used is the very prominent one called *specific speed*. The specific speed of a pump N_S is defined in industrial practice as

$$N_S = \frac{N\sqrt{\text{G.P.M}}}{H^{3/4}} \qquad (8\text{–}13)$$

where N is the revolutions per minute of the pump shaft, G.P.M. is the capacity in gallons per minute, and H is the pump head per stage in feet. The head and capacity values are those at the point of maximum efficiency for the pump-shaft speed employed.

Specific speed is useful as an index of the type of pump; specific speed provides a parameter that is useful in determining what combinations of head, speed, and capacity are possible and desirable.

Figure 8–19 shows a plot of approximate relative impeller shapes and efficiencies as a function of specific speed. The usual radial-flow impeller has a low specific speed whereas the usual axial-flow runner has a high specific speed. A specific speed less than 500 is not considered practicable or desirable.

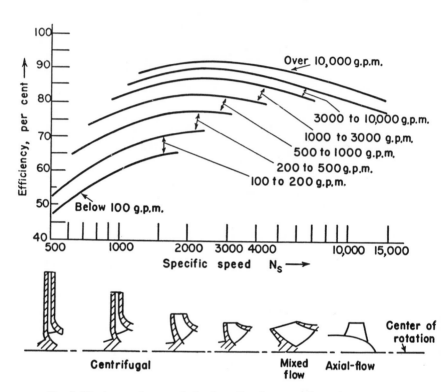

Fig. 8–19. Approximate relative impeller shapes and good average efficiencies for commercial pumps as a function of specific speed. (*Courtesy Worthington Pump & Machinery Corp.*)

The specific speed of a fan could be defined in various ways. One definition is the following

$$N_S = \frac{N\sqrt{\text{C.F.M.}}}{H^{3/4}} \qquad (8\text{–}14)$$

where N is the angular speed of the runner in revolutions per minute, C.F.M. is the capacity in cubic feet per minute, and H is the fan head in feet of fluid flowing. For narrow straight-blade centrifugal fans, N_S varies from about 140 to 900. For curved-blade centrifugal fans N_S varies from about 1000 to 3000. For axial-flow fans N_S varies from about 4000 to 8000.

8-13. Cavitation

If a liquid flows through a machine or a stationary passage, there may be an unfavorable change in performance as operating conditions vary. There may be two serious undesirable effects: (1) a marked drop in efficiency; and (2) a dangerous erosion or pitting of some of the metal parts. The operation of a centrifugal pump handling water will be used as a specific illustration. For a certain inlet head, if the discharge valve is opened, the efficiency rises to some peak as represented in Fig. 8-20. The sound emitted by the pump is normal. As the discharge valve is opened further, a condition may be reached at which the efficiency (and total head) drops markedly. This condition is sometimes called the *cutoff* point. As this condition is approached, the sound emitted by the pump changes. At first it sounds as if sand were passing through the pump (with clear water entering). Then the sound or noise may change (as the discharge is increased) to give the impression of rocks passing through the pump, or a machine-gun barrage. If the pump is operated for any length of time at these conditions, the impeller may be badly eroded and pitted.

These two effects, of efficiency drop and pitting, have been discovered in the operation of water pumps, water turbines, marine propellers (on surface

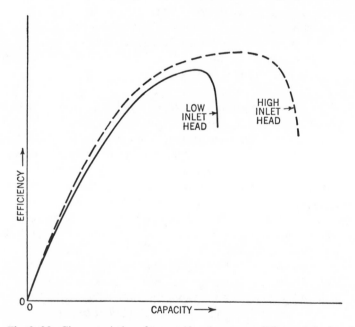

Fig. 8-20. Characteristics of a centrifugal pump at different inlet heads.

ships particularly), and in diverging channels, as in venturi tubes. In many cases these effects have been traced and ascribed to a phenomenon called *cavitation*. The word "cavitation" itself implies a cavity or a void. If, at some point in the liquid flow, the existing fluid pressure equals the vapor pressure at the particular temperature, then the liquid will vaporize—a cavity or void will form. If the fluid pressure fluctuates slightly above and below the vapor pressure (this fluctuation is common), there will be an alternate formation and collapse of the vapor bubbles. Evidence shows that this alternate collapse and formation of bubbles is responsible for the marked drop in efficiency and the pitting of the metal parts.

The violent collapse (taking place in a very short time) of vapor bubbles can force liquid at high velocity into the vapor-filled pores of the metal. The sudden stoppage at the bottom of the pore can produce surge pressures of high intensity on small areas. The process might be called an explosion or implosion. These surge pressures can exceed the tensile strength of the metal, and progressively blast out particles and give the metal a spongy appearance. Rapid pitting takes place, often eating holes through metal vanes and dangerously weakening the structure. Figure 8–21 shows the results of cavitation on some turbine runners.

It is to be noted that two phases of a substance, liquid and vapor, are involved in the cavitation process. This double character makes the phenomenon difficult to analyze completely. Furthermore, in actual applications water has a small percentage of absorbed and dissolved air, oxygen, and nitrogen. These gases help the bubble formation. Solid matter may act as a catalyzer in the formation of bubbles. There is evidence to show that cavitation effects are purely mechanical. Some chemical action, however, may be present to supplement cavitation and to accelerate pitting.

Although much theoretical and experimental work has been done on cavitation, the mechanism of cavitation has not been definitely or completely established yet on a quantitative basis for all fluid machines. Various parameters, such as specific speed and dimensionless ratios including the vapor pressure or head, have been used to advantage in correlating experimental data on cavitation limits. It is difficult to calculate exactly the pressure in some of the complicated flows involved, as at a certain point on a pump impeller or a marine propeller. It is frequently necessary to investigate experimentally the performance of a machine, or its model, to determine upper limits below which cavitation effects do not occur.

In some cases, as for the diverging channel in a venturi tube, it may not be difficult to calculate the liquid pressure at various points. A comparison between the liquid pressure and the vapor pressure at the particular temperature should give some indication as to the possibility of cavitation occurring. Values of vapor pressure for water at different temperatures are given in Fig. 8–22.

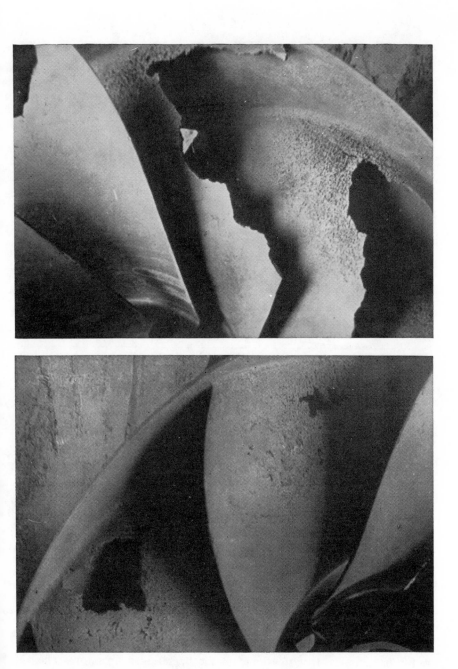

Fig. 8–21. Hydraulic turbine runners showing the results of cavitation. (*Courtesy Baldwin Southwark Division, Baldwin Locomotive Works.*)

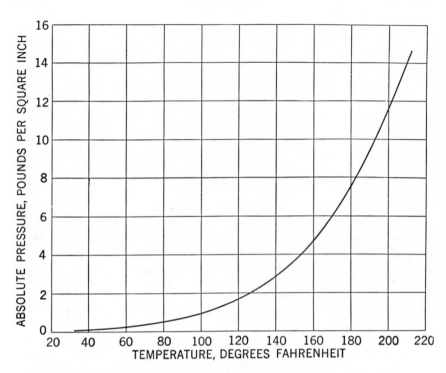

Fig. 8–22. Vapor pressure plotted against temperature for water. (Data adapted from J. H. Keenan and F. G. Keyes, *Thermodynamic Properties of Steam*, Wiley, 1936.)

For centrifugal pumps, particular care should be taken to have a high enough inlet pressure to avoid cavitation. In general, as the pump inlet pressure is reduced, the capacity at which cavitation occurs is reduced.

8–14. Net positive suction head

Operation of a pump with a suction or inlet head close to the vapor pressure of the liquid may cause cavitation within the pump impeller. In practice a certain term NPSH, or net positive suction head, has been established as an aid in evaluating the possibility of cavitation.

Let p_a represent the absolute static pressure at pump inlet, γ the specific weight of the liquid flowing, V the average velocity at inlet, and h_v the absolute vapor pressure head of the liquid at the operating temperature. The net positive suction head is defined as

$$\text{NPSH} = \frac{p_a}{\gamma} + \frac{V^2}{2g} - h_v \qquad (8\text{–}15)$$

The sum of the two terms

$$\frac{p_a}{\gamma} + \frac{V^2}{2g}$$

represents the total suction or inlet head with the pump centerline taken as the datum ($z = 0$ at pump centerline). The NPSH thus represents the total suction head of liquid above the vapor pressure head. It is desirable to avoid low values of NPSH which may cause cavitation.

8–15. Types of turbines

Except for the direction of the energy conversion, centrifugal pumps and one common class of turbines (reaction) have many features in common. The efficiency of each type of machine depends on the shape of the blades. In principle it is possible to use the same blading for either. Knapp,[5] for example, reported results of experiments on a high-head, high efficiency centrifugal pump. This machine was operated both as a turbine and as a pump. High efficiencies of the same order of magnitude were found in both the normal pump and turbine regions of operation.

Turbines are commonly divided into two general types: (*a*) impulse and (*b*) reaction. In both types the momentum of a stream of fluid is changed by passing it across some sort of wheel, or runner, with buckets or vanes. In both types the force acting on the buckets or vanes rotates the runner, which performs useful work, and the fluid falls away with reduced energy. The terms "impulse" and "reaction" alone do not provide a complete distinction between the action in the two types of machines. The presence or absence of a free jet is one feature which distinguishes between the two types.

8–16. Impulse turbines

In turbines of the impulse or Pelton type, the fluid energy, first in the potential form, is next converted wholly into the kinetic form by means of a free jet in one or two nozzles. In the jet the static pressure is practically that of the atmosphere in which the jet is moving. As indicated in Fig. 8–23 this circular free jet strikes buckets on a rotating wheel. In practice these buckets (see Fig. 8–24) are usually spoon-shaped, with a central ridge splitting the impinging jet into two halves which are deflected backward through an angle of about 165 degrees. A complete reversal of 180 degrees would be desirable. This is not possible because the fluid must be thrown to one side

[5] "Complete Characteristics of Centrifugal Pumps and Their Use in the Prediction of Transient Behavior," by R. T. Knapp. *A.S.M.E. Transactions*, November, 1937, page 683.

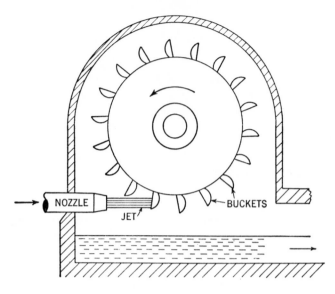

Fig. 8–23. Impulse turbine.

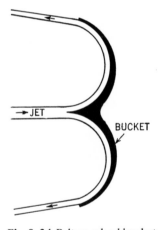

Fig. 8–24. Pelton-wheel bucket.

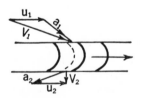

Fig. 8–25. Velocity diagram for an impulse-turbine wheel.

in order to clear the following bucket. The actual energy transfer from jet to wheel is wholly by means of a reaction, that is, by changing the momentum of the stream.

The action on a single vane of an impulse turbine is indicated in Fig. 8–25. The subscript 1 refers to entrance and the subscript 2 refers to exit. Let u represent the peripheral velocity of the wheel, and a the relative velocity of the fluid with respect to the vane. The absolute velocity of the fluid V is the vector sum of a and u. The dotted curve in Fig. 8–25 represents the free edge of the jet at atmospheric pressure. The passage between the vanes is not completely filled with fluid. Under favorable circumstances the exit velocity V_2 is nearly at right angles to the plane of rotation, and is very small. If the friction is very small, or neglected, then a_1 equals a_2 in magnitude. If the entrance point is at the same radius as the exit point, then u_1 equals u_2. The peripheral velocity of the wheel for maximum efficiency is slightly less than one-half the absolute velocity of the jet.

8–17. Reaction turbines

There is no formation of a free jet in the reaction turbine. In this type there is a casing fitted with guide vanes completely surrounding the runner. The runner is a wheel provided with vanes, and fluid enters it completely around the periphery. In the reaction turbine only a moderate amount of the available fluid energy is transformed into kinetic energy before entrance to the runner, so that the fluid enters the runner with an excess pressure. The fluid completely fills the vane passages. Although the path of the fluid through the runner is tortuous, the essential features of the diagram in Fig. 8–26 apply. The relative velocity of the fluid is increased as it flows in the narrowing passages between the blades. Under favorable circumstances the absolute velocity after passing through the runner is nearly at right angles to the plane of rotation, and is very small.

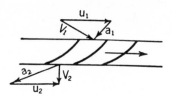

Fig. 8–26. Velocity diagram for a reaction-turbine runner.

Impulse turbines of modern design usually have horizontal shafts. Reaction turbines can be of either the vertical- or horizontal-shaft type.

8–18. Turbine runners

In a manner which is somewhat similar to that for centrifugal and axial-flow pumps, turbines are sometimes classified as (*a*) tangential (Pelton); (*b*) radial- and mixed-flow (Francis or American); and (*c*) axial-flow (Kaplan). The Pelton wheel is called a *tangential* wheel because the center line of the jet is tangent to the path of the centers of the buckets.

As represented in Fig. 8–27, fluid under pressure enters a spiral casing or housing which encompasses the runner. After flowing through adjustable

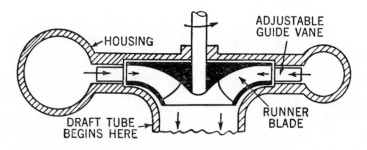

Fig. 8–27. Radial-flow turbine (Francis).

guide vanes, the fluid passes through the rotating runner in a plane practically normal to the axis of rotation; the flow is largely radially inward. This
machine is frequently called a *radial-flow* or Francis turbine.

The flow is partly radial and partly axial in the runner shown in Fig.
8–28. This is known as *mixed flow*. This type of turbine is sometimes called
the American type, although the name Francis is frequently extended to
designate all inward-flow types of turbines. A section of an actual installation
is shown in Fig. 8–29. The turbine runner is connected directly to the vertical
generator shaft.

Turbines of the type shown in Fig. 8–28 are variously called *axial-flow*

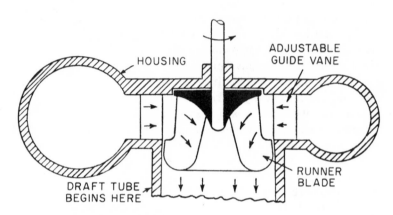

Fig. 8–28. Combined radial-axial-flow turbine.

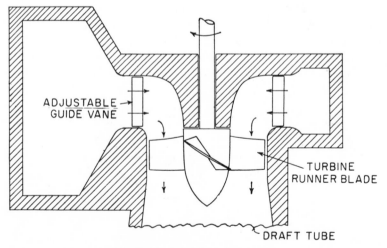

Fig. 8–29. Axial-flow turbine.

or *propeller type*. The arrangement of guide vanes is similar to that of the Francis turbine. The guide vanes give a spiral inward motion to the fluid. The fluid then passes axially through the runner. If there is not much variation in available head and if the load is fairly constant, the *fixed blade* axial-flow type of turbine runner is most economical. High efficiency can be maintained with a minimum number of mechanical parts. In some installations, however, the available head or the load may be variable. The Kaplan adjustable-blade axial-flow type of runner has been employed for such services. In this runner the pitch of each blade can be controlled to meet different conditions.

In many installations the amount of fluid entering the runner must be regulated in order to adapt the turbine to suit the load at any time. This regulation is usually done automatically in such a way that the rotational speed of the turbine is kept constant. In the Pelton wheel, speed regulation is sometimes obtained by means of a regulating needle or spear in the jet. In reaction turbines, the adjustable guide vanes are rotated so that the channels are widened or narrowed.

8-19. Some performance features of water turbines

Generally speaking, the tangential or impulse wheel is used for high heads and relatively small volumetric rates of flow. The radial- and mixed-flow turbine is used for intermediate heads and intermediate rates of flow. The axial-flow turbine is used for low heads at high rotational speeds and large rates of flow.

Impulse turbines are generally employed for heads above 800 feet. The efficiency of an impulse turbine may be more than 85 per cent; an average value is about 82 per cent.

Reaction turbines are generally regarded as not suited for heads greater than 800 feet, and are commonly employed under considerably lower heads. Intermediate heads between 15 and 750 feet have been used in economical installations. An average efficiency for reaction turbines might be placed at 90 per cent. Values slightly above this have been attained in large plants.

Units of the axial-flow type operate under heads up to 100 feet. Axial-flow turbines have been built with efficiencies above 90 per cent in units up to 60,000 horsepower.

The factor *specific speed* is used in turbine practice. The specific speed of a turbine N_S is defined as

$$N_S = \frac{N\sqrt{\text{B.hp.}}}{h^{5/4}} \tag{8-16}$$

where B.hp. is the brake horsepower, or power output at the rotating turbine shaft, h is the total available head in feet, and N is the runner speed in revolutions per minute, at which the maximum efficiency under a given head is attained. Specific speed is a characteristic index number; it serves to classify a turbine and to indicate its type. Considering average values, for impulse wheels N_S ranges from 0 to 4.5, for reaction turbines N_S varies from 10 to 100, whereas for axial-flow turbines N_S varies from 80 to 200.

8–20. Power plants

The turbine is but one part of a system. A turbine, with oil as the working fluid, is part of a fluid coupling and a fluid torque converter. Steam and water turbines are used in power-generating plants. The choice as to whether a thermal or a water-power plant should be used depends upon a careful evaluation of many factors. Thermal and water plants are combined in some power systems. The present discussion is not intended to cover all the economic and social aspects involved in power generation, but only to illustrate very briefly some of the technical features of water-power plants.

A hydroelectric power plant consists mainly of means for delivering available water under pressure to a site, machinery and equipment to produce the power, the necessary turbine setting or foundation, and a building. A storage reservoir may be used in order to hold enough water to operate the plant for some length of time. In some plants there is no such reservoir. A dam of some sort may be used to develop most or a portion of the available head. The intake equipment usually includes racks or screens to prevent trash from entering the runners, and a head gate to shut off the water flow if necessary.

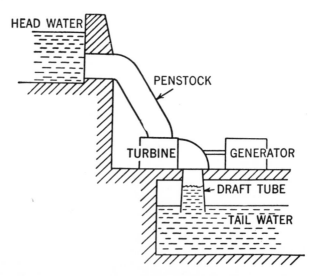

Fig. 8–30. Diagrammatic sketch of a water-power plant.

The water is conducted toward the power house by a conduit which may be an open channel, called a canal or flume, or a tunnel. In high head plants, as a rule, water must be conducted considerable distances. A forebay is often placed at the end of the conduit. The forebay is a small reservoir to equalize the flow. The water is led from the forebay directly to the turbine through pressure pipes or penstocks. The turbine with its case or pit, and possibly a draft tube, comprises the setting. The turbine discharges into a body of water called the tail water. The channel conducting the water away is the tail race.

As indicated in Fig. 8–30, draft tubes are used in connection with reaction turbines. If a draft tube is employed, the entire available head can be used, and the turbine can still be set high enough above the tailwater level to allow inspection and maintenance without draining the tail bay. Practically, the draft head should not be greater than 12 to 15 feet, to prevent the water column from breaking.

8–21. Fluid couplings

Various methods can be employed for coupling or joining two rotating shafts, as for connecting an engine or motor to its load. Purely mechanical, rigid, or flexible couplings are used in many applications. In some applications, however, service requirements are best met by some form of fluid connector.

A fluid coupling is simply a combination of a centrifugal pump and a turbine. Oil is usually employed in commercial units for the working fluid because of its lubricating properties, stability, and availability. Some recommendations call for a straight mineral oil having a maximum Saybolt reading of about 180 seconds at 130 degrees Fahrenheit. The principle of the fluid coupling can be easily demonstrated by means of two ordinary electric fans which are set facing each other. One fan, connected to an electric outlet, is put into motion by turning on the electric current. As its blades rotate, the air current which they develop turns the blades of the other fan, which is not receiving any electric current.

As indicated in Fig. 8–31, the rotating input, or primary, shaft A drives the pump impeller, which usually has straight radial vanes. Kinetic energy is added to the fluid as the pump builds up speed, and the fluid flows outward. After sufficient energy has been developed, the fluid rotates the turbine runner and output, or secondary, shaft B. The fluid moves through a closed path shaped like a vortex ring. There is *no* mechanical or rigid connection between shafts A and B. The connection is solely by means of a fluid, which has cushioning properties.

Two rotating shafts could be coupled by an arrangement consisting of a separate pump and a separate turbine joined by intermediate piping. The fluid coupling, however, provides a considerable saving in weight and space

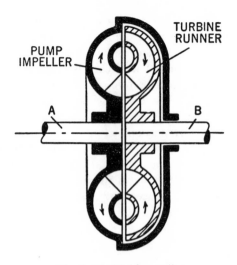

Fig. 8–31. Fluid coupling.

because of its concentric arrangement of impeller and runner. Further, the fluid coupling eliminates the friction loss in the intermediate piping.

Since there are no torque reacting elements in the fluid coupling besides the impeller and runner, under steady operating conditions the output torque (shaft *B*) always equals the input torque (shaft *A*)—hence the term "fluid coupling." The speed of shaft *A* always exceeds that of shaft *B*. At the beginning of operation, shaft *A* rotates while shaft *B* does not; the *slip* is 100 per cent. At rated speeds the slip may be reduced to from 1 to 4 per cent. At normal speeds and loads, the efficiency of a fluid coupling is high, and may be 96 to 99 per cent. The acceleration of the output shaft is smooth under all conditions of operation. The load may be stalled completely without stalling the driver.

Torsional vibrations or shocks on either shaft of the fluid coupling are damped out by the fluid. In some cases it is very important to eliminate the transmission of such vibrations or shocks. The fluid coupling does not replace the transmission in the usual automotive application. The extra cost of the coupling, plus the power loss through it, are offset by such factors as reduction in wear on the parts behind the coupling, reduction of vibrations, and improved operation of the vehicle.

8–22. Fluid torque converters

It is to be emphasized that a fluid torque converter is different from a fluid coupling. As indicated in Fig. 8–32, the fluid (usually oil) discharged from the pump flows through the turbine runner and a series of fixed guide vanes.

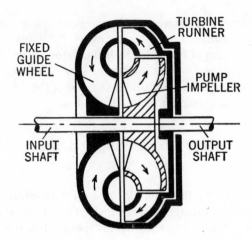

Fig. 8–32. Torque converter.

The stationary vanes change the direction of the fluid, thereby making possible a torque and speed transformation. Since the stationary vanes take some reaction (carrying it to the foundation), the turbine (or secondary) torque does not equal the pump (or primary) torque—hence the name "torque converter." Different arrangements are possible, with several stages of turbine runners and stationary vanes. The converter provides for smooth starting of the load and absorbs torsional vibrations and shocks.

In some converters the starting output torque is about five times the input torque. In general, the efficiency of a torque converter is high at low speeds. Converters have been built with peak efficiencies around 85 to 87

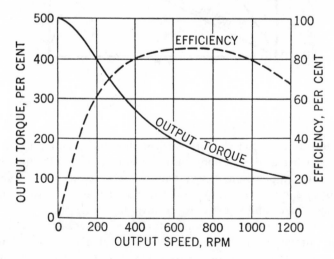

Fig. 8–33. Characteristics of a torque converter.

per cent. Some typical characteristics of a torque converter are illustrated in Fig. 8–33. The decreasing torque-speed characteristic is an advantage for the starting and acceleration of heavy loads. If the output shaft of the converter is stalled during operation, owing to extreme loads, the driver will not stall but will continue running. In some units the speed of the driver is practically constant regardless of the speed of the driven unit.

8–23. Combination of fluid coupling and torque converter

The "speed ratio" of a fluid coupling or a torque converter will be taken as the ratio of the output (turbine) speed divided by the input (pump) speed. Figure 8–34 shows plots of efficiency versus speed ratio; one curve is for a coupling and the other is for a torque converter. The efficiency of a fluid coupling starts at zero, rises in a straight line until it almost attains 100 per cent, and then drops sharply to zero. The efficiency of a torque converter, on the other hand, has a different trend. The efficiency of a torque converter starts at zero, rises above that of the fluid coupling curve to a maximum, and then gradually drops off to zero.

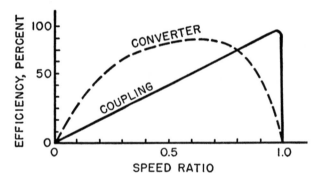

Fig. 8–34. Efficiency curves for a fluid coupling and a fluid torque converter.

The torque converter is more efficient than the coupling at the lower speed ratios, whereas the coupling has a higher efficiency at the higher speed ratios. In some applications (as automotive), it is desirable to combine a fluid coupling and a torque converter in such a manner as to avoid the inefficient range of each and to use only the advantages of each.

A possible method of combining features is to arrange a torque converter with guide vanes or reaction vanes as a one-way clutch or free-wheeling unit; this unit permits motion only in the direction of the driver and prevents motion in the reverse direction. The guide vanes are held stationary at the lower speed ratios; the assembly then operates as a torque converter. The guide vanes are arranged so that they rotate at the higher speed ratios; the assembly then operates as a fluid coupling. Thus it is possible to devise a fluid unit which has the advantages of both the coupling and the torque converter.

8–24. General relations

Consider steady flow in a torque converter. The fluid recirculates in a closed path. The total change in angular momentum of the fluid about the shaft axis in a complete circulation is zero. Therefore the total torque acting on the fluid is zero, because torque equals the time rate of change in angular momentum. Let T_P represent the torque of the pump acting on the fluid, T_S the torque of the turbine runner on the fluid, and T_F the torque of the fixed guide vanes acting on the fluid. Then for equilibrium

$$T_P + T_F - T_S = 0$$
$$T_S = T_P + T_F \tag{8–17}$$

The output or delivery torque is equal to the pump torque plus the torque applied to the fixed guide vanes.

Let ω_P represent the pump angular speed, and ω_S represent the turbine angular speed. The power delivered from the machine at the output shaft is $T_S\omega_S$. The power input to the machine is $T_P\omega_P$. The efficiency E of the machine is the ratio of power output divided by power input, or

$$E = \frac{T_S\omega_S}{T_P\omega_P} = \frac{(T_P + T_F)\omega_S}{T_P\omega_P} \tag{8–18}$$

8–25 Special relations for fluid couplings

In a fluid coupling there are no fixed guide vanes, $T_F = 0$, and thus $T_S = T_P$.

For a fluid coupling the relation for efficiency reduces to

$$E = \frac{\omega_S}{\omega_P} \tag{8–19}$$

The "slip" is given by the relation $\omega_S = (1 - \text{slip})\omega_P$. Then the efficiency for a fluid coupling is

$$E = \frac{(1 - \text{slip})\omega_P}{\omega_P} = 1 - \text{slip} \tag{8–20}$$

The efficiency of a fluid coupling varies linearly with slip.

Consider a runner as a turbine or pump runner. Let Q represent the volume rate of flow through this runner, ρ the density of the fluid flowing through, and γ the specific weight of the fluid. Equation (8–8) shows that the torque T is given by the relation

$$T = \rho Q(R_2 V_{2U} - R_1 V_{1U}) \qquad (8\text{–}21)$$

Fluid couplings usually have straight radial blades. Let us analyze a fluid coupling in which the flow relative to the blades at inlet and exit is radial. Assume that the pump impeller is identical with the turbine runner. Let us calculate the torque for the pump runner. In this case the fluid leaving the runner has a radial relative velocity which means that

$$V_{2U} = R_2 \omega_P$$

Because the fluid enters with a relative radial velocity,

$$V_{1U} = R_1 \omega_S$$

Thus Equation (8–21) becomes

$$T = \rho Q(R_2^2 \omega_P - R_1^2 \omega_S) \qquad (8\text{–}22)$$

The product $T\omega_P$ represents input power. The work input per unit weight of fluid is $T\omega_P/\gamma Q$. Thus the head loss h, or the energy loss in the coupling per unit weight of fluid flowing, is given by the expression

$$\frac{T(\omega_P - \omega_S)}{\gamma Q} = h \qquad (8\text{–}23)$$

Eliminating the torque from Equations (8–22) and (8–23) gives

$$\frac{1}{g}(R_2^2 \omega_P - R_1^2 \omega_S)(\omega_P - \omega_S) = h \qquad (8\text{–}24)$$

8–26. Maximum propulsive efficiency

There are various bodies and craft which are propelled through an expanse of fluid. Examples are aircraft, marine craft, and rockets. Sometimes it is felt that the basic concept behind a certain propulsion system is quite new. Actually, some of the basic ideas have been established for centuries. Galileo and Newton, for example, formulated some relations which are basic in modern propulsion systems.

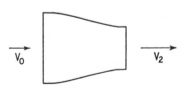

Fig. 8–35. Notation for propulsion system.

The maximum propulsive efficiency is a factor which is useful in evaluating the performance of a propulsion system. Consider the case in which fluid enters any device or structure with a velocity V_0 (relative to the body) and leaves with the velocity V_2 (relative to the body). The notation is shown in Fig. 8–35. If the device is moving through stationary fluid, then V_0 is the velocity of the body with respect to that stationary fluid. Let M represent the mass of fluid passing through the device per unit time, and let F represent the thrust or the force along the direction of V_0. From the basic relation between force, mass, and acceleration for steady flow,

$$F = M(V_2 - V_0) \tag{8-25}$$

The thrust times the velocity V_0 gives the useful power developed, or power output,

$$\text{power output} = M(V_2 - V_0)V_0 \tag{8-26}$$

The lost power due to lost kinetic energy in the exit jet is

$$\text{lost power} = \frac{M(V_2 - V_0)^2}{2} \tag{8-27}$$

The power input is the sum of power output plus lost power. The maximum propulsive efficiency E is the ratio of the power output divided by the power input:

$$E = \frac{M(V_2 - V_0)V_0}{M(V_2 - V_0)V_0 + \dfrac{M(V_2 - V_0)^2}{2}}$$

$$E = \frac{2}{1 + \dfrac{V_2}{V_0}} \tag{8-28}$$

Equation (8–28) gives the maximum efficiency as a function of the velocity ratio; a plot of these two factors is given in Fig. 8–36. Note that the efficiency decreases as the velocity ratio V_2/V_0 increases. When $V_2 = V_0$, the efficiency is 100 per cent; this limiting condition gives no thrust because of zero velocity change.

A propeller, such as that on air or marine craft, moves with the craft velocity V_0. The same relative motion between fluid and propeller exists if the blade screw rotates in a fixed plane and the fluid approaches some distance ahead with the velocity V_0, as shown in Fig. 8–37. The stream of fluid passing through the propeller disk is called the *slipstream* or the *propeller race*. The fluid velocity increases as it approaches and leaves the propeller. At the propeller the velocity is V_1; at some distance behind the propeller the velocity is V_2 and the pressure is p_0. As

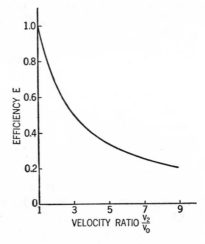

Fig. 8–36. Maximum propulsive efficiency as a function of velocity ratio.

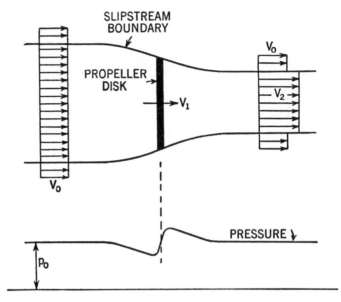

Fig. 8-37. Propeller slipstream.

indicated diagrammatically in the lower portion of Fig. 8–37, the pressure decreases from the value p_0, rises at the blade, and then drops to the value p_0. Momentum has been added to the fluid by the axial thrust T of the propeller. It will be assumed that all fluid elements passing through the screw disk have their pressures raised by exactly the same amount. The rotational motion actually imparted to the fluid will be neglected.

The mass of fluid passing through any section per unit time is $(\pi/4)D^2\rho V_1$. Since force equals the time rate of change of linear momentum,

$$T = \frac{\pi D^2}{4}\rho V_1(V_2 - V_0) \qquad (8\text{--}29)$$

The relation between the velocities can be found by applying the energy equation. Consider an accounting for unit weight of fluid flowing in the slipstream. Then

$$\frac{p_0}{\gamma} + \frac{V_0^2}{2g} + \text{work} = \frac{p_0}{\gamma} + \frac{V_2^2}{2g} \qquad (8\text{--}30)$$

where the "work" term represents the work done on unit weight of fluid by the propeller. The total work done on the fluid per unit time is TV_1. The work done per unit weight is

$$\frac{TV_1}{\dfrac{\pi D^2}{4}V_1\gamma}$$

The energy equation then becomes

$$\frac{V_0^2}{2g} + \frac{4T}{\pi D^2 \gamma} = \frac{V_2^2}{2g} \quad \text{or} \quad T = \frac{\pi D^2}{8}\rho(V_2^2 - V_0^2) \qquad (8\text{-}31)$$

A comparison of Equations (11-5) and (11-7) shows that V_1 is the arithmetic mean of V_0 and V_2:

$$V_1 = \frac{V_0 + V_2}{2} \qquad (8\text{-}32)$$

The useful power output, in driving the craft, is TV_0. The maximum efficiency E is defined as the ratio of power output to power input, or

$$\text{maximum efficiency} = E = \frac{TV_0}{TV_1} = \frac{V_0}{V_1} \qquad (8\text{-}33)$$

The actual propeller efficiencies obtained throughout the working range for good designs are usually from 80 to 88 per cent of the maximum efficiency. Equations (8-29), (8-32), and (8-33) can be arranged to obtain the following relation:

$$E = \frac{1}{1 + \dfrac{2T}{\pi D^2 \rho V_1 V_0}} \qquad (8\text{-}34)$$

Equation (8-34) shows that the maximum efficiency increases with (a) increase in propeller diameter, (b) increase in density, (c) increase in forward velocity, and (d) decrease in thrust. One practical conclusion shows that the propeller diameter should be as large as possible.

One limit on the diameter of propellers operating in air is set by the effect of shockwaves, which occur when the propeller tip speed equals or exceeds the velocity of sound. One limit on the diameter of propellers operating in water is set by *cavitation*. Cavitation is a phenomenon in which the fluid pressure equals the vapor pressure at the existing temperature. Vapor bubbles alternately form and collapse; this action may cause serious pitting of the blade screw and a marked decrease in efficiency. Cavitation requires serious consideration in connection with marine propellers.

8-27. Propeller characteristics

In order to illustrate how propeller characteristics are studied, specific reference will be made to aircraft propellers.

In a wind-tunnel test of a propeller, the propeller rotates in a fixed plane, and air is pushed toward the propeller. The following main measurements are made: the incoming air speed V, the angular speed of the propeller n (as revolutions per unit time), the outside diameter of the propeller D, the thrust T developed by the propeller, the power input to the propeller P, and the density of the air ρ. Thus there are six variables in this problem and there are three primary units. There are different possible groupings of the variables. In practice, however, a certain set of coordinates has become established. They are

$$\frac{V}{nD}, \qquad C_T = \frac{T}{\rho n^2 D^4}, \qquad C_P = \frac{P}{\rho n^3 D^5}$$

Each ratio is dimensionless, and any set of consistent units could be used. For the ratio V/nD, for example, velocity could be expressed in feet per second, diameter in feet, and angular speed in revolutions per second. The factor C_T is a *thrust coefficient*, and C_P is a *power coefficient*. Experimental data are usually expressed in a form with C_T and C_P as functions of the ratio V/nD. The actual overall efficiency of the propeller is defined as the ratio of the useful power TV and the input power P; $E = TV/P$. Since the propeller

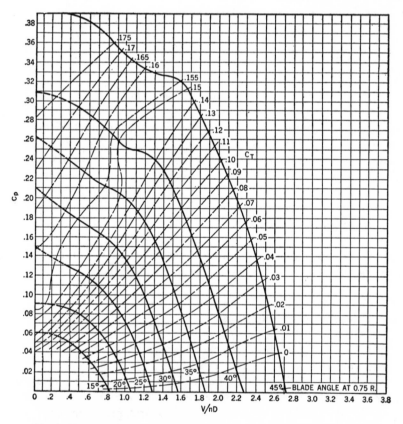

Fig. 8–38. Propeller characteristics for propeller 5869–9, Clark Y section, 3 blades. (From N.A.C.A. *Technical Report* No. 640.)

blade is twisted, the blade angle for a propeller varies in magnitude from the hub to the tip. The nominal blade angle for the entire propeller is taken as the blade angle at the position which is three-fourths of the radius of the propeller.

Figure 8-38 shows the characteristics of a particular propeller as determined from actual tests. Each solid line refers to the relation between C_P and V/nD for a particular blade angle. The dashed lines are lines of constant C_T. Consider a propeller operating at a certain angular speed with a certain forward speed V at a certain blade angle. For this particular value of V/nD and blade angle, reference to the chart will give the power coefficient along the vertical scale and the thrust coefficient from the constant C_T lines. Then the thrust and power are calculated from the relations

$$T = C_T \rho n^2 D^4$$
$$P = C_P \rho n^3 D^5$$

The analysis of the performance of a blade screw based on an integration of the action of the individual elements is frequently called the blade-element theory. The following gives some discussion of this theory.

Let n be the rotational speed of the blade screw, in revolutions per unit time. The peripheral velocity (in the plane of rotation) of the element at A in Fig. 8-39 is $V_P = \pi n d$. Let V be the velocity of the propeller along its axis, or the forward velocity of the craft to which the propeller is attached. Figure 8-40 shows the velocity and force components at a blade element.

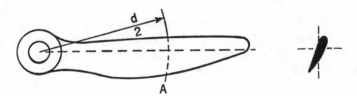

Fig. 8-39. Blade of a propeller.

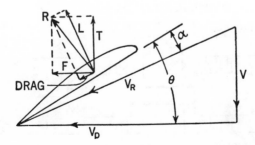

Fig. 8-40. Forces and velocity components at a blade element of a propeller.

This blade element follows a helical path in space. The resultant velocity V_R is the vector sum of V and V_P. The resultant force R acting on the blade element can be broken up into two components, a lift L perpendicular to V_R, and a drag parallel to V_R.

Propeller study is concerned with two other components of R. One component of R is T, a thrust force along the axis of rotation for moving the craft; T does useful work. The other component of R, parallel to the plane of rotation, is the torque force F produced by the power unit. α is the angle of attack of the lifting-vane element, and θ is the blade angle at the section. $\alpha = \theta - \tan^{-1}(V/V_P)$. Since V_P varies with the blade radius, the blade angle changes with radius; the blade is twisted.

8–28. Turbojet engines

Figure 8–41 illustrates a turbojet engine. Atmospheric air enters the left opening or inlet to the compressor. The compressor may be a centrifugal or axial-flow machine. The air from the compressor outlet passes through a chamber where fuel, such as kerosene or gasoline, is injected. The combustion products pass through the turbine; the turbine drives the compressor. The fluid leaves the plant through the tailpipe with a velocity higher than that entering.

For normal operation at steady flight the process is one of steady flow through the power plant. The motion of the parts is one of rotation, not reciprocation. As the fluid passes through the unit there is a large increase in specific volume, and therefore an increase in velocity. The unit pushes backward on the fluid; the fluid pushes forward on the craft.

Figure 8–42 illustrates the pressure-specific volume diagram. The sequence of events is variously called the Brayton cycle, the constant pressure cycle, or the gas turbine cycle.

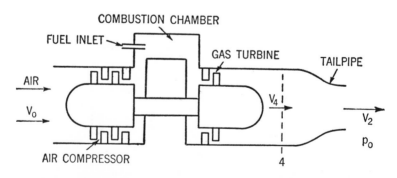

Fig. 8–41. Diagrammatic sketch of a turbojet.

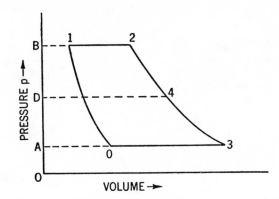

Fig. 8–42. Cycle for turbojet.

Atmospheric air enters at point 0 and is compressed to point 1. Frequently, the process from 1 to 2 is regarded as a constant-pressure process; actually there is some pressure drop as the gases flow through the combustion chamber. In the process from 1 to 2 the fuel is injected, the fuel is evaporated, the fuel combines with the oxygen in the air, and combustion takes place. The increase in specific volume during 1 to 2 is the combined effect of increase in mass, change in chemical composition, and thermal expansion at practically constant pressure.

The products of combustion enter the turbine at point 2 and leave the turbine at point 4. The fluid leaving the turbine can expand further to atmospheric pressure through the tailpipe and the exit region just outside the tailpipe. One important feature is the increase in specific volume of the fluid as it passes through the engine.

8–29. Rocket engines

The screw-type propeller is used on marine craft and some aircraft. There are other types of propulsion systems. In one kind, an air-breathing system, oxygen from the air is used for combustion; the turbojet (Fig. 8–28) is an example. Another kind of propulsion system is the rocket, which carries its own oxygen for combustion. Craft or vehicles which operate during part of their powered flight at an altitude where the air density is insufficient for propellers or air-breathing propulsion must carry the propellants aboard the craft. A common rocket engine is the chemical type in which hot exhaust gases are developed by chemical combustion. The chemicals, or propellants, are of two types, fuel and oxidizer.

Figure 8–43 illustrates a solid-fuel type of rocket. The fuel and oxidizer are intimately mixed together and cast into a solid mass, called a grain, in

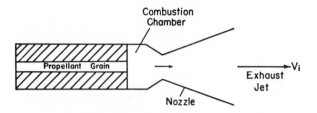

Fig. 8-43. Diagram of solid-propellant rocket.

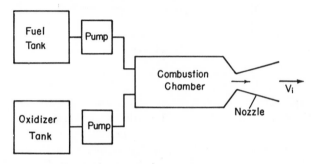

Fig. 8-44. Diagram of liquid-propellant rocket.

the combustion chamber. The propellant is cemented to the inside of the case. The propellant is usually cast with a hole or perforation down the center; this hole may be shaped in different ways, as a star, gear, or some other outline. The propellant and the perforation size and shape influence the burning rate and thus the engine thrust. The propellant may be ignited by an electrical impulse. When the propellant is ignited, the propellant grain burns on the inside surface of the perforation. The hot gas, at a high pressure, then flows back through the converging-diverging nozzle.

Figure 8-44 illustrates a liquid-propellant rocket. The fuel and oxidizer are in separate tanks. The fuel and oxidizer are pumped into the combustion chamber, where a reaction takes place to develop the propellant gas at a high pressure. The gas then flows back through the converging-diverging nozzle.

Consider linear velocities with respect to the rocket. The propellant velocity increases from zero in the combustion chamber to the velocity V_j, which is the final jet velocity of discharged gas. Let M represent the mass of propellant discharged per unit time. The thrust F is given by the relation

$$F = MV_j$$

If M is in slugs per second and V_j in feet per second, then F is in pounds. If the propellant flow is expressed as W, weight per unit time (as pounds per second) then, since $W = gM$, the thrust is

$$F = \frac{WV_j}{g} \tag{8-36}$$

where g is gravitational acceleration.

The specific impulse I of a rocket is defined as the ratio of the pounds of thrust per W (pounds per second). Thus

$$I = \frac{F}{W} = \frac{V_j}{g} \tag{8-37}$$

The specific impulse I is one of the main parameters used in studying rocket propellants. The specific impulse is a measure of how efficiently the propellant is used. For example, in comparing two propellants, the one with the higher specific impulse requires a smaller weight of propellant (for the same thrust) and thus could handle a higher payload.

The term specific fuel consumption c is used in studying other propulsion systems, as the turbojet. Let us express c in pounds of fuel per second per pound of thrust. If we take I as pounds of thrust per weight rate, in pounds per second, then

$$I = \frac{1}{c} \tag{8-38}$$

A turbojet may have an I value of 3500 seconds, whereas a rocket may have an I value of 250 seconds. The large difference between these values of I is due to the fact that the turbojet carries fuel only, whereas the rocket carries both fuel and oxidizer. Solid and liquid propellant rockets may have specific impulse values of the order of magnitude of 250 seconds.

The basic relations for the flow in a converging-diverging nozzle are derived in Chapter Six. Here we will use some of the final results to discuss rocket performance directly. Let k = ratio of specific heats (specific heat at constant pressure divided by specific heat at constant volume), p_c the pressure in the combustion chamber, p_o the ambient pressure just outside the rocket nozzle, R_1 the universal gas constant, and m the molecular weight of the propellant gas after combustion. Then, for a frictionless adiabatic process through the nozzle, the ideal exhaust velocity is given by the relation

$$V_j = \sqrt{2g\left(\frac{k}{k-1}\right)\frac{R_1 T_c}{m}\left[1 - \left(\frac{p_o}{p_c}\right)^{\frac{k-1}{k}}\right]} \tag{8-39}$$

The factors T_c, m, and k depend on the propellant gas. The ratio k may vary from 1.20 to 1.40. Propellants of low molecular weight m and high combustion temperature T_c should give high jet velocity and a high specific impulse. The jet velocity and specific impulse should be high for a high combustion-chamber pressure. The combustion-chamber pressure may be 1000 pounds per square inch absolute for a solid rocket and 500 pounds per square inch absolute for a liquid rocket.

8-30. Motion of craft through a fluid-force study

For convenience, the word "craft" will be used to represent a vehicle, structure, missile, or other body moving through a fluid medium. The medium may be air, water, or some other fluid. First assume steady horizontal motion of the body. Assume that some vertical "lift" force acts on the body to over-come the weight of the body. This lift may be a dynamic lift or a static buoy-ant force. The fluid resistance or drag of the body is a force in the direction of motion. We might call the drag the "force-required" (as force required to overcome resistance).

Some propulsion device is necessary to move the craft. This may involve some gas propellant, a propeller driven by an internal-combustion engine, or a propeller driven by a turbine. The thrust developed by the propulsion device is in the direction of motion; this thrust force acting on the craft might be called the "force-available" (as force available to move the craft).

The force-required is a function of the velocity of the craft. For a parti-cular craft we can select velocities and compute the corresponding drag forces; this procedure involves a study of drag coefficients. Thus a characteris-tic curve of force-required versus velocity can be obtained as illustrated in Fig. 8–45. On the same coordinates can be plotted a curve of force-available

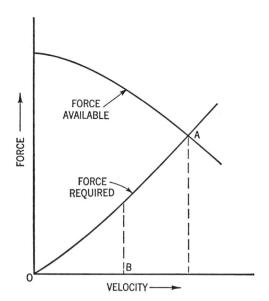

Fig. 8–45. Curves for force study of craft motion.

as a function of velocity; this is determined from a separate study of the propulsion system itself. The point of intersection, point *A*, represents the maximum speed of the craft. Above this speed the force-required is greater than the force the propulsion system can provide. If steady uniform motion at a lower speed, as at *B*, is desired, then it is necessary to reduce either the force-available (as by throttling the engine) or by increasing the force-required (as by braking).

8–31. Motion of craft through a fluid-power study

The question of vertical motion is involved with some craft. Refer to Fig. 8–46 and consider the forces acting *on* the body. The motion of the craft is at some angle θ with the horizontal. Let L represent the lift on the craft, a force perpendicular to the direction of motion. Let W represent the weight

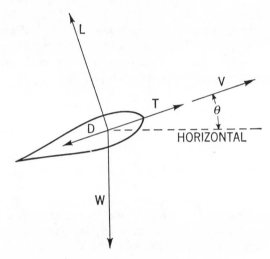

Fig. 8–46. Forces acting on some craft.

of the craft, a vertical force downwards. Let T represent the thrust produced by the propulsion system and D the drag of the craft; each of these two forces is in the direction of motion. For steady unaccelerated motion of the craft, a balance of forces in the direction along the motion, and a balance of forces perpendicular to the motion gives the two relations

$$L = W \cos \theta \qquad (8\text{-}40)$$
$$T = D + W \sin \theta \qquad (8\text{-}41)$$

The force method presented in the previous article is direct and simple. With some propulsion systems it is not easy or convenient to measure thrust directly, but it is more convenient to measure power. An example is the system of propeller and conventional internal-combustion engine. It is very convenient to measure the power of an internal engine by means of a dynamometer. Each term of Equation (8–41) can be multiplied by V to give

$$TV = DV + WV \sin \theta \qquad (8\text{-}42)$$
$$WV \sin \theta = TV - DV \qquad (8\text{-}43)$$
$$WV \sin \theta = P_A - P_R \qquad (8\text{-}44)$$

The product TV or P_A is frequently called the "power-available," and DV or P_R is called the "power-required." A diagram of power versus velocity could have two curves, one for power-required and one for power-available. Let z represent the vertical height above some arbitrary horizontal datum. The quotient $C = dz/dt$ represents a vertical velocity or "climb rate." Since $V \sin \theta = C$, Equation (8–44) can be written in the form

$$C = \frac{V(T - D)}{W} = \frac{P_A - P_R}{W}$$

For speeds below the maximum there is a range in which the force-available T is greater than the drag D. For such a condition a vertical climb is possible. When the force-available equals the force-required, then the power-available equals the power-required, and no climb is possible.

PROBLEMS

8-1. A pump discharges 2000 gallons of brine (specific gravity = 1.2) per minute. The pump inlet, of 12 inch diameter, is at the same level as the outlet, of 8 inch diameter. At inlet the vacuum is 6 inches of mercury. The center of the pressure gage connected to the pump discharge flange is 4 feet above the discharge-flange center. This gage reads 20 pounds per square inch gage. For a pump efficiency of 82 per cent, what is the power output of the motor?

8-2. A water pump develops a total head of 200 feet. The pump efficiency is 80 per cent and the motor efficiency is 87.5 per cent. If the power rate is 1.5 cents per kilowatt-hour, what is the power cost for pumping 1000 gallons? (1 horsepower = 0.746 kilowatts.)

8-3. A test on a centrifugal pump operating at 1150 revolutions per minute showed a total head of 37.6 feet at a capacity of 800 gallons per minute. Estimate the total head and capacity if this pump were operated at 1750 revolutions per minute. Assume normal operation at point of maximum efficiency in each case.

8-4. A test on a centrifugal pump operating at 1150 revolutions per minute showed a total head of 40.0 feet at a capacity of 600 gallons per minute. The impeller diameter is 10.5 inches. Estimate the total head and capacity of a geometrically similar pump at 1150 revolutions per minute with an impeller diameter of 10.0 inches.

8-5. A centrifugal pump operating at 1800 revolutions per minute develops a total head of 200 feet at a capacity of 2500 gallons per minute. What is its specific speed?

8-6. Prove that the specific speed for the same impeller does not change with the impeller speed. (*Hint:* Express the head and capacity at one speed in terms of the head and capacity at another speed.)

8-7. For a certain system it is required to select a pump that will deliver 2400 gallons per minute at a total head of 360 feet, and a pump shaft speed of 2900 revolutions per minute. What type of pump would you suggest?

8-8. For a certain system it is required to select a pump that will deliver 2400 gallons per minute at a total head of 20 feet, and at a pump shaft speed of 2600 revolutions per minute. What type of pump would you suggest?

8-9. It is desired to pump 1000 gallons of gasoline (specific gravity = 0.85) per hour from the bottom of one tank to the top of another. The level in the inlet tank is 4 feet above the pump center line, and the level in the discharge tank is 96 feet above the pump center line. The inlet-pipe diameter equals the discharge-pipe diameter of 2 inches. The dynamic viscosity of the gasoline is 0.80 centipoise. The total length of commercial steel pipe is 125 feet. If the pump efficiency is 80 per cent, what power output of the motor is required?

8-10. In a pumping system handling water (at 59 degrees Fahrenheit) the level in the suction tank is 10 feet below the pump-shaft center line, and the level in the discharge tank is 70 feet above the pump-shaft center line. The inlet piping is 3 inches in diameter and together with its valves and fittings is equivalent to 84 feet of straight steel pipe. The discharge line, $2\frac{1}{2}$ inches in diameter, with its valves and fittings is equivalent to 235 feet of straight steel pipe. If the motor delivers 9.5 horsepower to the pump shaft, what is the pump efficiency for a flow of 200 gallons per minute? Take the relative roughness for each pipe as 0.001.

8-11. Consider that the following quantities are involved in the performance of a centrifugal pump: capacity Q, shaft speed N, impeller diameter D, fluid density ρ, dynamic viscosity μ, and the quantity $E = gH$, where H is total head. E is energy per unit mass. By dimensional analysis, determine a convenient set of coordinates for organizing data.

8-12. Consider two geometrically similar fans or pumps. Using Equations (8-11) and (8-12), show that similar flow conditions exist when the specific speed for one machine equals that for the other machine.

8-13. A prototype pump is to deliver 1000 gallons of water per minute at a head of 70 feet with a runner speed of 1200 revolutions per minute. A model pump is to be constructed for a flow of 200 gallons per minute and a power input of 15 horsepower. Assume that each pump has an efficiency of 75 per cent, and that the flow patterns are similar. What would be the speed of the model runner?

8-14. Consider two geometrically similar machines of different sizes operating at the same shaft speed. For similar flow patterns, show that the capacity varies directly as the diameter cubed, the head varies directly as the diameter squared, and the fluid power varies directly as the fifth power of the diameter.

8-15. Consider two identical machines or the same machine. For similar flow patterns, show that the capacity varies directly as the speed, the head varies directly as the square of the speed, and the fluid power varies directly as the cube of the speed.

8-16. Let P represent shaft input power to a machine, N shaft speed, and D runner diameter. Assume constant efficiency and similar flow patterns for geometrically similar machines. Show that the ratio P/N^3D^5 is a constant.

8-17. Let T represent shaft input torque to a machine, N shaft speed, and D runner diameter. Assume constant efficiency and similar flow patterns for two geometrically similar machines. Show that the ratio T/N^2D^5 is a constant.

8-18. A pump impeller 18 inches in diameter rotates at 1500 revolutions per minute. Water leaves the pump impeller with a relative velocity of 30 feet per second in a backward direction (opposite to the direction of rotation) at 45 degrees with respect to the impeller tangent. What is the magnitude of the absolute or resultant exit velocity?

8-19. A centrifugal fan handling standard air (assume incompressible) has a resultant inlet velocity in an axial direction. The impeller rotates at 1500 revolutions per minute, has an outer diameter of 16 inches, and a width of 2 inches at the outer diameter. The volume rate of flow is 18 cubic feet per second. Fan head is equivalent to 4 inches of water. Efficiency is 70 per cent. Is the blade curved forward or backward? What is the angle between the relative velocity at impeller exit and the impeller tangent?

8-20. An axial-flow fan moving standard air (assume incompressible) consists only of a runner in a cylindrical casing. Air enters with a resultant axial velocity of 100 feet per second. The runner operates at 1400 revolutions per minute. Consider flow at a radius of 12 inches. As the fluid passes through the blades the relative velocity is changed by 25 degrees. What is the magnitude of the velocity at the fan exit?

8-21. Water enters the runner of a centrifugal pump at a radius of 5 inches. The velocity of the fluid, with respect to the runner, is 20 feet per second at 45 degrees with the tangent; the velocity is in the general direction of rotation. Water leaves the runner at a radius of 13 inches with a velocity of 100 feet per second at an angle of 30 degrees with the tangent; the velocity is in the general direction of rotation. Volume rate of flow is 5.0 cubic feet per second. Runner speed is 1750 revolutions per minute. What power is transferred from runner to fluid?

8-22. Water at 150 degrees Fahrenheit flows up a vertical pipe with an average velocity of 10 feet per second. At section A in this pipe the pressure is maintained at 1.0 pounds per square inch gage. At what distance above point A would cavitation start?

8-23. Standard air enters a series of identical fan blades with a relative velocity of 100 feet per second at an angle of 45 degrees with the leading edges of the blades. Assume two-dimensional, incompressible frictionless flow, and a constant depth of one inch perpendicular to the plane of flow. The blades bend the flow by an angle of 20 degrees toward the normal to the line of blade edges. What is the pressure change across the blades?

8-24. Standard air enters a series of identical fan blades with a relative velocity of 140 feet per second at an angle of 45 degrees with the leading edges of the blades. Assume two-dimensional incompressible flow, and unit depth perpendicular to the plane of flow. The blades bend the flow by an angle of 20 degrees toward the normal to the line of blade edges. The actual rise in static pressure across the blades (in the direction of flow) is 0.30 pound per square inch. What is the head loss across the blades?

8-25. A venturi meter installed in a pipe line of 12-inch diameter has a throat diameter of 6 inches. The pressure at inlet is 20 pounds per square inch gage. At what velocity in the main line will cavitation begin if the fluid is water at 104 degrees Fahrenheit?

8-26. The relative velocity of the water at a certain point on the blade of a marine propeller is always 3.6 times the velocity of the boat which it propels. If this point is 10 feet below the water surface, what will be the velocity of the boat when cavitation begins at the propeller, if the water temperature is 68 degrees Fahrenheit?

8-27. The total suction lift at the inlet to a water pump is -15.0 feet (elevation above datum $z = 0$); this means that the total suction head is 15 feet of water below atmospheric pressure. If the water temperature is 85 degrees Fahrenheit, what is the NPSH?

8-28. A single-acting piston pump makes 1600 strokes per minute, has a piston face area of 1.8 square inches, and a length of stroke of 2.5 inches. Neglecting leakage, what is the volume rate of flow in gallons per minute?

8-29. A positive-displacement pump has a displacement of 1.2 cubic inches per revolution, a speed of 1500 revolutions per minute, and a slip of 0.1 gallon per minute for each 100 pounds per square inch. What is the ratio between the actual discharge and the ideal discharge when the load pressure is 900 pounds per square inch?

8-30. In an industrial process, water is available at 500 pounds per square inch. It is proposed to use this water to operate a turbine at 1750 revolutions per minute against a back pressure of 30 pounds per square inch. What type of turbine would you suggest if it is estimated that a power output of 250 horsepower could be obtained?

8-31. By what combination of g and specific weight γ could the specific speed of a turbine be multiplied in order to give a dimensionless ratio?

8-32. A supply of water is available at a head of 48 feet. It is proposed to build a turbine to operate at 75 revolutions per minute to develop 42,000 horsepower. What type of turbine would you suggest?

8-33. In a fluid coupling the pump impeller is identical with the turbine runner. The blades are radial and the fluid is an oil with a specific gravity of 0.85. The pump impeller rotates at 2200 revolutions per minute, the pump inlet radius is 8 inches, the pump outlet radius is 12 inches, the turbine speed is 1400 revolutions per minute, and the pump torque is 40 pound-feet. What is the head loss as one pound of fluid makes a complete circuit through the coupling?

8-34. Consider two different fluid couplings, each with the same efficiency. Let N represent revolutions per unit time for the driver, Q volume rate of flow, and T torque. For similar flow conditions, show that the factor NQ^5/T^3 is a constant.

8-35. A 14-inch diameter propeller drives a torpedo through fresh water at 21.7 knots. Ideal propeller efficiency is 75 per cent. What is the useful power output? What power does the propeller add to the fluid?

8-36. An airplane propeller 8 feet in diameter develops a thrust of 1400 pounds when flying at 200 miles per hour. What is the ideal propeller efficiency? What is the theoretical value of the power absorbed by the propeller?

8-37. A helicopter weighing 500 pounds is hovering at a certain level in standard air. The diameter of the propeller is 10 feet. What is the effective or useful work done? What is the power delivered by the propeller to the fluid?

8-38. Starting with Equations (8-29), (8-32), and (8-33), derive Equation (8-34).

8-39. The drag of a plane can be approximated by the relation $C_D = 0.0254 + 0.047C_L^2$. Consider flight in standard sea-level air, assume that lift equals weight equals 18,000 pounds, and an effective wing area of 550 square feet. Plot a "polar diagram" of lift coefficient C_L versus drag coefficient C_D for the range of lift coefficient from 0 to 1.2. Plot a curve of horsepower-required versus air speed for a range from 100 to 400 miles per hour.

8-40. The drag equation for a craft is $C_D = 0.020 + 0.061C_L^2$ where C_D is the drag coefficient and C_L is the lift coefficient. The craft weighs 15,000 pounds and has an effective wing area of 300 square feet, the air density is 0.80 times that of standard sea-level air, and the craft speed is 450 miles per hour. What is the power required for steady horizontal flight?

8-41. A craft weighing 8000 pounds is traveling 375 miles per hour through air. The propeller develops a thrust of 840 pounds, and the engine delivers 1050 horsepower. Horsepower-required is 540. What are the propeller efficiency and the rate of climb?

8-42. The jet leaving a rocket has a diameter of 6 inches, a velocity of 1500 feet per second relative to the rocket, and a gas density of 0.0015 slug per cubic foot. The drag coefficient of the rocket is 0.60 based upon the projected frontal area of 0.80 square foot. What is the velocity of the rocket if it travels steadily horizontally through standard air?

8-43. A jet airplane is flying through standard air at 360 miles per hour. The hot gases leave the 8-inch diameter jet with a velocity of 1500 feet per second relative to the plane. Assume that the hot gases have the same properties as air at 1200 degrees Fahrenheit and 14.7 pounds per square inch absolute. What is the thrust on the plane?

8-44. A plane with a thermal jet engine is flying at 450 miles per hour at an altitude of 25,000 feet (pressure is 5.45 pounds per square inch absolute, temperature is −30 degrees Fahrenheit). The engine takes in 100,000 cubic feet per minute. The products of combustion leave the jet engine with a specific volume of 40 cubic feet per pound; the exit area is 250 square inches. Neglect the weight of fuel added to the air entering the engine. What is the useful power of the propulsion system?

8-45. A propeller of 7-inch diameter drives a water craft at a uniform velocity. The propeller develops a thrust of 20 pounds, has an ideal efficiency of 80 per cent, and rotates at 1200 revolutions per minute. What is the speed of the craft? If the angle of attack at 75 per cent of the radius is 3 degrees, what is the blade angle at this radius?

8-46. An airplane is equipped with two propellers, each 9.5 feet in diameter, with characteristics given in Fig. 8-38. Each engine delivers 800 horsepower at 2500 revolutions per minute of engine shaft through a gear reduction of 3 to 2 to the propeller (propeller rotates slower than engine). Craft speed is 300 miles per hour, and air density is 0.69 times that of standard air. What are the total power delivered to the air, the blade angle setting, and the propeller efficiency?

8-47. Standard air approaches a propeller in an axial direction with a velocity of 60 feet per second. The propeller rotates at 1200 revolutions per minute. Picture flow at a radius of 11 inches, where the blade section is an N.A.C.A. 2418 airfoil (Fig. 5-35); the chord is 3 inches. Consider a span or length of blade in the radial direction as one inch. The blade angle is 30 degrees. Neglecting drag, what is the torque force on this piece of blade?

8-48. An airplane has a propeller 9 feet in diameter, with characteristics given in Fig. 8-38. The engine delivers 700 horsepower to the propeller; the propeller speed is 1500 revolutions per minute. The fluid is standard air. Just before the plane starts its take-off, what thrust does the propeller exert on the air?

CHAPTER NINE

Unsteady
Flow-Galloping,
Water Hammer

"They were obliged to have him with them," the Mock Turtle said.
"No wise fish would go anywhere without a porpoise."
"Wouldn't it, really?" said Alice, in a tone of great surprise.
"Of course not," said the Mock Turtle. "Why, if a fish came to me,
and told me he was going on a journey, I should say, "With what
porpoise?"
"Don't you mean 'purpose'?" said Alice.
"I mean what I say," the Mock Turtle said, in an offended tone.
— LEWIS CARROLL[1]

Steady flow is defined as flow in which the velocity at a point does not change with time. There are numerous cases which involve steady flow. There are other important cases, however, in which the flow is unsteady. Some discussion of unsteady flow is useful in drawing a contrast with steady flow, in detailing a method of analysis, and in indicating possible undesirable motions. The following discusses two major topics: first, galloping; and second, water hammer.

In various cases a structure in a fluid stream can show a definite vibration. Examples are suspension bridges, suspended telephone wires and cables, electric power transmission lines, smokestacks, marine craft hydrofoils and submarine periscopes. In some cases the vibration has been disastrous.

Two general types of vibration can be identified: (1) an *aoelian vibration*, with a relatively small amplitude normal to the approach flow (as a fraction of an inch), and (2) a *galloping* or *flutter vibration*, with a relatively large amplitude normal to the approach flow (as 6 inches to 15 feet). The aoelian vibration involves a circular cross-section, whereas galloping requires a noncircular cross-section.

An *aoelian harp*, a box having strings which produces musical tones

---

[1] *Alice's Adventures in Wonderland* by Lewis Carroll.

when placed in the wind, had been known for some time. Other examples are the "singing" of circular telephone wires. On some electric power transmission wires, a damper, consisting of a spring (elasticity) and a mass (inertia), is installed for the purpose of reducing the aoelian vibration and the alternating stress action which may fracture the wire. The damper is normally installed on the line close to the tower clamp where the alternate bending action is severe.

9-1. The galloping system

Figure 9-1 shows a schematic of the system to be analyzed. The noncircular mass (with inertia) in the fluid stream can rotate about an axis and can also move in a vertical direction. There is a rectilinear spring; the motion of this spring is in a vertical direction, with the spring force being linearly proportional to the spring deflection. The mass is also connected to a torsional

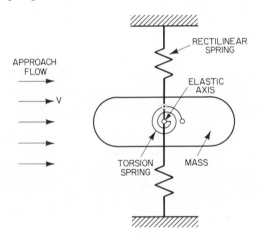

Fig. 9-1. Schematic of system.

spring; the mass can execute motion in a torsional, rotational, or pitch manner. The torsional spring is linear, in that the spring torque (about the axis of rotation) is directly proportional to the spring angular displacement. One end of each spring is fastened to a fixed frame; the other end is fastened to the elastic axis on the moving body. The elastic axis is at the center of gravity.

The mass in the incompressible fluid stream can rotate about the elastic axis and can move in a vertical direction. Thus, the mass has two degrees of freedom; two independent coordinates are needed to specify the mass position. Figure 9-2 illustrates the notation. The lift force acting on the body is perpendicular to the undisturbed approach flow. The angle of attack between

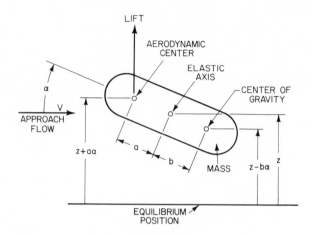

Fig. 9–2. Notation.

the approach flow and the body centerline is α. The center of gravity is determined by the body mass distribution. The *aerodynamic center* is defined as the point, in a body, about which the pitch moment does not change with angle of attack. The aerodynamic center, the lift coefficient, and other flow characteristics can be determined by a model test in a wind or water tunnel.

The distance between the aerodynamic center and the center of gravity is a (as feet), the rectilinear vertical displacement of the center of gravity from the equilibrium position is z (as feet), and the angular displacement is α (as radians). Let M represent the mass of the body (as slugs), K_z the rectilinear vertical spring rate (as pounds per foot), let K_α be the torsional spring rate (as foot-pounds per radian), and let R be the effective radius of gyration of the mass (as feet). It will be assumed that the structure has the same cross-section perpendicular to the plane of motion. In other words, two-dimensional flow is assumed.

9–2. General analytical relations

The general procedure is to set up two ordinary differential equations of motion. Each is a linear differential equation with constant coefficients. One is a force equation; the other is a moment or torque equation. The two equations will be solved simultaneously to determine possible galloping action. Attention will be focused on varying the approach flow velocity to see if a critical condition for galloping can be determined.

Let L represent lift force (as pounds), A represent a reference area (as square feet), V linear fluid approach velocity (as feet per second), ρ constant fluid density (as slugs per cubic foot), q the approach stream dynamic pressure $\frac{1}{2}\rho V^2$, and C_L the dimensionless lift coefficient defined as

$$C_L = \frac{L}{qA} \qquad (9\text{–}1)$$

As indicated in Fig. 5–35, we assume a linear relation between lift coefficient C_L and angle of attack α. Let $C_{L\alpha} = \partial C_L / \partial \alpha$ be the slope of the curve of lift coefficient versus angle of attack in the linear region. The net aerodynamic lift force and the net aerodynamic moment become

$$\text{Lift} = qA\alpha C_{L\alpha} \qquad (9\text{–}2)$$
$$\text{Moment} = qA\alpha a C_{L\alpha} \qquad (9\text{–}3)$$

The two differential equations of motion for the lumped-element system are

$$M\frac{d^2z}{dt^2} + K_z z = \text{Force} = (qA C_{L\alpha})\alpha \qquad (9\text{–}4)$$

$$R^2 M\frac{d^2\alpha}{dt^2} + K_\alpha \alpha = \text{Moment} = qA\alpha a C_{L\alpha} = (qAa C_{L\alpha})\alpha \qquad (9\text{–}5)$$

It is convenient to define two natural frequencies, translational or rectilinear ω_z and torsional ω_α, by the relations

$$\omega_z = \sqrt{\frac{K_z}{M}} \qquad \omega_\alpha = \sqrt{\frac{K_\alpha}{MR^2}} \qquad (9\text{–}6)$$

Each frequency has the dimensions of radians per second. Then the basic differential equations become

$$\frac{d^2z}{dt^2} + \omega_z^2 z - \left(q\frac{A C_{L\alpha}}{M}\right)\alpha = 0 \qquad (9\text{–}7)$$

$$\frac{d^2\alpha}{dt^2} + \omega_\alpha^2 \alpha - \left(q\frac{Aa C_{L\alpha}}{MR^2}\right)\alpha = 0 \qquad (9\text{–}8)$$

Equations (9–7) and (9–8) are linear differential equations with constant coefficients. The next step is to solve these two equations simultaneously. Let each displacement and acceleration be expressed in the general form

$$Z = De^{st} \qquad \alpha = Ee^{st} \qquad (9\text{–}9)$$
$$\frac{d^2z}{dt^2} = Ds^2 e^{st} \qquad \frac{d^2\alpha}{dt^2} = Es^2 e^{st} \qquad (9\text{–}9\text{A})$$

where D and E are each an amplitude constant, t is time (as seconds), and s is a root (dimensions of radians per second). Substituting the relations in Equation (9–9) into Equations (9–7) and (9–8) gives the algebraic relations

$$(s^2 + \omega_z^2)De^{st} - \left(q\frac{A C_{L\alpha}}{M}\right)Ee^{st} = 0 \qquad (9\text{–}10)$$

$$\left(s^2 + \omega_\alpha^2 - q\frac{A C_{L\alpha}}{MR^2}\right)Ee^{st} = 0 \qquad (9\text{–}11)$$

Equations (9–10) and (9–11) are each a linear relation. In order to obtain a simultaneous solution of these equations which will not be trivial and which will not vanish identically, it is necessary that the determinant of the coefficients be equal to zero.[2] Thus

$$\begin{vmatrix} (s^2 + \omega_z^2) & -\left(q\dfrac{AC_{L\alpha}}{MR^2}\right) \\ & \left(s^2 + \omega_\alpha^2 - q\dfrac{AC_{L\alpha}}{MR^2}\right) \end{vmatrix} = 0$$

$$(s^2 + \omega_z^2)\left(s^2 + \omega_\alpha^2 - q\dfrac{AaC_{L\alpha}}{MR^2}\right) = 0 \tag{9–12}$$

The characteristic Equation (9–12) has the general form

$$s^4 + s^2 B_1 + B_2 = 0 \tag{9–13}$$

with four roots.

In the analysis of flutter or galloping a prime interest is the influence of the approach velocity V. It is helpful to develop a plot of components of the roots each as a function of a parameter y, where y is defined as

$$y = \sqrt{q\dfrac{AaC_{L\alpha}}{MR^2}} \tag{9–14}$$

The parameter y is proportional to the approach velocity V. Then Equation (9–12) can be cast in the compact form

$$(s^2 + \omega_z^2)(s^2 + \omega_\alpha^2 - y^2) = 0 \tag{9–15}$$
$$s^2 + \omega_z^2 = 0 \tag{9–15A}$$
$$s^2 + \omega_\alpha^2 - y^2 = 0 \tag{9–15B}$$

9–3. Equations of motion

Let n be a complex number given by the relation

$$n = f + ig \tag{9–16}$$

where f is a real number, g alone is a real number, and $i = \sqrt{-1}$. The factor i can be regarded as a convenient mathematical operator. The term ig is called an *imaginary number*. In general, a complex number has a real part and an imaginary part. Equation (9–15A) can be written in the form

$$s = \pm i\omega_z \tag{9–17}$$

which gives a pair of imaginary roots with no real component. Figure 9–3 shows a plot of the general variation of the variable angular frequency ω

[2] Various mathematical references on the simultaneous solution of linear differential equations can be consulted, for example, *Advanced Engineering Mathematics* by C. R. Wylie, McGraw-Hill Book Co.

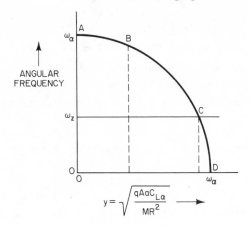

Fig. 9–3. Variation of ω as a function of y.

as a function of y. In Fig. 9–3 the line through $\omega = \omega_z$ represents the positive, constant, undamped, vertical natural frequency over a range of y values (which are proportional to the approach velocity V).

Consider only values of y equal to or less than ω_α. Then Equation (9–15B) can be recast in the form

$$s^2 = y^2 - \omega^2\alpha = -\omega^2$$

$$y^2 + \omega^2 = \omega\alpha^2 \tag{9–18}$$

$$s = \pm i\sqrt{\omega^2 - y^2} \tag{9–18A}$$

Equation (9–18) is the equation of a circle of radius ω_α. For only positive values of y and ω, in Fig. 9–3 the curve ABCD represents one-quarter of a circle. At $\omega = 0$, the value of y is ω_α. At the intersection, point C, let the value of y be y_C.

Refer to Fig. 9–3, and consider only values of y less than that at y_C. For example, consider a value of y_B for point B. At y_B there are two frequencies, one ω_z and the other $\sqrt{\omega_\alpha^2 - y_B^2}$. Then the integrated equations of motion become

$$z = A_1 \sin [\sqrt{\omega_\alpha^2 - y_B^2}\, t + \phi_1] + A_2 \sin [\omega_z t + \phi_2] \tag{9–19}$$

$$\alpha = A_3 \sin [\sqrt{\omega_\alpha^2 - y_B^2}\, t + \phi_1] + A_4 \sin [\omega_z t + \phi_2] \tag{9–20}$$

where the A's are amplitude coefficients, and ϕ_1 and ϕ_2 are phase angles. The motion is limited by the amplitude coefficients A_1, A_2, A_3, and A_4.

9–4. Conditions for galloping

Refer to Fig. 9–3 and consider the y value at y_C. At this value there are two identical roots or two identical frequencies $\omega_z = \omega_\alpha$. Thus the solution of the differential equations takes the form

$$z = A_1 \sin [\omega_z t + \phi_1] + t A_2 \sin [\omega_z t + \phi_2] \tag{9-21}$$

$$\alpha = A_3 \sin [\omega_z t + \phi_1] + t A_4 \sin [\omega_z t + \phi_2] \tag{9-22}$$

Because of the t term, there is an amplitude build-up with time. For overhead electrical transmission lines and other structures, as suspension bridges, this condition represents galloping or flutter. The basic differential equations do not include friction or damping. In a practical case some damping would restrict the displacement amplitude to a finite limit or a failure might take place. Note that the condition for galloping involves two identical frequencies $\omega_z = \omega_\alpha$.

As the approach velocity V increases from zero to some higher value, an important question is the velocity at which galloping starts. Referring to Fig. 9–3, galloping starts at point C, with the y value

$$y = \sqrt{\omega_\alpha^2 - \omega_z^2} = \sqrt{q \frac{A a C_{L\alpha}}{M R^2}} \tag{9-23}$$

Since $q = \frac{1}{2}\rho V^2$, the corresponding galloping velocity V_G (at point C) is given by the relation

$$V_G = \sqrt{\frac{2 M R^2 (\omega_\alpha^2 - \omega_z^2)}{\rho A a C_{L\alpha}}} \tag{9-24}$$

The simple closed form solution given by Equation (9–24) points out the factors to consider in studies of possible undesirable vibrations. To develop a high V_G, attention could be given to increasing the torsional frequency ω_α, to increasing the radius of gyration R, to reducing the area A, and to reducing the distance a. In an analysis there is a question as to the possible values of the lift coefficient slope $C_{L\alpha}$. In order to be on the safe side and calculate the lowest V_G, one might select a high value for $C_{L\alpha}$. For example, for a flat plate $C_{L\alpha}$ is about 2π.

9–5. Experimental observations

Probably there are various cases, as aircraft, missiles, marine craft, and other structures in a fluid stream, in which the center of gravity is at or close to the elastic axis. Some pertinent data are available for overhead electrical transmission lines; thus these data will be discussed briefly to provide a check on the main features of the foregoing analysis and to illustrate what might happen in dynamically similar systems.

Under certain conditions the symmetrical circular section of an overhead transmission line may accumulate sleet or ice to provide an unsymmetrical section to the wind. The line can vibrate in torsion and it can vibrate vertically in translation. In certain winds the transmission line may gallop. Evidence indicates that the center of gravity is very close to the elastic axis.

One investigator (1)³ reported measurements of a particular energized transmission line in Canada. During a freezing rain with wind, ice was deposited on the conductor and the line galloped. Measurements were made of the vibration characteristics of the line. The natural frequency in torsion ω_α was measured as 15.7 radians per second, and the natural translation frequency ω_z was measured as 7.8 radians per second. Observation showed that the line galloped at the natural frequency in translation. The motion was simple harmonic. The torsional frequency was in phase with the translational vibration. This matching of frequencies checks point C in Fig. 9-3, and Equations (9-21) and (9-22). The line galloped in winds varying from 20 to 30 feet per second. A value of 2π was taken for $C_{L\alpha}$ and the aerodynamic center was taken at the quarter-point (aft from the leading edge). From the data reported it is difficult to make a precise calculation of the radius of gyration R due to the ice formation. Making some reasonable calculations as to ice formation, the use of Equation (9-24) indicates a galloping velocity in the region from 25 to 32 feet per second; this theoretical calculation checks with the measurements.

For a series of tests with another transmission line. the investigator gave the following conclusions: (a) galloping took place at the natural frequency of the conductor in translation; (b) both the torsional and the translation motion at any point of the galloping conductor are basically simple harmonic; (c) galloping is a phenomenon of an extremely critical nature; and (d) the fundamental frequency of the torsional motion during galloping is the same as that of the translation motion.

A number of other investigations (laboratory, wind tunnel, and field) were conducted (2). Different prototype field test lines were erected. Semicircular wooden foils were fastened along the length of each conductor to provide an unsymmetrical section normal to a horizontal wind. Small scale models of a transmission line, with unsymmetrical sections along an elastic line, were arranged in a laboratory wind tunnel.

Field prototype tests, with natural winds, showed that the galloping action was very critical. Galloping started only at a certain approach wind velocity. For example, one field test line galloped only when the normal wind reached 12 feet per second. One field line did not gallop with the normal prevailing wind velocities. Referring to Fig. 9-3, the torsional frequency ω_α was so high that the condition at point C was not reached. Torsional inertia was added to the lines by means of cross bars; this lowered the torsional natural frequency so that the lines galloped with normal winds. This confirmed the indication given in Fig. 9-3.

It may be possible that a transmission line without ice could have a very high torsional frequency. An accumulation of ice may add enough torsional

---

³ A number in parentheses refers to a reference at the end of this chapter.

inertia to lower the torsional frequency sufficiently so that the line would gallop with prevailing winds.

At different times, over 20 cases of natural galloping, first mode, were observed with the prototype field test lines. The natural frequency in translation was measured during periods of no galloping. In each case of natural galloping the frequency was equal to the natural translational frequency (about 2 radians per second for one set of spans). In some cases the vertical double displacement of the line was about 12 feet. For each case of natural galloping the natural torsional frequency was definitely higher than the natural translational frequency.

Observations of both field test lines and wind tunnel models showed two distinct stages in the galloping action: a small irregular vertical displacement random motion at the start; and a fully developed large vertical displacement periodic motion following the irregular random start. The initial start resembled a motion due to random excitation from vertical fluctuations in the approach stream. A mode or modes of the transmission line can be excited by the random vertical fluctuations in the approach horizontal flow; the vertical displacement may be small. If the unsymmetrical conductor is displaced sufficiently in a vertical direction, it will start to twist. Sooner or later a significant twisting of the line develops and the galloping starts.

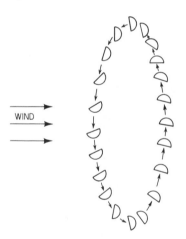

WIND

An example of a measured fully developed orbit of field line galloping is indicated in Fig. 9–4; the double vertical displacement amplitude was about 10 feet. Various modifications of this orbit have been observed. The line twists back and forth as it moves up and down. There is a sharp twist or snap at the top and bottom of the orbit. If the conductor starts and continues in an orbit of the type shown in Fig. 9–4, it can build up to a high steady state displacement amplitude.

Fig. 9–4. Measured orbit of galloping field line.

Let z_0 represent the vertical displacement amplitude, z the instantaneous vertical displacement. Figure 9–5 shows a typical plot of the measured vertical displacement ratio z/z_0 versus time. This plot was drawn for a case in which the double vertical displacement amplitude was about 6 feet; the motion is close to simple harmonic. Figure 9–6 shows the corresponding angular displacement α versus time; the motion is close to simple harmonic. The phase difference between vertical displacement and the angular displacement is very small.

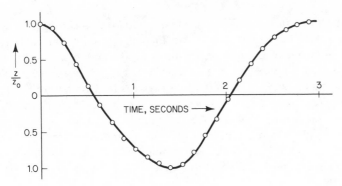

Fig. 9–5. Typical plot of measured displacement ratio z/z_0 versus time.

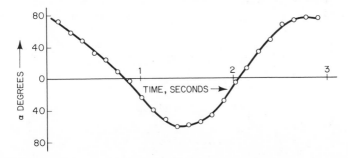

Fig. 9–6. Typical plot of angular measured displacement α versus time.

9–6. Water hammer

Water hammer is defined as the change in pressure in closed conduits above and below normal caused by sudden changes in the steady flow. Water hammer is an unsteady flow action. When a liquid flows through a pipe there is a certain amount of energy in the liquid. If the flow is stopped suddenly, as by a valve, the energy of the liquid is used up in compressing the liquid and stretching the pipe. There may be violent changes in pressure, a hammering effect or action. A liquid machine, as a centrifugal pump or a water turbine, may be in a piping system subject to water hammer. Water hammer is of practical importance because of the danger of possible damage to machines and piping. The same action may occur in a fuel injection system.

The following discussion covers basic relations and an introduction to the graphical method of solving water-hammer problems. One feature of prime importance is the pressure-versus-time curve during the transient surge action. If we know this curve we can investigate the extent of possible danger. The following discussion brings out the role of an elastic pipe.

As illustrated in Fig. 9–7 imagine first liquid flowing through a long pipe with a steady velocity V_0. With this flow there is a certain "normal" pressure or head at the end of the pipe A. Picture next the action as a valve at A is suddenly closed. Liquid next to the end is first brought to rest; then adjacent layers, in turn, are successively brought to rest. After a time interval the fluid in the pipe from A to B is at rest whereas the fluid in the pipe from C to B is still moving with the velocity V_0. The fluid pressure in A to B is "super-normal" or higher than the normal pressure with steady flow. The fluid in the pipe from A to B has been compressed while the elastic pipe has been stretched. The stopping of fluid continues toward C; a super-normal pressure wave travels from A to C.

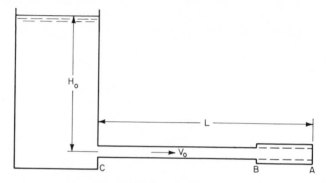

Fig. 9–7. Notation for long pipe line.

When this pressure wave reaches C, the fluid in the entire pipe is at rest, but at a pressure higher than that in the reservoir. Then a period of unloading starts; a wave of pressure decrease passes from C to A; when this wave reaches A the liquid is at the normal pressure. During this unloading the elastic pipe contracts and liquid flows toward the reservoir. After a time this flow toward the reservoir is stopped and the fluid pressure becomes "sub-normal" or drops below the normal value (for steady flow). Thus a sub-normal wave travels from A to C and the liquid expands. This cycle is repeated until frictional effects dampen the surging.

9–7. Velocity of pressure wave

An important factor in water hammer calculations is the velocity of pressure-wave travel through the liquid; let this velocity be represented by a. The velocity of pressure propagation through a fluid c is given by the expression

$$c = \sqrt{E/\rho}$$

where E is the bulk modulus of the liquid alone, as in a rigid pipe. Let K represent the effective bulk modulus of the liquid as modified by the pipe expansion. Then

$$a = \sqrt{K/\rho}$$

The next step is to work out the relation between E and K.

Figure 9–8a illustrates a prism of liquid $ABCD$ at the end of the pipe just before closing of the valve. The initial length is $dx + dx'$. As illustrated in Fig. 9–8b the length after closing of the valve is dx; the fluid is compressed by the excess pressure dp.

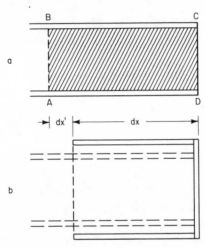

Fig. 9–8. Notation for wave velocity.

Let e be the thickness of the pipe wall, s the tension in the wall induced by the excess pressure dp, E_1 the elastic modulus of the wall material, R the initial inner radius of the pipe, dR the increase in radius, and dl the increase in circumference of the wall. The elastic modulus is given by the relation

$$E_1 = \frac{s}{dl/2\pi R}$$

The stress is given by the relation $s = dpR/e$. The change in length dl is equal to $2\pi dR$. Thus

$$E_1 = \frac{dp2\pi R^2}{e2\pi dR} \qquad dR = \frac{dpR^2}{eE_1} \tag{9-25}$$

The increase in volume Δv_1 resulting from pipe stretching is

$$\Delta v_1 = 2\pi R \, dR \, dx = \frac{2\pi R^3 \, dp \, dx}{eE_1} \tag{9-26}$$

Even if the pipe did not stretch, there would be a change in the fluid volume due to the increase in pressure dp; let this volume increase be Δv_2. Since E is the bulk modulus of the liquid, the magnitude of Δv_2 is

$$\Delta v_2 = \frac{dp\pi R^2(dx + dx')}{E} \qquad (9\text{--}27)$$

The magnitude of the effective bulk modulus K is the ratio of the pressure change divided by the net volumetric strain. Let Δv_3 be the net volumetric change in the fluid. Then the magnitude of K is given by the relation

$$K = \frac{dp(dx + dx')\pi R^2}{\Delta v_3} \qquad (9\text{--}28)$$

The change Δv_3 equals the net of Δv_1 and Δv_2. Thus

$$\Delta v_3 = \frac{dp(dx + dx')\pi R^2}{K} = \frac{dp\pi R^2(dx + dx')}{E} + \frac{2\pi R^3 \, dp \, dx}{eE_1}$$

It is common practice to neglect dx' with respect to dx; doing this yields the final relation

$$\frac{1}{K} = \frac{1}{E} + \frac{2R}{eE_1} \qquad \text{or} \qquad K = \frac{E}{1 + 2RE/eE_1} \qquad (9\text{--}29)$$

Thus the wave velocity a becomes

$$a = \sqrt{\frac{E}{\rho}} \sqrt{\frac{1}{1 + 2RE/eE_1}} \qquad (9\text{--}30)$$

Note that the elasticity of the pipe reduces the wave velocity as compared with that of liquid in a rigid pipe. For steel, E_1 is commonly taken as 30,000,000 pounds per square inch. For water, E is commonly taken as 300,000 pounds per square inch.

9–8. Rapid or sudden valve closure

Refer to Fig. 9–7. The liquid flows through the pipe of length L under the head H_0 with the steady velocity V_0. If the pressure wave velocity is a, then the time of travel of a wave from A to C is L/a units of time. For example, if L is in feet and a is in feet per second, then L/a is in seconds. The time of travel of a wave from A to C and back to A is $2L/a$ units of time. If the valve is closed within $2L/a$ units of time or closed before the pressure wave makes the round trip back to the valve, we will say that the closure is "rapid" or "sudden." If the valve is closed in a time interval greater then $2L/a$ units, we will say that the closure is "slow." This section will deal with rapid closure, and the next section will deal with slow closure.

Considering rapid closure, if the fluid is brought to rest, the maximum pressure rise above normal is Δp. The change in kinetic energy per unit weight of fluid is $V_0^2/2g$. The bulk modulus $K = -\Delta pv/\Delta v$, where Δv is the change in the initial specific volume v. The work done in elastic compression is the average stress $\Delta p/2$ times the volume change Δv or

$$\frac{\Delta p \Delta v}{2} = \frac{\Delta p^2 v}{2K}$$

Assuming no friction loss, the elastic work equals the change in kinetic energy. Since $K = a^2 \rho$, we get the relations

$$\frac{\Delta p^2 v}{2K} = \frac{V_0^2}{2g}$$

$$\Delta p = \rho a V_0 \qquad (9\text{–}31)$$

The maximum pressure rise for rapid closure is given by Equation (9–31). Sometimes this relation is written in terms of a maximum pressure-head rise h_0, such that $\Delta p = w h_0$. This gives

$$h_0 = \frac{a V_0}{g} \qquad (9\text{–}32)$$

A plot of pressure head at the valve versus time after valve closure is given in Fig. 9–9.

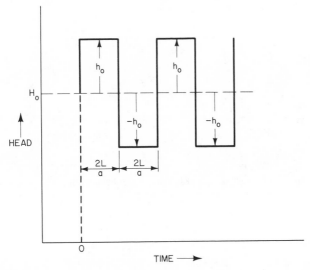

Fig. 9–9. Plot of pressure head at valve versus time, for rapid valve closure.

9–9. Slow valve closure

If the valve closes in a time interval longer than $2L/a$ units of time, it is necessary to make a detailed study of reflected waves. Let n represent the number of $2L/a$ units of time after the valve starts closing. For example, a value $n = 2$ means a time interval of $4L/a$.

As illustrated in Fig. 9–7, H_0 is the pressure-head at the valve for steady flow before valve closing starts, and V_0 is the steady-flow velocity. Let V_1 represent the velocity and H_1 the maximum pressure-head at $2L/a$ units of time. In the time interval $2L/a$ the velocity has been reduced from V_0 to V_1. In this time interval the length of liquid column slowed down is $2L$. The mass of this column is $2L\pi R^2 w/g$, where w is specific weight. Since force equals mass rate of flow times velocity change, we can write

$$w(H_1 - H_0)\pi R^2 = \frac{2L\pi R^2 w}{g2L/a}(V_0 - V_1)$$

$$H_1 - H_0 = \frac{a}{g}(V_0 - V_1)$$

(9–33)

Imagine next the state at $4L/a$ units of time, or when $n = 2$. At the end of this time the velocity has been reduced from V_1 to V_2. The increased pressure-head $(H_1 - H_0)$ has traveled up toward the reservoir and has been reflected as a decreased pressure-head or negative wave. At $n = 2$ there is a positive pressure-head $(H_2 - H_0)$ traveling toward the reservoir. Thus the total pressure-head decelerating the flow is $(H_1 - H_0) + (H_2 - H_0)$. The dynamic equation thus gives

$$w[(H_2 - H_0) + (H_1 - H_0)]\pi R^2 = \frac{2L\pi R^2 w}{g2L/a}(V_1 - V_2)$$

$$H_2 - H_0 = \frac{a}{g}(V_1 - V_2) - (H_1 - H_0)$$

(9–34)

We could carry out the calculations for $n = 3$, $n = 4$, and higher values; the general result is

$$H_n - H_0 = \frac{a}{g}(V_{n-1} - V_n) - (H_{n-1} - H_0)$$

(9–35)

Equation (9–35) is a basic equation in studying water hammer for slow valve closure. One more basic relation is necessary before we can calculate pressure-time curves. Considering steady flow, the velocity V_0 in the pipe can be expressed in the form

$$V_0 = \frac{A_v}{A}C\sqrt{2gH_0}$$

(9–36)

where A_v is the effective valve-opening area, A is the pipe area, and C is the discharge coefficient. Let $B_0 = A_v C\sqrt{2g}/A$. Then

$$V_0 = B_0\sqrt{H_0}$$

(9–37)

For the sake of simplicity, assume that the valve opening is linear with respect to time. Let V_t be the velocity in the pipe at any time t, T the time of valve closing, and h_t the increased pressure-head above H_0. Then the pipe velocity is

$$V_t = \left(1 - \frac{t}{T}\right)B_0\sqrt{H_0 + h_t}$$

(9–38)

As an example, let $n = 1$, then $t = 2L/a$, and Equation (9–38) becomes

$$V_1 = \left(1 - \frac{2L}{aT}\right)B_0\sqrt{H_1} \qquad (9\text{–}39)$$

which includes the valve-opening and discharge coefficient for this particular steady-flow state.

A solution of the water-hammer problem involves a simultaneous solution of Equations (9–35) and (9–38).

$$H_n - H_0 = \frac{a}{g}(V_{n-1} - V_n) - (H_{n-1} - H_0) \qquad (9\text{–}35)$$

$$V_t = \left(1 - \frac{t}{T}\right)B_0\sqrt{H_0 + h_t} \qquad (9\text{–}38)$$

For example, if we wish to find H_1, we must get a simultaneous solution of Equations (9–33) and (9–39).

$$H_1 - H_0 = \frac{a}{g}(V_0 - V_1) \qquad (9\text{–}33)$$

$$V_1 = \left(1 - \frac{2L}{aT}\right)B_0\sqrt{H_1} \qquad (9\text{–}39)$$

Various methods of solution have been proposed. From the engineering viewpoint, one of the simplest and most convenient is the graphical method.

9–10. Graphical Method of Solution

The graphical method of solution involves the construction of lines on a plot of pressure-head H at valve versus velocity as illustrated in Fig. 9–10. The curve marked B_0 is a parabola passing through the points H_0 and V_0. Assume that the time of valve closing T is given. Then curve B_1 can be constructed, taking different values of H_1 and calculating the corresponding values of V_1 by Equation (9–39). We do not know yet what point on curve B_1 satisfies the other Equation (9–33). We can find this point by drawing a line with a slope $-a/g$. Thus we obtain the point of intersection which gives the pressure-head H_1 at the time $2L/a$. The reason for this construction can be indicated by rearranging Equation (9–38) in the form

$$\frac{H_1 - H_0}{V_1 - V_0} = -\frac{a}{g}$$

For the next step, the curve marked B_2 can be calculated from the relation

$$V_2 = \left(1 - \frac{4L}{aT}\right)B_2\sqrt{H_2} \qquad (9\text{–}40)$$

taking different values of H_2 and calculating the corresponding values of V_2. We wish a simultaneous solution of Equations (9–34) and (9–40). Starting

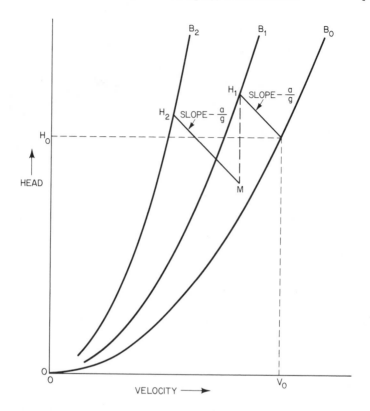

Fig. 9-10. Illustration of the graphical solution of a water-hammer problem.

with point H_1 on Fig. 9-10, measure below H_0 a distance equal to $H_1 - H_0$ to give point M. From point M draw a line with the slope $-a/g$. The point of intersection with B_2 gives the head H_2. The reason for this construction can be indicated by rearranging Equation (9-34) in the form

$$\frac{(H_2 - H_0) + (H_1 - H_0)}{V_2 - V_1} = -\frac{a}{g}$$

This method of construction can be used to give the points H_3 and H_4 as indicated. From this solution can be obtained a curve of pressure-head, or pressure, versus time. From such a curve one can study possible danger of damage to the system.

REFERENCES

(1) "Vibration Characteristics of a Conductor Which Had Previously Galloped" by A. Madeyshi, *Research Division Report No. 57-317*, Hydro-Electric Power Commission of Ontario, Canada, August 1957.

"Galloping Conductor Test on the Experimental Line" by A. Madeyshi, *Research Division Report No. 57-325*, Hydro-Electric Power Commission of Ontario, Canada, August 1957.

(2) "The Flutter or Galloping of Certain Structures in a Fluid Stream" by Raymond C. Binder, *The Shock and Vibration Bulletin No. 39*, Part 3, January 1969, Shock and Vibration Information Center, Naval Research Laboratory, Washington D.C., January 1969, p. 171.

PROBLEMS

9-1. Water fills a pipe 42 inches inside diameter having a wall thickness of $\frac{1}{4}$ inch. Modulus of elasticity for the steel in the pipe is 30,000,000 pounds per square inch. The bulk modulus for the water alone is 300,000 pounds per square inch. Considering the elasticity of the pipe, what is the velocity of the pressure wave in the water?

9-2. Water flows through a steel pipe 42 inches inside diameter having a wall thickness of $\frac{1}{4}$ inch. The steady flow velocity is 6 feet per second. For a sudden valve closure, what is the maximum pressure rise? What would be the maximum pressure rise if the pipe were assumed rigid?

9-3. Water from a reservoir flows steadily through a horizontal pipe 1100 feet long. There is a valve at the end of the line. The steady flow average velocity V_0 is 10 feet per second, and the corresponding pressure head H_0 is 500 feet. The valve closure is slow, linear, and takes 2.1 seconds. Friction is neglected. By the graphical method, determine the pressure head, feet, above the steady flow head H_0, at intermediate stages during the valve closure. Assume the wave velocity a is 3140 feet per second.

CHAPTER TEN

Flow of
Liquids in
Open Channels

Vigorous writing is concise. A sentence should contain no unnecessary words, a paragraph no unnecessary sentences, for the same reason that a drawing should have no unnecessary lines and a machine no unnecessary parts.

—WILLIAM STRUNK, JR.[1]

An open channel is defined as one in which the liquid stream is not completely enclosed by solid boundaries and thus has a free surface subjected only to atmospheric pressure. The flow in an open channel depends upon the slope of the channel bottom and the slope of the liquid surface. Natural streams, rivers, artificial canals, sewers, tunnels, and pipes not completely filled with liquids are examples of open channels. An accurate solution of problems of open-channel flow is much more difficult than those of pipe flow. One distinguishing feature of the flow of liquids in open channels, as contrasted with pipe flow, is that the cross-sectional area is free to change instead of being fixed. The presence of a free surface introduces complexities of a primary magnitude. It is difficult to obtain reliable experimental data on the flow in open channels. A relatively smaller range of conditions is found in pipe flow. Most pipes are round, but open-channel sections vary from circular to the irregular forms of natural rivers; channel surface conditions range from the smoothness of timber to the roughness and irregularity in the beds of some rivers.

The treatment of open-channel flow is therefore somewhat more empirical than that of pipe flow. The empirical treatment, however, is the best available, and can yield results of practical value if cautiously applied.

[1] *The Elements of Style*. The Macmillan Co., New York, 1959.

324

10-1. Definitions

Sometimes the word "stage" is used in place of the word "depth." Either word refers to the vertical height of liquid in a channel. *Steady flow* refers to a condition in which the flow characteristics at any *point*, such as velocity and pressure, do not change with time. The flow of a liquid in an open channel is said to be *uniform* when the depth, velocity, liquid cross-sectional area, slope, and other such flow elements remain *constant from section to section*. Steady flow should not be confused with uniform flow. As indicated in Fig. 10-1, in uniform flow the liquid-surface line is parallel to the channel-bottom line. The surface angle α equals the bottom angle α_0, or $\sin \alpha = \sin \alpha_0$.

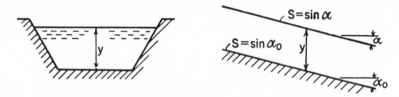

Fig. 10-1. Uniform flow in an open channel.

The flow is said to be *nonuniform* or *varied* whenever the depth and other features of flow, such as the liquid cross-sectional area, the velocity, and the slope, vary from section to section. Two examples of nonuniform flow are given in Fig. 10-2, in which the water surface is not parallel to the bottom surface. The flow in an open channel may be uniform in one length and nonuniform in another length.

The term "hydraulic mean depth" or "hydraulic radius" is customarily used in open-channel studies. This term is defined as

$$\text{hydraulic mean depth} = R = \frac{\text{cross-sectional area}}{\text{wetted perimeter}}$$

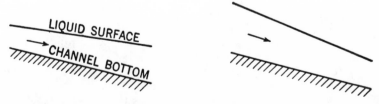

Fig. 10-2. Nonuniform or varied flow in an open channel.

The cross-sectional area in Fig. 10-3 is the area $ABCD$. The wetted perimeter is the length of solid boundary in contact with the liquid, that is, the distance $ABCD$. The wetted perimeter does *not*

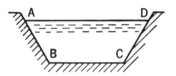

include the distance across the free surface. For a circular pipe flowing full, the hydraulic mean depth equals $D/4$, where D is the pipe diameter.

Fig. 10–3. Cross section of an open channel.

The following treatment applies only to channels with small slopes. Strictly speaking, the slope S should be taken as the sine of the angle of inclination. The following relations apply only to small slopes for which the sine and the tangent are very nearly equal. The diagrams in this chapter exaggerate the slope for the sake of illustration.

10–2. Energy relation for steady, uniform flow

The case will be taken in which the velocity V is the same at all depths. In the notation shown in Fig. 10-4, the energy equation between any two points 1 and 2, for steady flow in any stream tube, becomes

$$z_1 + \frac{p_1}{\gamma} + \frac{V_1^2}{2g} = z_2 + \frac{p_2}{\gamma} + \frac{V_2^2}{2g} + h \qquad (10\text{-}1)$$

where h is the lost head. All the streamlines are parallel to the bottom of the

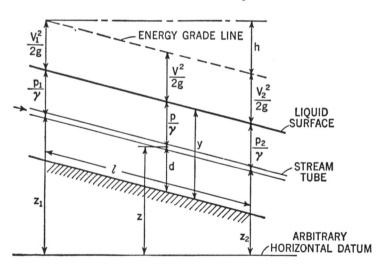

Fig. 10–4. Profile of channel.

channel for uniform flow. Thus $p_1 = p_2$, and $V_1 = V_2$. Then

$$\text{lost head} = h = z_1 - z_2$$

The energy required to maintain the flow in the channel is obtained at the expense of potential energy. In uniform flow the entire change in elevation is charged to the maintenance of the flow. The slope S of the energy grade line is h/l. The slope of the energy grade line equals the slope of the water surface equals the slope of the bottom only for the special case of steady uniform flow.

10–3. Friction relation for steady, uniform flow

It is customary to express the head loss for flow in a circular pipe completely filled as

$$h = f\frac{l}{D} \cdot \frac{V^2}{2g} \tag{10–2}$$

The hydraulic mean depth R for a circular pipe is $D/4$. Replacing D, substituting $S = h/l$, and solving for V, gives

$$V = \sqrt{\frac{8g}{f}}\sqrt{RS} \tag{10–3}$$

For open-channel flow, it is common practice to express Equation (10-3) in the form

$$V = C\sqrt{RS} \tag{10–4}$$

where C is a dimensional coefficient, with the dimensions $L^{1/2}T^{-1}$. Equation (10-4) is commonly known as the Chezy equation.

The pipe friction coefficient f is a function of the Reynolds number and the pipe roughness. Since $C = \sqrt{8g/f}$, it follows that C is a function of the same factors; that is, C is a function of the average velocity, hydraulic mean depth, kinematic viscosity, and wall roughness. Because applications and experiments are generally concerned with water at ordinary atmospheric temperatures, the viscosity variation is small; the viscosity has not been explicitly included in the friction relations. Apparently the available data on friction for open-channel flow are based on experiments with fairly well-developed turbulent flow in rough conduits. The application of such data to flow with small depths and low velocities is questionable.

Most open channels are relatively large as compared with pipes, and have surfaces which are rougher than pipes. The effect of roughness becomes prominent in pipe flow at high Reynolds numbers. Since C is a function of f, it appears that for usual applications the coefficient C depends largely on the character of the surface and the cross section of the channel.

A wide variety of empirical formulas for C have been presented, among them being those by Kutter, Ganguillet, Manning, and Bazin. For a general introductory study, as is given in this chapter, the Manning equation is preferable. Manning's equation states that

$$C = \frac{1.49}{n} R^{1/6} \qquad (10\text{–}5)$$

where n is a roughness factor. Some average values of n are listed in Table 10-1, with the foot taken as the length unit.

TABLE 10-1

SOME AVERAGE VALUES OF n FOR USE
IN MANNING'S EQUATION

Nature of surface	n	Nature of surface	n
Planed wood	0.012	Earth, good condition	0.025
Unplaned wood	0.013	Earth, with stones or weeds	0.035
Finished concrete	0.012	Gravel	0.028
Unfinished concrete	0.014	Vitrified sewer pipe	0.013
Brick	0.016	Cast iron	0.014
Rubble	0.025	Riveted steel	0.015

Example. A rectangular channel lined with finished concrete is 10 feet wide. The bottom slope is 0.002, and the depth is 7 feet. Compute the rate of discharge by the Manning equation for steady uniform flow.

$$\text{Hydraulic mean depth} = R = \frac{70}{24} \text{ feet}$$

$$V = C\sqrt{RS} = \frac{1.49}{n} R^{1/6}\sqrt{RS} = \frac{1.49}{n} R^{2/3} S^{1/2}$$

$$Q = AV = \frac{70(1.49)}{0.012} \left(\frac{70}{24}\right)^{2/3} (0.002)^{1/2}$$

The rate of discharge $Q = 794$ cubic-feet per second.

10–4. Velocity distribution in open channels

The velocity distribution in an open channel is influenced by the walls and also by the free surface. The distribution of velocity is quite irregular in channels of variable section, particularly in natural streams. The velocity is usually highest at the point or points least affected by the solid boundaries and the free surface. Figure 10-5 shows some representative velocity profiles in a vertical plane. Experiments have shown that the maximum velocity in a straight section of an open channel may occur at a point below the water surface. This maximum velocity frequently does not occur at the center. No

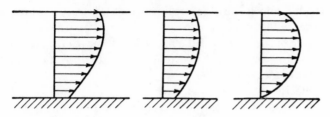

Fig. 10–5. Some vertical velocity profiles in open channels.

completely rational explanation of these observations has been established at the present time. Surface tension and wind effects alone cannot account for the relatively pronounced action retarding the water layers above the level of maximum velocity.

Some measurements indicate that the depth of the maximum velocity increases with depth of the stream. The thread of maximum velocity lies very close to the surface in some shallow streams having rough beds. In some channels the vertical velocity distribution approximates a parabola whose axis is horizontal and passes through the point of maximum velocity. The velocity distribution can be represented by velocity contours, or lines of equal velocity, as shown in Fig. 10-6. The velocity within the area enclosed by a contour curve is higher than that at a point of the curve. The velocity outside the enclosed area is less than that on the curve. Figure 10-6 shows that the velocity varies from top to bottom and from side to side. A bend in the channel or an irregular bed (as in a natural stream) may give very irregular and distorted velocity contour lines.

The measurement of the rate of discharge of a stream is frequently called *stream gaging*. A common procedure is to measure the velocity at various stations in the stream by means of a *current meter*. A current meter in the stream may be suspended from a cable or attached to a rod. Measurements may be made from a bridge, a car suspended from a cable, or a boat. The actuating element of a current meter is a wheel, consisting of a series of vanes

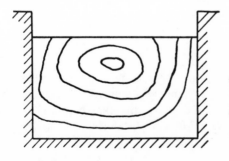

Fig. 10–6. Velocity contours in the straight length of an open channel.

or cups, which is impelled by the stream current. The rate at which the wheel revolves varies with the water velocity. An indication of the rate of rotation gives a measure of the velocity if the meter is suitably calibrated. One procedure for calibrating or "rating" a current meter is to draw it through stationary water at a known speed.

If the velocity distribution in a stream varies, the actual kinetic energy of the stream is greater than that computed on the basis of average velocity. It is general practice, however, to express the kinetic energy term in the energy equation as $V^2/2g$, where V is the average velocity. This use of the average velocity, then, introduces an error. This error is sometimes compensated by empirical coefficients. The use of the average velocity V would be correct if we had one-dimensional flow. In very accurate studies, as for laminar flow, it is necessary to take the velocity variation into account.

10–5. Specific energy

The concept of *specific energy* has proved very fruitful in giving a simple and clear explanation of various phenomena of open-channel flow. An introduction to this concept, the specific-energy diagram, and some applications will be given in the following paragraphs.

There is a difference between *total head* and *specific energy*. The total head (total energy per unit weight of liquid) at any point in the stream shown in Fig. 10-4 is

$$\text{total head} = z + \frac{p}{\gamma} + \frac{V^2}{2g}$$

where z is taken with respect to some *arbitrary horizontal datum plane*. The specific energy is defined as the energy per unit weight of the flowing liquid with respect to a line passing through the *bottom of the channel*. The specific energy E at any point, then, is

$$\text{specific energy} = E = d + \frac{p}{\gamma} + \frac{V^2}{2g} \tag{10–6}$$

Since $d + p/\gamma = y$,

$$E = y + \frac{V^2}{2g} \tag{10–7}$$

The specific energy in open-channel flow is simply the sum of the depth and the velocity head in the channel. In uniform open-channel flow the total head is decreased as flow takes place, but the specific energy remains constant. In nonuniform or varied flow the total head is continually decreased, but the specific energy may be increased or decreased.

It is most convenient and instructive to illustrate principles by dealing with the two-dimensional flow in a channel of rectangular cross section.

In such a channel, with vertical side walls, the flow in parallel vertical planes is identical. The study of flow in a rectangular channel gives simple equations and helps to bring out fundamentals clearly. The same general method of attack can be applied to channels of other shapes, but the resulting relations are more complicated.

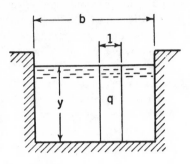

The total rate of discharge Q of the entire channel in Fig. 10-7 equals Vby. The rate of discharge q per unit width is simply Vy. For this two-dimensional flow, since $V = q/y$, the specific energy becomes

$$E = y + \frac{V^2}{2g} = y + \frac{q^2}{2gy^2} \qquad (10\text{–}8)$$

Fig. 10–7. Rectangular cross section of an open channel.

The foregoing equation involves essentially three variables. The physical aspects of Equation (10-8) can be brought out by treating separately two cases: one in which q is constant but E and y vary, and another in which E is constant but q and y vary.

Figure 10-8 shows a plot of depth against specific energy for a constant q. The curve is asymptotic to the horizontal axis and the 45 degree line marked "potential energy." The 45 degree line is simply a plot of the first term of Equation (10-8), whereas that part of the abscissa between the potential-energy line and the specific-energy curve represents the kinetic energy term $V^2/2g$ of Equation (10-8). The specific-energy curve has a point of minimum specific energy at a depth which is called the critical depth y_c. Considering values of depth other than y_c, for every specific energy there are two alternate depths at which the flow may take place—one below and one above the

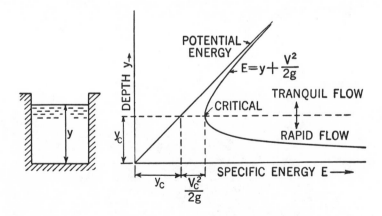

Fig. 10–8. Specific energy diagram, with constant rate of discharge q.

critical depth. The flow is said to be *tranquil* if the actual depth y is greater than y_c. The flow is said to be *rapid* if the actual depth y is less than y_c.

For flow in the region of the critical depth, a depth change may correspond to a very small change in specific energy. Several depths may exist for practically the same specific energy. This feature offers some explanation of the fact that liquid flowing in the region of the critical depth may have an unstable wavy surface. Figure 10-8 shows that a loss of specific energy is accompanied by a reduction of depth in tranquil flow, whereas a loss of specific energy is accompanied by an increase of depth in rapid flow.

Critical depth relations can be established by differentiating Equation (10-8) with respect to y and setting the result equal to zero. E has a minimum value when $dE/dy = 0$. The subscript c will be used for critical depth values. For example, E_c represents the specific energy at the critical depth y_c. Then

$$\frac{dE}{dy} = \frac{d}{dy}\left(y + \frac{q^2}{2gy^2}\right) = 1 - \frac{q^2}{gy^3} \qquad (10\text{--}9)$$

Setting Equation (10-9) equal to zero gives

$$q^2 = gy_c^3 \qquad \text{or} \qquad y_c = \sqrt[3]{\frac{q^2}{g}} \qquad (10\text{--}10)$$

Eliminating q from Equation (10-8) shows that

$$E_c = y_c + \frac{gy_c^3}{2gy_c^2} = \frac{3}{2}y_c \qquad (10\text{--}11)$$

Figure 10-9 shows a plot of depth y versus q for a constant value of

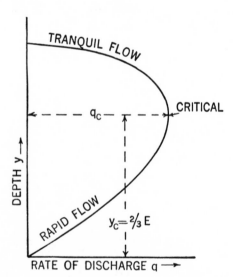

Fig. 10–9. Plot of depth against rate of discharge for a constant specific energy E.

specific energy E, as obtained by using Equation (10-8). Equation (10-8) can be rearranged as follows:

$$q^2 = 2g(Ey^2 - y^3) \tag{10–12}$$

The term dq/dy equals zero at the maximum value of q for a given E. Differentiating Equation (10-12) and setting $dq/dy = 0$ gives

$$\frac{dq}{dy} = \frac{g}{q}(2Ey - 3y^2) \qquad y = \frac{2}{3}E \tag{10–13}$$

A comparison of Equations (10-11) and (10-13) shows that maximum rate of discharge and minimum specific energy both refer to conditions of critical flow.

The critical velocity V_c at the critical depth can be found from Equation (10-10) (noting that $q = yV$) to be

$$V_c = \sqrt{gy_c} \tag{10–14}$$

An inspection of Equation (10-8) and Fig. 10-8 shows that at depths less than the critical the velocity $V > \sqrt{gy}$, whereas at depths greater than the critical the velocity $V < \sqrt{gy}$.

10–6. Nonuniform flow in channels of constant shape

The foregoing articles have dealt with uniform flow, a condition in which the depth, velocity, liquid cross-sectional area, slope, and other characteristics remain constant from section to section. In uniform flow the liquid surface is parallel to the bottom of the channel. Uniform flow might be regarded as a limit that is approached in some cases, as in very long channels. There is an unlimited number of ways in which a liquid might flow steadily through a certain channel. It might flow with a variable depth and with a surface slope different from the bottom slope. In the most general case the energy line, the liquid surface, and the channel bottom have different slopes.

The following discussion is limited to nonuniform flow in a *prismatic* channel; the cross-sectional shape is constant along the channel axis in a prismatic channel. It will be assumed that the slope of the streamlines is constant or varies so gradually that centrifugal forces are negligible. It will be assumed that the slope of the streamlines is so small that the pressure head does not differ appreciably from the depth. Such flow is sometimes called *gradually varied flow*. Because of a lack of a more satisfactory method, calculations are usually based on the assumption that the rate of energy loss in nonuniform flow is equivalent to that which would occur if the flow were uniform under identical conditions of depth, rate of discharge, and boundary roughness. This assumption has not been fully confirmed by experiment. Errors arising from such an assumption are likely to be small in comparison with those involved in the selection of roughness and friction coefficients.

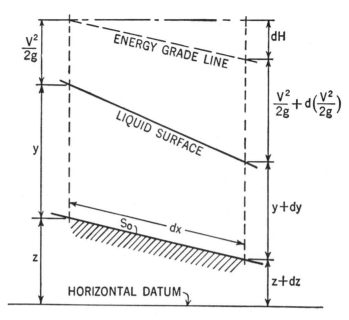

Fig. 10–10. Notation for nonuniform flow.

The total head H at any point in the stream (see Fig. 10-10) is

$$H = z + y + \frac{V^2}{2g} \qquad (10\text{–}15)$$

where z is now the elevation of the channel floor. Differentiating with respect to distance in the direction of flow yields the corresponding rate of change of each of the variables:

$$\frac{dH}{dx} = \frac{dz}{dx} + \frac{dy}{dx} + \frac{d}{dx}\left(\frac{V^2}{2g}\right) \qquad (10\text{–}16)$$

The term dH/dx represents the rate of change of total head. This quantity is always negative in the direction of flow. Using the Chezy notation gives

$$\frac{dH}{dx} = -\frac{V^2}{C^2 R}$$

The quantity dz/dx represents the slope of the bottom of the channel. Following the usual arbitrary custom regarding signs (downward slope as positive), dz/dx will be taken as equal to $-S_0$.

The quantity dy/dx is the rate of change of depth with distance. It is to be recalled that $V = Q/A$, where A is the cross-sectional area of the liquid. If infinitesimals of order higher than the first are neglected, then an increment of cross-sectional area dA equals the surface width b multiplied by the depth increment dy (see Fig. 10-11). The last term of Equation (10-16) thus becomes

$$\frac{d}{dx}\left(\frac{V^2}{2g}\right) = \frac{V}{g}\cdot\frac{dV}{dx} = -\frac{Q^2}{A^3g}\cdot\frac{dA}{dx} = -\frac{Q^2b}{A^3g}\cdot\frac{dy}{dx}$$

Inserting these several equivalent terms in Equation (10-16) and solving for dy/dx gives, finally,

$$\frac{dy}{dx} = \frac{S_0 - \dfrac{Q^2}{A^2C^2R}}{1 - \dfrac{Q^2b}{A^3g}} \qquad (10\text{–}17)$$

Equation (10-17) is the basic differential equation for the steady nonuniform flow in a prismatic channel. Equation (10-17) can represent a large number of surface profile forms. An excellent treatment of methods for solving such equations is given by von Kármán and Biot.[2]

If the denominator of Equation (10-17) is not zero, then $dy/dx = 0$ (uniform flow) when

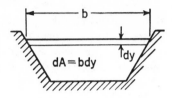

Fig. 10–11. Cross section of open channel.

$$S_0 = \frac{Q^2}{A^2C^2R} \qquad \text{or} \qquad A^2 = \frac{Q^2}{S_0C^2R} \qquad (10\text{–}18)$$

Therefore, for a particular form of channel with a certain quantity and slope, there is only one depth at which the flow will be uniform. This depth is called the *normal depth*. When the actual bed slope is greater than the slope for uniform flow at a given depth, the numerator in Equation (10-17) is positive in sign, and vice versa. If the numerator in Equation (10-17) is not zero, then the slope dy/dx approaches infinity as the denominator of Equation (10-17) approaches zero. The limit is reached at a critical depth y_c when

$$\frac{bQ^2}{gA^3} = 1 \qquad (10\text{–}19)$$

For a rectangular channel, $Q = byV$. A rearrangement of Equation (10-19) shows that the velocity V_c at the critical depth y_c is

$$V_c = \sqrt{gy_c}$$

Equation (10-20) is identical with Equation (10-14).

10–7. Hydraulic jump

The *hydraulic jump* is an example of nonuniform flow in an open channel which will be studied by principles previously discussed. Under suitable circumstances the flow at a depth less than the critical may suddenly change

[2] *Mathematical Methods in Engineering*, by T. von Kármán and M. A. Biot. McGraw-Hill Book Co., Inc., New York, 1940.

to a flow at a certain depth greater than the critical. With this increase in depth there is a corresponding reduction in velocity. This local phenomenon, in which flow passes abruptly from a *rapid* to a *tranquil* state, is called the hydraulic jump. The hydraulic jump is somewhat similar to the sudden expansion in a pipe, in which there is a rapid increase in flow area, a corresponding decrease in velocity, and a loss of available mechanical energy.

The hydraulic jump may occur in different forms which may be classified broadly as (*a*) direct and (*b*) undular. As illustrated in Fig. 10-12(*a*), in the *direct* jump the upper depth is reached practically by one continuous rise of the surface. There is a surface roll and an eddying region at the face of the jump. The direct form is characteristic of jumps of relatively large height. In the *undular* form, Fig. 10-12(*b*), the surface is wavy but not broken, and there is no back roll. The undular form is typical of jumps of relatively low height.

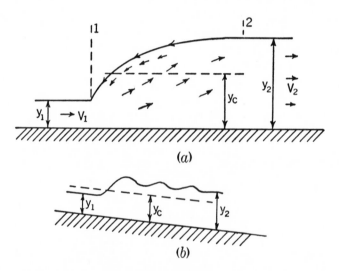

Fig. 10–12. Forms of hydraulic jump: (*a*) direct, (*b*) undular.

Figure 10-13(*a*) indicates the profile of a jump, and Fig. 10-13(*b*) shows the corresponding plot of depth against specific energy. The subscript 1 refers to conditions before the jump, and the subscript 2 refers to conditions after the jump. At the depth $y_1 (<y_c)$, the specific energy is E_1. At the depth y_2 $(>y_c)$, the specific energy is $E_2 \cdot E_1 - E_2$ is the loss of specific energy inherent in the jump. This loss may be quite large under some circumstances.

In many problems in fluid mechanics and rigid-body mechanics an energy accounting alone does not provide a completely quantitative description of the phenomenon. Particularly in cases where the internal detailed mechanism is difficult to formulate, as when shock or impact is involved, it

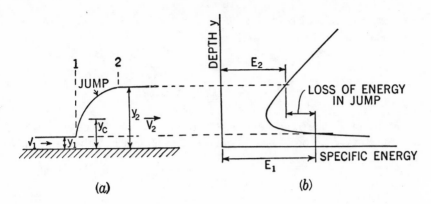

Fig. 10–13. Physical interpretation of the hydraulic jump by means of the specific-energy diagram.

is very helpful to make a momentum study—a particular study or accounting based on conditions at the boundaries. The relation between the depths and velocities upstream and downstream from the hydraulic jump can be developed most conveniently by means of a momentum analysis.

Picture the body of liquid in the jump, Fig. 10-14, as a free or isolated body, with no tangential forces acting on the horizontal channel bottom and the side walls. It seems reasonable to neglect the tangential wall forces if the length of the jump is short or the distance between sections is short and the shock losses are relatively large. Then the net force in the direction of flow acting on the body of liquid is $F_1 - F_2$. Since force equals the time rate of change of linear momentum (Chapter Three), then

$$F_1 - F_2 = \frac{Q\gamma}{g}(V_2 - V_1) \qquad (10\text{–}21)$$

In order to demonstrate methods and principles with a minimum of mathematical complications, the case of two-dimensional flow will be treated, with a unit width of a rectangular channel, as shown in Fig. 10-7. Then, from principles of fluid statics,

$$F_1 = \frac{\gamma y_1}{2}, \qquad A_1 = \frac{\gamma y_1^2}{2}, \qquad F_2 = \frac{\gamma y_2^2}{2}$$

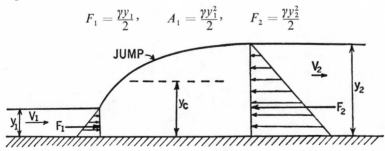

Fig. 10–14. Notation for momentum study of hydraulic jump.

From the equation of continuity, $V_1 y_1 = V_2 y_2 = q$. Substituting the foregoing relations in Equation (10-21) gives

$$\frac{y_1^2 - y_2^2}{2} = \frac{q^2}{g}\left(\frac{1}{y_2} - \frac{1}{y_1}\right)$$

$$\frac{y_1^2}{2} + \frac{q^2}{gy_1} = \frac{y_2^2}{2} + \frac{q^2}{gy_2} \tag{10-22}$$

If the rate of discharge and one depth are known, the other depth can be calculated by Equation (10-22). Equation (10-22) can be solved explicitly, for y_1 and y_2, to give

$$y_2 = \frac{y_1}{2}\left(-1 + \sqrt{1 + \frac{8q^2}{gy_1^3}}\right)$$

$$y_1 = \frac{y_2}{2}\left(-1 + \sqrt{1 + \frac{8q^2}{gy_2^3}}\right) \tag{10-23}$$

As an example, if the rate of discharge and y_1 are known, then y_2 can be calculated from the foregoing momentum equation. Then the loss of energy, per unit weight, in the jump can be computed by the energy equation, as

$$\text{lost energy} = \left(y_1 + \frac{V_1^2}{2g}\right) - \left(y_2 + \frac{V_2^2}{2g}\right) \tag{10-24}$$

10–8. Flow over a fall

Figure 10-15 illustrates another example of nonuniform flow, that of the flow over a fall. With a horizontal channel bottom, the motion takes place entirely at the expense of the specific energy stored in the liquid. The passage from section 1 to section 2 on the falling liquid surface curve corresponds to a shift down on the upper branch (tranquil) of the specific-energy curve, with a loss in specific energy. This loss in specific energy is accompanied by a lowering of the depth. The surface of the moving liquid cannot drop below the critical depth. The critical depth corresponds to the lowest possible

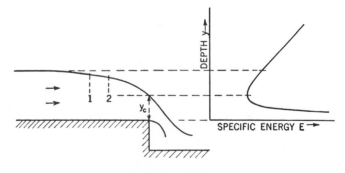

Fig. 10–15. Flow over a fall.

energy content of the falling liquid. Any further lowering of the liquid surface below y_c would mean a change of the movement into the lower branch of the specific-energy curve, and this change could be possible only if energy were added from outside. Therefore the critical depth is the lowest depth which the surface can attain in the natural process of dissipating energy.

10–9. Supercritical, critical, and subcritical flows

Imagine liquid at a depth y, as illustrated in Fig. 10-16(*a*). Next, picture a surface disturbance of small amplitude traveling with a wave velocity c_w as indicated in Fig. 10-16(*b*). The next step is to calculate this wave velocity.

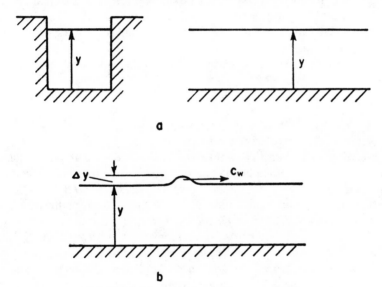

a

b

Fig. 10–16. Notation for surface wave action.

The equation of continuity for flow through a channel having a simple, rectangular cross section is

$$Vy = \text{constant} \tag{10–25}$$

The energy equation, with no friction, is

$$y + \frac{V^2}{2g} = \text{constant} \tag{10–26}$$

Differentiating Equations (10-25) and (10-26)

$$V\,dy + y\,dV = 0 \tag{10–27}$$

$$dy + \frac{V\,dV}{g} = 0 \tag{10–28}$$

Applying Equations (10-27) and (10-28) to the case of a surface wave disturbance as shown in Fig. 10-16(b) gives

$$c_w = \sqrt{gy} \tag{10–29}$$

For example, if a stone were dropped into still water, waves would move with the velocity c_w. Equation (10-29), however, should not be applied to ocean waves or waves in very deep channels.

It should be noted that the wave velocity c_w equals the velocity of open-channel flow at the critical depth y_c as indicated by Equations (10-14) and (10-20). For "critical" flow, the liquid velocity equals the velocity of propagation of a small surface wave. If a wave were started, it could not progress upstream because the two velocities are equal. In tranquil or "subcritical" flow, the liquid velocity is less than c_w. In subcritical flow the small surface waves would travel upstream. In rapid or "supercritical" flow the liquid velocity is greater than c_w. In supercritical flow small surface waves would be swept downstream.

There is an analogy between open channel flow and the flow of a compressible fluid. This analogy is discussed in Sec. 6-17.

10–10. Laminar flow with a free surface

Preceding articles in this chapter have presented what might be classed as a one-dimensional method of analysis. It was pointed out in Sec. 10-4 than in some studies, as for laminar flow, it is necessary to take the velocity variation into account. The following treats one case of fully developed laminar flow, and illustrates a general method of approach.

A fundamental relation for laminar flow will be derived first. Figure 10-17 shows the notation; u is the velocity at a distance y from some boundary; and x is a distance measured in the direction of flow. Consider the forces acting on a small element of height dy, of length dx, and of unit depth

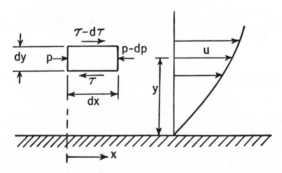

Fig. 10–17. Notation for two-dimensional laminar flow.

perpendicular to the plane of motion. The only forces acting on this element are the pressure and viscous shear forces. For steady flow the pressure forces on this element balance the viscous forces. Therefore

$$dp\,dy = d\tau\,dx$$

$$\frac{dp}{dx} = \frac{d\tau}{dy}$$

where $d\tau$ is the increment of shear stress τ. Since $\tau = \mu\,du/dy$ the pressure gradient dp/dx becomes

$$\frac{dp}{dx} = \mu\frac{d^2u}{dy^2} \tag{10–30}$$

The pressure gradient will be regarded as constant in the direction of y. Integration of the foregoing fundamental equation gives

$$u = \frac{1}{\mu}\left(\frac{dp}{dx}\right)\frac{y^2}{2} + C_1 y + C_2 \tag{10–31}$$

The constants C_1 and C_2 are to be evaluated from boundary conditions.

The next step is to apply the fundamental Equation (10-31) to the particular case of laminar flow with a free surface. Figure 10-18 illustrates the notation. The depth of liquid is h; at the surface the velocity is v.

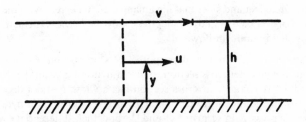

Fig. 10–18. Laminar flow with a free surface.

When $y = 0$, the velocity $u = 0$. When $y = h$, the velocity u is v. Thus we can evaluate the constants C_1 and C_2. The final expression for the variable velocity u becomes

$$u = \frac{1}{\mu}\cdot\frac{dp}{dx}\cdot\frac{y^2}{2} + \left(\frac{v}{h} - \frac{1}{\mu}\cdot\frac{dp}{dx}\cdot\frac{h}{2}\right)y \tag{10–32}$$

Equation (10-32) shows a parabolic velocity distribution. The volume rate of flow through a channel of unit width is q. This volume rate can be found by integration

$$q = \int_0^h u\,dy \tag{10–33}$$

We can define an average velocity V as

$$V = \frac{q}{\text{area}} = \frac{q}{h} \tag{10–34}$$

Using Equation (10-33) we find that the average velocity V is $\frac{2}{3}$ of the surface velocity v. Also, the pressure gradient can be expressed in the final form

$$\frac{dp}{dx} = \frac{3V\mu}{h^2} \tag{10-35}$$

Noting that $p = \gamma h$, where γ is specific weight, we find that the slope of the free surface is given by the relation

$$\frac{dh}{dx} = \frac{3\mu V}{\gamma h^2} \tag{10-36}$$

PROBLEMS

10-1. A circular brick-lined conduit, 5 feet in diameter, has a slope of 1 in 1000. Compute the rate of discharge for steady uniform flow if the conduit flows half full.

10-2. A rectangular channel 25 feet wide and 6 feet deep is lined with unplaned wood. The rate of discharge is 400 cubic feet per second. What is the slope if the flow is steady and uniform?

10-3. A canal lined with unfinished concrete is of trapezoidal section, with a bottom width of 10 feet and sides making an angle of 60 degrees with the horizontal. The bottom slope is 0.0015 and the depth of flow is 7 feet. Calculate the rate of discharge for steady uniform flow.

10-4. Sometimes for practical purposes it is important to proportion the dimensions of the channel cross section to give a hydraulic mean depth which is as large as possible. A rectangular cross section has a width b and a depth D. For a given area A, find the ratio between b and D to give a maximum hydraulic mean depth. (*Hint:* Express R in terms of A and D, and then differentiate with respect to D).

10-5. A trapezoidal canal has a bottom width of 20 feet and side slopes of 1 to 1. The rate of discharge is 364 cubic feet per second at a depth of 3.60 feet. What is the specific energy?

10-6. If 600 cubic feet per second at a depth of 4.1 feet flows in a rectangular channel 20 feet wide, is the flow rapid or tranquil?

10-7. What is the maximum rate of flow which could occur in a rectangular channel 16 feet wide for a specific energy of 6.8 feet?

10-8. Starting with Equation (10-8), for a constant rate of discharge derive the equation

$$\frac{E}{y_c} = \frac{1}{2}\left(\frac{y_c}{y}\right)^2 + \frac{y}{y_c}$$

in which all the ratios are dimensionless. Plot a dimensionless specific energy diagram with y/y_c as ordinate and E/y_c as abscissa.

10-9. Starting with Equation (10-22), for a constant specific energy derive the relation

$$\frac{q}{q_c} = \sqrt{3\left(\frac{y}{y_c}\right)^2 - 2\left(\frac{y}{y_c}\right)^3}$$

where q_c is the critical discharge, and each ratio is dimensionless. Plot a dimensionless diagram with y/y_c as ordinates and q/q_c as abscissas.

10-10. Derive Equation (10-23) from Equation (10-22).

10-11. Water flows 3 feet deep in a rectangular channel with an average velocity of 4 feet per second. Can a jump be formed downstream?

10-12. Water flows at a rate of 360 cubic feet per second in a rectangular channel 18 feet wide at a depth of 1.0 foot. What is the total power loss in a hydraulic jump which has occurred from this flow?

10-13. With a discharge of 400 cubic feet per second in a rectangular channel 20 feet wide, determine the depth y_1 which will sustain a jump resulting in $y_2 = 4.0$ feet. What portion of the initial energy is lost in the jump?

10-14. The average velocity of flow in a rectangular channel is 6 feet per second at a section which is 10 feet deep. What is the volume rate of flow per unit width?

10-15. There is a flow of 100 cubic feet per second in a rectangular channel 5 feet wide and 5 feet deep. Assuming the same specific energy, at what other depth can this flow take place in the same channel?

10-16. Consider a symmetrical trapezoidal cross section. Assume a given area, but different liquid depths. For the maximum hydraulic mean depth, show that the hydraulic mean depth equals one-half the liquid depth.

10-17. How fast would a surface wave travel in water 14 inches deep?

10-18. The width of a rectangular channel is reduced from 6 to 5 feet and the bottom is raised 1 foot vertically at a given section. The depth of the approach flow is 4 feet. At the contracted section there is a drop of 3 inches in the surface elevation. Neglecting friction, what is the steady rate of flow?

10-19. Rain water flows down a roof which has a slope of 30 degrees with the horizontal. Assume laminar flow. If the depth is 0.03 inch, what is the average velocity?

10-20. Castor oil is to flow steadily down a flat sloping surface with a thickness of 0.20 inch with an average velocity of 0.1 foot per second. Assume laminar flow. What is the slope?

10-21. Consider two-dimensional laminar flow between two parallel, fixed plates. Show that the pressure drop per length equals $12\mu V/h^2$, where μ is dynamic viscosity, V is average velocity, and h is the distance between plates.

10-22. Consider two-dimensional laminar flow between two fixed parallel plates. Show that the average velocity is $\frac{2}{3}$ of the maximum velocity.

CHAPTER ELEVEN

General Analytical Relations for Frictionless and Viscous Flow

Some of the foregoing relations have been presented in a simplified form, primarily to answer practical problems and to bring out physical aspects with a minimum of complexity. For example, in some cases a one-dimensional study of flow is sufficiently accurate for certain purposes, and thus the simplified equations are quite satisfactory. In other cases, however, a one-dimensional analysis is not sufficiently accurate, and it is necessary to consider other variables. The following presents a discussion of some more general relations.

11–1. Velocity and acceleration

Let x, y, and z represent the space coordinates, and t time. Let V represent the resultant velocity at any point in a body of fluid, u the x-component of the resultant velocity, v the y-component, and w the z-component. The notation is illustrated in Fig. 11–1. As an example of the Eulerian method, the velocity can be expressed symbolically as

$$u = f_1(x, y, z, t)$$
$$v = f_2(x, y, z, t) \qquad (11\text{–}1)$$
$$w = f_3(x, y, z, t)$$

The first relation in Equation (11–1) can be stated in words as: the velocity component u is some function of the space co-ordinates x, y, z, and the time t. For a certain value of x, of y, of z, and of t, there is a corresponding value for u. Equation (11–1) is a general statement of the velocity relation. If the flow is steady, the velocity at a point is not a function of time, and the fore-

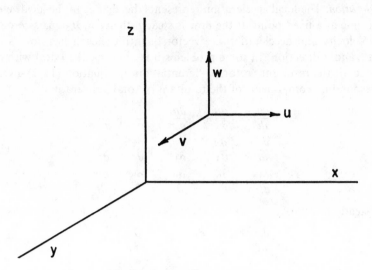

Fig. 11–1. Coordinate system.

going relations become

$$u = f_1(x, y, z)$$
$$v = f_2(x, y, z) \qquad \text{(11–2)}$$
$$w = f_3(x, y, z)$$

Similar expressions could be written for acceleration, pressure, and other flow features.

In the most general case of fluid movement the resultant velocity V is a function of both the distance s along a streamline and the time t, that is,

$$V = f(s, t) \qquad \text{(11–3)}$$

The velocity changes from point to point in space in one instant of time and also from moment to moment of time at any one point in space. Thus

$$dV = \frac{\partial V}{\partial s} ds + \frac{\partial V}{\partial t} dt \qquad \text{(11–4)}$$

It is to be recalled that the partial differential quotient is taken with respect to one variable only, keeping all other variables constant. For example, $\partial V/\partial s$ is the derivative of V with respect to s only, with time constant.

Acceleration is the time rate of change of velocity. Thus the total acceleration dV/dt is

$$\frac{dV}{dt} = \frac{\partial V}{\partial s}\frac{ds}{dt} + \frac{\partial V}{\partial t} = V\frac{\partial V}{\partial s} + \frac{\partial V}{\partial t} \qquad \text{(11–5)}$$

The term $\partial V/\partial t$ is the *local acceleration*, and the term $V(\partial V/\partial s)$ the *convective acceleration*. The local acceleration represents the change in the local velocity with time at a fixed point. If the flow is steady, then $\partial V/\partial t$ equals zero.

Velocity and acceleration are vector quantities; each has both a magnitude and a direction. In some problems it is convenient to deal with components of the resultant vector. Differentiation of Equation (11-1) gives the corresponding components of the resultant or total acceleration

$$\frac{du}{dt} = u\frac{\partial u}{\partial x} + v\frac{\partial u}{\partial y} + w\frac{\partial u}{\partial z} + \frac{\partial u}{\partial t}$$

$$\frac{dv}{dt} = u\frac{\partial v}{\partial x} + v\frac{\partial v}{\partial y} + w\frac{\partial v}{\partial z} + \frac{\partial v}{\partial t} \qquad (11-6)$$

$$\frac{dw}{dt} = u\frac{\partial w}{\partial x} + v\frac{\partial w}{\partial y} + w\frac{\partial w}{\partial z} + \frac{\partial w}{\partial t}$$

For steady motion

$$\frac{\partial u}{\partial t} = 0 \qquad \frac{\partial v}{\partial t} = 0 \qquad \frac{\partial w}{\partial t} = 0$$

11-2. Continuity equation in three dimensions

In Chapter Three the equation of continuity was given for steady flow in one dimension. The equation of continuity for the most general case of three dimensions will now be derived. Figure 11-2 shows an infinitesimal parallelepiped in a body of fluid; the sides have the lengths dx, dy, and dz. Consider first the mass flow per unit time in the x-direction. The mass per

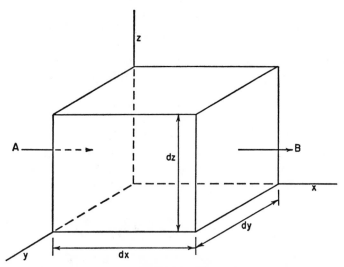

Fig. 11-2. Infinitesimal parallelepiped of fluid, for a continuity study.

unit time entering the left face is $\rho u \, dy dz$; this flow is indicated by vector A in Fig. 11-2. As indicated by the vector B in Fig. 11-2, the mass rate leaving the opposite face is

$$\left(\rho u + \frac{\partial}{\partial x}(\rho u) \, dx \right) dy dz$$

The net gain in the x-direction is

$$\frac{\partial}{\partial x}(\rho u) \, dx dy dz$$

Calculating in a similar manner for the other directions gives the gain

$$\left[\frac{\partial}{\partial x}(\rho u) + \frac{\partial}{\partial y}(\rho v) + \frac{\partial}{\partial z}(\rho w) \right] dx dy dz$$

The foregoing considers only a space variation. In the most general case there is also a time gain

$$\frac{\partial}{\partial t}(\rho) \, dx dy dz$$

The flow of fluid across the closed boundary surface from outward to inward must equal that from inward to outward if there is to be no creation of mass. If inflow is taken as positive and outflow as negative, then the total mass rate of flow across the closed boundary must be zero. Thus the equation of continuity in three dimensions becomes

$$\frac{\partial \rho}{\partial t} + \frac{\partial}{\partial x}(\rho u) + \frac{\partial}{\partial y}(\rho v) + \frac{\partial}{\partial z}(\rho w) = 0 \qquad (11\text{-}7)$$

If the fluid is incompressible (ρ is constant), then the equation of continuity takes the special form

$$\frac{\partial u}{\partial x} + \frac{\partial v}{\partial y} + \frac{\partial w}{\partial z} = 0 \qquad (11\text{-}8)$$

11-3. Dynamic equation for a frictionless fluid

The foregoing relations are kinematic; that is, they refer to the geometry of motion without regard to the forces causing that motion. The next relation to be developed is simply a statement in mathematical form that the resultant force acting on a small element of fluid equals the product of the mass of the element and the acceleration produced. Note that both force and acceleration are vector quantities. The following treatment is limited to flow with no friction or viscous forces acting (the dynamic viscosity being equal to zero). The general relation with viscous forces is given in Sec. 11-8.

Figure 11-3 shows an infinitesimal parallelepiped in a body of fluid; the sides have the lengths dx, dy, and dz. Picture first the forces acting on this element in the x-direction. The pressure on the left face is p; the pressure force

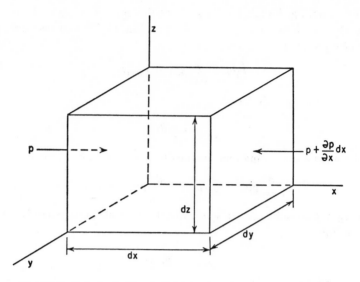

Fig. 11-3. Infinitesimal parallelepiped of fluid, for a force study.

on the left face is $p\,dydz$. The pressure on the right face is $p + (\partial p/\partial x)\,dx$; the pressure force on the right face is

$$\left(p + \frac{\partial p}{\partial x}\,dx\right)dydz$$

The net pressure force acting on the element in the positive direction of x is

$$-\frac{\partial p}{\partial x}\,dxdydz$$

Assume that some force other than that due to pressure is acting on the fluid; such a force might be a gravity force. The attraction of the earth is an example of a *body* force. A *body* force is proportional to the volume or mass, whereas a *surface* force, such as pressure, is proportional to the area. Let P, Q, and R represent the x-, y-, and z-components, respectively, of the body force per unit mass. Then the total force acting on the element in the positive x-direction is

$$\left(P\rho - \frac{\partial p}{\partial x}\right)dxdydz$$

Since force equals mass times acceleration, the dynamical relation for the x-direction becomes

$$\left(P\rho - \frac{\partial p}{\partial x}\right)dxdydz = \rho\,dxdydz\,\frac{du}{dt}$$

$$P\rho - \frac{\partial p}{\partial x} = \rho\,\frac{du}{dt} \tag{11-9}$$

where du/dt is the x-component of the total acceleration. Equations for the

y- and z-directions can be obtained in a similar fashion. The final set of dynamical equations for the three directions is thus

$$P\rho - \frac{\partial p}{\partial x} = \rho \frac{du}{dt} \qquad (11\text{-}10)$$

$$Q\rho - \frac{\partial p}{\partial y} = \rho \frac{dv}{dt} \qquad (11\text{-}11)$$

$$R\rho - \frac{\partial p}{\partial z} = \rho \frac{dw}{dt} \qquad (11\text{-}12)$$

The foregoing equations are called Euler's hydrodynamical equations. The total acceleration in each relation in Equation (11-10) can be expressed in the form given by Equation (11-6).

11-4. Simple case of viscous flow

There are cases that involve viscous flow directly. Examples are lubrication, some pumps (as rotary, vane, and other positive-displacement types), laminar flow in pipes, and laminar flow around bodies. There are cases that involve viscous flow directly and indirectly, as those under the general heading of resistance. Examples are skin-friction drag, drag of bodies with eddying flow, and eddying flow in channels.

In many of these cases the simplified relations previously presented are sufficient. In other cases it is necessary to make some study of the more general analytical relations for viscous flow. The purpose of the following is to present an introduction to these relations for incompressible fluids.

As illustrated in Fig. 11-4, imagine laminar flow in one direction only, parallel to the x-axis. Let u represent velocity, consider a variation in the y-direction only, and let T represent shear or tangential stress. Then the absolute or dynamic viscosity μ is defined as

$$\mu = \frac{\text{shear stress}}{\text{rate of shearing strain}}$$

$$T = \mu \frac{du}{dy} \qquad (11\text{-}13)$$

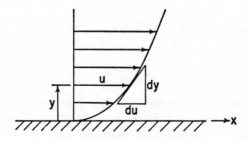

Fig. 11-4. Velocity distribution for simple cases of flow.

The rate of shearing strain is an angular change per unit time. One object of this discussion is to extend the foregoing concept to the more general case of variable flow in three dimensions.

11–5. Stress components for the general case

Figure 11–5 shows three faces of an infinitesimal parallelepiped of fluid and illustrates the notation. Let S_x represent a direct stress along the x-axis, S_y a direct stress along the y-axis, and S_z a direct stress along the z-axis. Each direct stress is considered as a positive stress when acting in the positive direction. A compressive direct stress on a surface would be negative. Let T_{xy} represent a shear or tangential stress on a surface normal to the x-axis and in a direction parallel to the y-axis. Similarly, T_{zx} is a shear stress in a surface perpendicular to the z-axis and in a direction parallel to the x-axis.

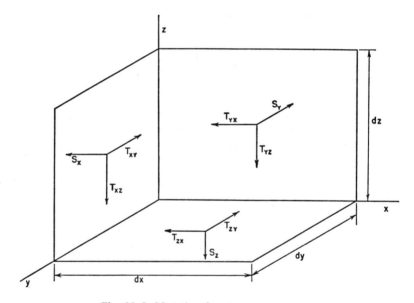

Fig. 11–5. Notation for stress components.

In the general case there are nine stress components, as follows:

$$
\begin{matrix}
S_x & T_{xy} & T_{xz} \\
T_{yx} & S_y & T_{yz} \\
T_{zx} & T_{zy} & S_z
\end{matrix}
$$

Certain quantities are expressed with reference to a coordinate system. The word "tensor" is a general name applied to these quantities which transform according to certain laws as the coordinate system is changed. Sometimes a tensor is called a "matrix."

Consider only a Cartesian coordinate system and only such changes as rotations about the origin. A scalar quantity is a tensor of "rank zero"; the scalar remains unchanged as the coordinate system is rotated. The number of coordinates is three; the number of quantities necessary to specify the scalar is one. Three raised to the "zero" power is one; thus the rank is zero. A vector is considered as a tensor of "rank one." A vector can be completely specified by its three scalar components. When the coordinate system is rotated about the origin, the vector does not change, but the three coordinates or components of the vector will change. Three raised to the "one" power is three; thus the rank is one.

The nine stress components illustrated in Fig. 11-5 can be considered as a tensor of "rank two." There are nine components (three raised to the "two" power), which can change as the coordinate system is rotated.

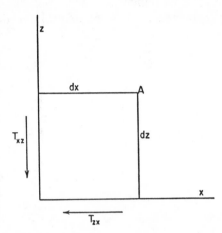

Figure 11-6 shows an end view of the parallelepiped looking in the y-direction at the face with the sides dx and dz. The stress T_{xz} times the area $dydz$ gives a shear force. The stress T_{zx} times the area $dxdy$ gives a shear force. Consider the equilibrium of the small parallelepiped with respect to rotations. Let us first take

Fig. 11-6. End view of parallelepiped, looking in y-direction at face with sides dx and dz.

moments of the forces about a line through point A parallel to the y-axis. Then, for equilibrium

$$T_{xz}dxdydz = T_{zx}dxdydz$$
$$T_{xz} = T_{zx}$$

Treating the other shear forces in a similar fashion gives the final results

$$T_{xy} = T_{yx}$$
$$T_{yz} = T_{zy}$$

There are six shear terms in the following stress tensor:

$$
\begin{array}{ccc}
S_x & T_{xy} & T_{xz} \\
T_{yx} & S_y & T_{yz} \\
T_{zx} & T_{zy} & S_z
\end{array}
$$

There are three particular pairs of two each; each in a pair has the same magnitude as the other. The tensor is "symmetric" because the shear stresses are symmetrical with respect to the diagonal line of direct stresses.

11–6. Deformation rate components for the general case

Let V represent the resultant velocity at any point in a body of fluid, u the x-component of the resultant velocity, v the y-component, and w the z-component. The notation is illustrated in Fig. 11–7.

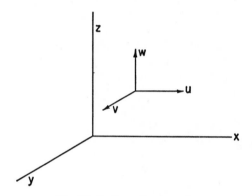

Fig. 11–7. Coordinate system.

Figure 11–4 illustrates a simple case of deformation. There is a velocity variation only in the y-direction. The rate of shearing strain, or the angular deformation per unit time, is du/dy. The shear stress is only in the x-direction.

Consider next a more general case of viscous flow in which there is a shear stress in both the x- and y-directions. Figure 11–8(a) shows the element before deformation. Figure 11–8(b) illustrates the angular changes. The angular change A_1 per unit time is $\partial u/\partial y$. The angular change A_2 per unit time is $\partial v/\partial x$.

Besides angular deformation rates, there can be rates of linear deformation extension, such as $\partial u/\partial x$. In the most general case of three dimensions, there are nine differential coefficients. Each coefficient involves a rate of change of shape. These coefficients can be arranged as the deformation tensor

$$
\begin{array}{ccc}
\dfrac{\partial u}{\partial x} & \dfrac{\partial v}{\partial x} & \dfrac{\partial w}{\partial x} \\[2ex]
\dfrac{\partial u}{\partial y} & \dfrac{\partial v}{\partial y} & \dfrac{\partial w}{\partial y} \\[2ex]
\dfrac{\partial u}{\partial z} & \dfrac{\partial v}{\partial z} & \dfrac{\partial w}{\partial z}
\end{array}
$$

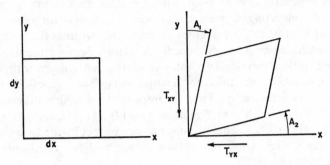

Fig. 11–8. Effect of shear stresses in x- and y-directions.

11–7. General relation between stresses and deformation rates

For the simple case discussed in Sec. 11–4, the relation between shear stress and rate of shearing strain is given by the relation

$$T = \mu \frac{du}{dy}$$

where the dynamic viscosity forms a connecting link between stress and rate of deformation.

Consider some other cases, and refer to Fig. 11–8(b). The relation between total shear stress and total rate of deformation is

$$T_{xy} = T_{yx} = \mu \left(\frac{\partial v}{\partial x} + \frac{\partial u}{\partial y} \right) \tag{11–14}$$

In a similar fashion we get the results

$$T_{yz} = T_{zy} = \mu \left(\frac{\partial w}{\partial y} + \frac{\partial v}{\partial z} \right) \tag{11–15}$$

$$T_{zx} = T_{xz} = \mu \left(\frac{\partial u}{\partial z} + \frac{\partial w}{\partial x} \right) \tag{11–16}$$

Equations (11–14), (11–15), and (11–16) take care of the shear stresses. The next step is to investigate the direct stresses.

Let p represent a pressure which is the average value of the three direct stresses. Thus, if the three direct stresses act,

$$p = -\tfrac{1}{3}(S_x + S_y + S_z) \tag{11–17}$$

If the fluid is frictionless, or has zero viscosity, then there are no shear forces; the forces acting on the element are normal and

$$S_x = S_y = S_z = -p \tag{11–18}$$

For the general case of viscous flow in three dimensions, there is the problem of connecting a stress tensor with a rate-of-deformation tensor. Note, however, that the stress tensor is symmetric, whereas the general rate-of-deformation tensor is not symmetric. A relation between the tensors can be worked out by forming two rate-of-deformation tensors with the differential coefficients in different arrangements. The rate-of-deformation tensor is made symmetric by adding a transposed tensor. Also, the generalized relation should be such that Equations (11–14), (11–15), (11–16), (11–17), and (11–18) are satisfied. These conditions can be met by the following relation between a stress tensor, a pressure tensor, and two transformed rate-of-deformation tensors:

$$
\begin{bmatrix} S_x & T_{xy} & T_{xz} \\ T_{xy} & S_y & T_{yz} \\ T_{xz} & T_{yz} & S_z \end{bmatrix} = - \begin{bmatrix} p & 0 & 0 \\ 0 & p & 0 \\ 0 & 0 & p \end{bmatrix} + \mu \begin{bmatrix} \dfrac{\partial u}{\partial x} & \dfrac{\partial v}{\partial x} & \dfrac{\partial w}{\partial x} \\ \dfrac{\partial u}{\partial y} & \dfrac{\partial v}{\partial y} & \dfrac{\partial w}{\partial y} \\ \dfrac{\partial u}{\partial z} & \dfrac{\partial v}{\partial z} & \dfrac{\partial w}{\partial z} \end{bmatrix} + \mu \begin{bmatrix} \dfrac{\partial u}{\partial x} & \dfrac{\partial u}{\partial y} & \dfrac{\partial u}{\partial z} \\ \dfrac{\partial v}{\partial x} & \dfrac{\partial v}{\partial y} & \dfrac{\partial v}{\partial z} \\ \dfrac{\partial w}{\partial x} & \dfrac{\partial w}{\partial y} & \dfrac{\partial w}{\partial z} \end{bmatrix}
$$

$$(11\text{–}19)$$

Equation (11–19) is a generalization of the simple special form given by Equation (11–13). The meaning of the tensor Equation (11–19) is that the corresponding equations in the ordinary sense hold between the quantities which have corresponding places in the various tensors. For example, the first place in the second line of the tensors gives the equation

$$
T_{xy} = \mu \left(\frac{\partial u}{\partial y} + \frac{\partial v}{\partial x} \right)
$$

which is identical with Equation (11–14). Other corresponding places give Equations (11–15) and (11–16). For a frictionless fluid, Equation (11–19) shows that $S_x = S_y = S_z = -p$. For a fluid with friction, Equation (11–19) shows that

$$
S_x = -p + 2\mu \frac{\partial u}{\partial x}
$$

$$
S_y = -p + 2\mu \frac{\partial v}{\partial y}
$$

$$
S_z = -p + 2\mu \frac{\partial w}{\partial z}
$$

$$
S_x + S_y + S_z = -3p + 2\mu \left(\frac{\partial u}{\partial x} + \frac{\partial v}{\partial y} + \frac{\partial w}{\partial z} \right)
$$

$$(11\text{–}20)$$

For an incompressible fluid, the equation of continuity gives

$$
\frac{\partial u}{\partial x} + \frac{\partial v}{\partial y} + \frac{\partial w}{\partial z} = 0
$$

and thus Equation (11–20) agrees with Equation (11–17). We might consider Equation (11–19) as a generalized definition of viscosity.

11–8. Dynamic equation with viscous forces

The basic dynamic relation is: Mass times acceleration equals the sum of the forces acting. In Article 11–3 is given the dynamic relations

$$\rho \frac{du}{dt} = P\rho - \frac{\partial p}{\partial x} \tag{11–10}$$

$$\rho \frac{dv}{dt} = Q\rho - \frac{\partial p}{\partial y} \tag{11–11}$$

$$\rho \frac{dw}{dt} = R\rho - \frac{\partial p}{\partial z} \tag{11–12}$$

where P, Q, and R represent the x-, y-, and z-components respectively, of the body force per unit mass. Equations (11–10), (11–11), and (11–12), however, do not include viscous forces. Note that each term in Equations (11–10), (11–11), and (11–12) has the dimensions of a force per unit volume.

Figure 11–9 illustrates the stresses acting on an infinitesimal parallelepiped with sides dx, dy, and dz. The stress S_x acts on the surface indicated. In the most general case the direct stress is a function of x. Thus, on the opposite face the stress is

$$S_x + \frac{\partial S_x}{\partial x} dx$$

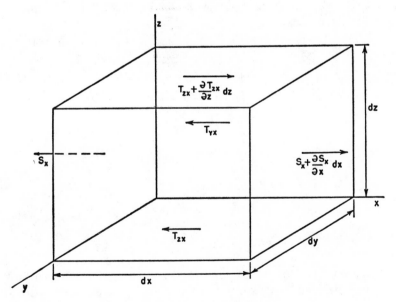

Fig. 11–9. Stress variations.

On the bottom face, with the area $dxdy$, there is a stress T_{zx}. On the top parallel surface the shear stress is

$$T_{zx} + \frac{\partial T_{zx}}{\partial z} dz$$

On the back vertical face of area $dxdz$ there is a shear stress T_{yz}. On the front parallel surface the shear stress is

$$T_{yx} + \frac{\partial T_{yx}}{\partial y} dy$$

Summing up all the forces in the x-direction gives the net force in the x-direction as

$$\left(\frac{\partial S_x}{\partial x} + \frac{\partial T_{yx}}{\partial y} + \frac{\partial T_{zx}}{\partial z} \right) dxdydz$$

The force per unit volume can be found by dividing the foregoing expression by $dxdydz$. The corresponding forces per unit volume for the x-, y-, and z-directions respectively are

$$\frac{\partial S_x}{\partial x} + \frac{\partial T_{yx}}{\partial y} + \frac{\partial T_{zx}}{\partial z}$$

$$\frac{\partial T_{xy}}{\partial x} + \frac{\partial S_y}{\partial y} + \frac{\partial T_{zy}}{\partial z}$$

$$\frac{\partial T_{xz}}{\partial x} + \frac{\partial T_{yz}}{\partial y} + \frac{\partial S_z}{\partial z}$$

Differentiating Equation (11–19) to get the three foregoing x-, y-, and z-forces gives the dynamic equations with viscous forces. For the x-direction the resulting equation is

$$\rho \frac{du}{dt} = P\rho - \frac{\partial p}{\partial x} + \mu \frac{\partial}{\partial x} \left(\frac{\partial u}{\partial x} + \frac{\partial v}{\partial y} + \frac{\partial w}{\partial z} \right) + \mu \left(\frac{\partial^2 u}{\partial x^2} + \frac{\partial^2 u}{\partial y^2} + \frac{\partial^2 u}{\partial z^2} \right)$$

$$(11–21)$$

For an incompressible fluid, the equation of continuity (see Eq. 11–8) gives

$$\frac{\partial u}{\partial x} + \frac{\partial v}{\partial y} + \frac{\partial w}{\partial z} = 0 \qquad (11–22)$$

and the basic differential equations with viscous forces become

$$\rho \frac{du}{dt} = P\rho - \frac{\partial p}{\partial x} + \mu \left(\frac{\partial^2 u}{\partial x^2} + \frac{\partial^2 u}{\partial y^2} + \frac{\partial^2 u}{\partial z^2} \right) \qquad (11–23)$$

$$\rho \frac{dv}{dt} = Q\rho - \frac{\partial p}{\partial y} + \mu \left(\frac{\partial^2 v}{\partial x^2} + \frac{\partial^2 v}{\partial y^2} + \frac{\partial^2 v}{\partial z^2} \right) \qquad (11–24)$$

$$\rho \frac{dw}{dt} = R\rho - \frac{\partial p}{\partial z} + \mu \left(\frac{\partial^2 w}{\partial x^2} + \frac{\partial^2 w}{\partial y^2} + \frac{\partial^2 w}{\partial z^2} \right) \qquad (11–25)$$

Equations (11–23), (11–24), and (11–25) are called the Navier-Stokes equations, after two early workers in this field. These equations were first presented in 1822. The mathematical difficulties in solving the Navier-Stokes equations are very great. As yet no general methods are established for solving the equations. A small number of special cases have been worked out.

11–9. Vector analysis in flow studies

A "scalar" quantity is one that has magnitude only; examples are mass, density, temperature, energy, volume, entropy, and heat. A "vector" quantity is one that has both magnitude and direction; examples are displacement, velocity, acceleration, force, and momentum.

Vector analysis is an analytical shorthand, or abbreviated language, which makes it possible to write physical relations in a compact form. The following discussion defines some of the basic terms and introduces some of the main concepts involved in vector analysis.

A scalar quantity can be represented by a single number, whereas a vector quantity requires a vector for its representation. A vector is a line whose length, to a certain scale, represents the magnitude of the quantity and whose direction is the direction of the quantity represented. In order to represent the direction completely, it is necessary to distingush between the origin and the end of a vector.

Two vectors can be added vectorially or geometrically by placing the origin of the second vector at the end of the first. The sum is then the single vector whose origin is the origin of the first and whose end is the end of the second. As an example, let **a** and **b** represent two vectors. The vector sum **r** can be written in the form

$$\mathbf{r} = \mathbf{a} + \mathbf{b} \qquad (11\text{–}26)$$

whereas the difference can be written as

$$\mathbf{r} = \mathbf{a} - \mathbf{b} \qquad (11\text{–}27)$$

These two operations are illustrated in Fig. 11–10. Multiplication of a vector by a scalar gives a vector whose direction is that of the original vector, but whose length is equal to that of the original vector multiplied by the scalar. For example, multiplication by -1 interchanges the origin and the end. Equation (11–27) can be considered as the sum of **a** plus $(-\mathbf{b})$.

Let **i**, **j**, and **k** each represent a unit vector along a positive direction of the **x**-, **y**-, and **z**-coordinate axes respec-

Fig. 11–10. Vector addition and subtraction.

tively. If **r** is the vector from the origin to some point in space whose co-ordinates are x, y, and z, the the vector can be written as

$$\mathbf{r} = \mathbf{i}x + \mathbf{j}y + \mathbf{k}z \tag{11-28}$$

These features are illustrated in Fig. 11–11. The combination $\mathbf{i}x$ is a vector in the x-direction with the scalar values or magnitude equal to x. A similar interpretation holds for the other components of **r**.

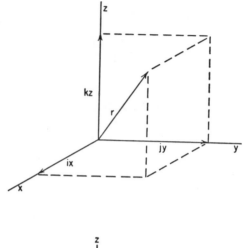

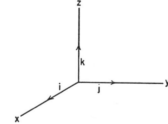

Fig. 11–11. Components of a vector.

11–10. Scalar and vector products

The "scalar" or "dot" product of two vectors is defined as

$$\mathbf{a} \cdot \mathbf{b} = ab \cos(\mathbf{a}, \mathbf{b})$$

where $(\mathbf{a}, \mathbf{b})$ represents the angle between the directions of **a** and **b**; this is angle θ in Fig. 11–12. The term a is the scalar value of **a**; the term b is the scalar value of **b**. The scalar product is a scalar quantity.

Work is an example of a scalar product. Referring to Fig. 11–13, the work

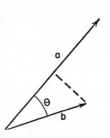

Fig. 11-12. Notation for scalar product.

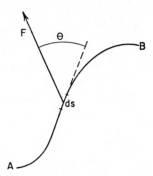

Fig. 11-13. Notation for line integral.

done by a force $\mathbf{F}$ in the distance or displacement $\mathbf{ds}$ is by definition equal to

$$\mathbf{F}\cos\theta\, ds = \mathbf{F} \cdot \mathbf{ds}$$

The force component tangent to the path has the magnitude $\mathbf{F}\cos\theta$. The work done along the path from A to B is the "line integral"

$$\text{Work} = \int_{A}^{B} \mathbf{F} \cdot \mathbf{ds}$$

Let the x-, y-, and z-components of $\mathbf{a}$ be a_x, a_y, and a_z respectively. Let the x-, y-, and z-components of $\mathbf{b}$ be b_x, b_y, and b_z respectively. Then the scalar product can be written as

$$\mathbf{a} \cdot \mathbf{b} = a_x b_x + a_y b_y + a_z b_z$$

The "vector" or "cross" product of two vectors is defined as

$$\mathbf{a} \times \mathbf{b} = \mathbf{e}\, ab \sin(\mathbf{a}, \mathbf{b})$$

where $\mathbf{e}$ is a unit vector perpendicular to the plane including vectors $\mathbf{a}$ and $\mathbf{b}$. The result of a vector multiplication is a vector and therefore has both direction and magnitude. The direction of $\mathbf{e}$ is such that a right-hand screw would advance in the positive direction of $\mathbf{e}$ if it were turned from $\mathbf{a}$ to $\mathbf{b}$.

The vector product can be written as

$$\mathbf{a} \times \mathbf{b} = (\mathbf{i}a_x + \mathbf{j}a_y + \mathbf{k}a_z) \times (\mathbf{i}b_x + \mathbf{j}b_y + \mathbf{k}b_z)$$

The vector product can be put in the form of a determinant

$$\mathbf{a} \times \mathbf{b} = \mathbf{i}(a_y b_z - a_z b_y) + \mathbf{j}(a_z b_x - a_x b_z) + \mathbf{k}(a_x b_y - a_y b_x)$$

$$\mathbf{a} \times \mathbf{b} = \begin{vmatrix} \mathbf{i} & \mathbf{j} & \mathbf{k} \\ a_x & a_y & a_z \\ b_x & b_y & b_z \end{vmatrix}$$

11–11. Vector calculus

If in a region of space there is a corresponding value of some physical property for every point, then this region is called a "field." If this physical property is a scalar quantity, then the region is a "scalar field." For example, in a temperature field the temperature would be defined for every point in a body; one could say that the scalar quantity temperature is a function of the position of each point. In a density field the density would be defined for all parts of a body.

If, on the other hand, the physical property is a vector, the region is a "vector field." For example, if there is a corresponding velocity for each point in a body of fluid, we would call this region a vector field. Another example of a vector field is the earth's gravitational field in which there is a corresponding force for each position in space. It is assumed that the functions under consideration in this chapter are continuous functions, are single-valued, and possess continuous first space-derivatives in the regions under consideration.

It is convenient in many studies of physical phenomena to use a certain differential operator. Let the symbol ∇ (called "del" or "nabla") represent the vector differential operator

$$\nabla = i\frac{\partial}{\partial x} + j\frac{\partial}{\partial y} + k\frac{\partial}{\partial z}$$

The quantity ∇P is the "gradient" of the scalar quantity P

$$\nabla P = \operatorname{grad} P = i\frac{\partial P}{\partial x} + j\frac{\partial P}{\partial y} + k\frac{\partial P}{\partial z}$$

Note that the gradient is a vector quantity.

The combination $\nabla \cdot P$ is the "divergence" of the vector P and is written

$$\nabla \cdot P = \operatorname{div} P = \left(i\frac{\partial}{\partial x} + j\frac{\partial}{\partial y} + k\frac{\partial}{\partial z}\right) \cdot (iP_x + jP_y + kP_z)$$

$$\nabla \cdot P = \frac{\partial P_x}{\partial x} + \frac{\partial P_y}{\partial y} + \frac{\partial P_z}{\partial z}$$

The divergence is a scalar quantity.

The combination $\nabla \times P$ is the "curl" or "rot" of the vector P and can be written

$$\operatorname{curl} P = \nabla \times P = \left(i\frac{\partial}{\partial x} + j\frac{\partial}{\partial y} + k\frac{\partial}{\partial x}\right) \times (iP_x + jP_y + kP_z)$$

$$\operatorname{curl} P = \begin{vmatrix} i & j & k \\ \dfrac{\partial}{\partial x} & \dfrac{\partial}{\partial y} & \dfrac{\partial}{\partial z} \\ P_x & P_y & P_z \end{vmatrix}$$

The curl is a vector quantity.

11-12. Examples of flow relations using vector analysis

A number of examples will be given to illustrate the use of vector analysis in flow studies.

Imagine a fluid flowing through a plane surface with the uniform velocity **V** as illustrated in Fig. 11-14. Let **n** represent a vector of unit magnitude normal to the plane. Then the volume of fluid passing through unit area of the plane per unit time is the scalar product **V · n**, because **V · n** = $V \cos A$, where A is the angle between **V** and **n**. $V \cos A$ is the velocity component perpendicular to the plane.

Circulation is formally defined as the line integral of the velocity along a closed contour line in the fluid. This line integral is the integral of the product of the contour element ds and the component of the velocity in the direction of ds. Figure 11-15 shows a closed contour arbitrarily drawn in a field of

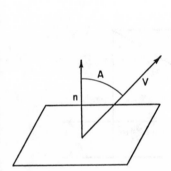

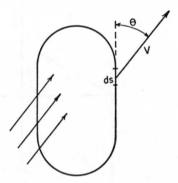

Fig. 11-14. Notation for flow rate calculation.

Fig. 11-15. Contour line in a field of flow.

flow. Let **V** be the velocity at any point and θ the angle between **V** and the contour element **ds**. The circulation Γ around the closed contour is represented by the following integral

$$\Gamma = \int_c V \cos \theta \, ds = \int_c \mathbf{V} \cdot \mathbf{ds}$$

An analogy might be drawn between work done on a body and circulation. Force corresponds to velocity, and the body displacement corresponds to the contour element ds.

Let **V** represent the resultant velocity at a point. Then the general equation of continuity [see Equation (13-7)] in vector form becomes

$$\frac{\partial \rho}{\partial t} + \mathbf{\nabla} \cdot (\rho \mathbf{V}) = 0$$

Let the x-, y-, and z-components of the resultant velocity be represented by u, v, and w respectively. Consider the special case of steady incompressible flow. The divergence of the velocity $\mathbf{V}$ can be written as

$$\mathbf{V} \cdot \mathbf{V} = \frac{\partial u}{\partial x} + \frac{\partial v}{\partial y} + \frac{\partial w}{\partial z} \tag{11-29}$$

Let us multiply both sides of Equation (11–29) by the volume of the element; this gives

$$(\mathbf{\nabla} \cdot \mathbf{V})\, dxdydz = \left(\frac{\partial u}{\partial x}\, dx\right) dydz + \left(\frac{\partial v}{\partial y}\, dy\right) dxdz + \left(\frac{\partial w}{\partial z}\right) dxdy \tag{11-30}$$

Refer to Fig. 11–16. The term $(\partial u / \partial x)\, dx$ represents a velocity increment. This velocity increment times $dydz$ represents an increment in volume rate of flow. A similar interpretation can be given to the other two terms on the right-hand side of Equation (11–30). Thus, we can picture the divergence of the velocity as the increment in volume rate of flow per unit volume.

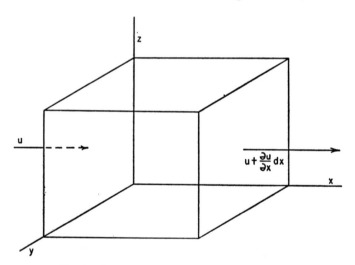

Fig. 11–16. Illustration for divergence of velocity.

For steady incompressible flow the equation of continuity shows that the divergence of the velocity is zero; there is no increase or decrease in the volume rate of flow per unit volume.

Let S be equal to $\mathbf{i}P + \mathbf{j}Q + \mathbf{k}R$, where P, Q, and R represent the x-, y-, and z-components, respectively, of the body forces per unit mass. Then the Euler dynamical equations [see Equation (11–10)] can be replaced by one vector relation

$$\rho \frac{d\mathbf{V}}{dt} = \rho \mathbf{S} - \mathbf{\nabla} p$$

The term ∇p represents the resultant pressure gradient. The x-component of the pressure gradient is $\partial p/\partial x$, a pressure change per unit length in the x-direction.

PROBLEMS

11-1. Find the scalar products: $\mathbf{i} \cdot \mathbf{i}$; $\mathbf{j} \cdot \mathbf{j}$; $\mathbf{i} \cdot \mathbf{j}$; and $\mathbf{j} \cdot \mathbf{k}$.

11-2. Find the vector products: $\mathbf{j} \times \mathbf{k}$; $\mathbf{k} \times \mathbf{i}$; and $\mathbf{i} \times \mathbf{j}$.

11-3. Show that the magnitude of the vector product $\mathbf{a} \times \mathbf{b}$ equals the area of the parallelogram of which $\mathbf{a}$ and $\mathbf{b}$ are the adjacent sides.

11-4. Find the gradient of V if $V = (xy^2z)$.

11-5. Find the divergence of V if $V = \mathbf{i}x + \mathbf{j}2y + \mathbf{k}z$.

11-6. In a mass of fluid the distribution of pressure p is given by the expression $p = k(x^2 + y^2)$ where k is a constant. What is the magnitude of the pressure gradient?

11-7. For two-dimensional flow in the x–y plane, what is the slope at each point along a streamline in terms of velocity components?

11-8. Show that the general equation of continuity can be written in the form

$$\frac{d\rho}{dt} + \rho \left(\frac{\partial u}{\partial x} + \frac{\partial v}{\partial y} + \frac{\partial w}{\partial z} \right) = 0$$

CHAPTER TWELVE

Steady
Two-Dimensional
Incompressible Flow

A general problem in fluid mechanics is to map out the velocity or pressure distribution in the field or region of a flowing fluid. If the velocity at each point in the fluid is known, an application of the dynamic equation yields the pressure at each point. Knowledge of the pressure distribution around a body is helpful in determining the total force of the fluid on the body. For example, in investigations of pumps, compressors, and turbines, the force acting between a fluid and a blade may be determined with the aid of a pressure-distribution study. The pressure distribution along a channel is useful in investigations of the energy conversion, in studies of separation, and in studies of possible boiling or cavitation in a liquid.

Various methods are used in determining the pressure distribution around a body or along a channel. One direct method is to make measurements on the actual body or channel, on the so-called *prototype*. Sometimes it is more convenient to make measurements on a geometrically similar *model* of the body or channel. A body may be tested in a wind tunnel or a water tunnel. In each case pressure openings along a surface can be connected to suitable manometers or pressure gages in order to find the pressure distribution.

In some cases it may be difficult and expensive to make an experimental investigation. The expense and difficulty of a test may not be warranted. Then there is this question: Can the velocity and the pressure distribution be determined theoretically?

In order to illustrate a general method, picture the flow of some actual fluid around a body. Assume that there is no separation or eddying wake. At the body surface the fluid velocity is zero. There is a boundary layer close to the body surface. In this boundary layer there is a velocity gradient and viscous shearing. Some distance away from the body, outside of the boundary layer, the flow is not influenced so much by the viscous action in the boundary layer. Thus, for purposes of analysis, it is convenient to divide the entire flow area into two regions. One region is the boundary layer; for this region we must take into account friction or the viscosity of the fluid. Outside of the boundary layer is the other region in which, for many practical cases, we can neglect the viscosity of the fluid. A purely analytical calculation of viscous flow is very difficult; a discussion of such flow is given in Chapters Five and Eleven. On the other hand, the calculation of frictionless flow is much simpler.

The remainder of this chapter will be devoted to two-dimensional steady flow of a frictionless incompressible fluid. In two-dimensional motion the flow is identical in parallel planes. The x-y plane will be taken as the plane of flow. The problem of frictionless flow of an incompressible fluid is more or less a problem of pure geometry.

12-1. Source and sink flow

Certain simple types of flow will be combined to give various resultant streamline patterns. In such a *potential* or *possible* flow the streamlines may be regarded as solid boundaries, since no fluid crosses a streamline. A streamline, then, may be replaced by a solid body without affecting the flow. Three

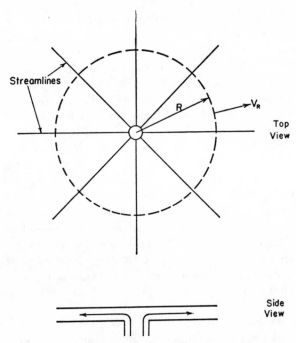

Fig. 12-1. Two-dimensional frictionless source flow.

simple types of flow will be employed; one in which the streamlines are rectilinear or parallel straight lines; one in which the streamlines are radial; and one in which the streamlines are concentric circles (vortex flow).

In Fig. 12-1 the fluid enters the opening and moves radially outward. Strictly speaking, the opening should be infinitesimal; at the exact center the velocity approaches infinity. These singular conditions will be avoided by confining attention to a radius some distance away from the center. The flow shown in Fig. 12-1 will be referred to as one due to a *source*. Flow in the opposite direction is a *sink*. In either case the streamlines are radial. Let Q be the volume rate of flow, V_R the radial velocity at radius R, and b the distance perpendicular to the plane of flow. Then, from Fig. 12-1,

$$Q = 2\pi R V_R b$$

$$V_R = \frac{Q}{2\pi R b} \tag{12-1}$$

In many treatments b is considered as unity, for convenience.

12-2. Vector addition of rectilinear, source, and sink flows

Figure 12-2(a) shows a uniform rectilinear flow with a velocity V_0; the streamlines are parallel and spaced at equal intervals. A source is placed at A. At any point, such as B, the velocity component V_0 is added vectorially to the radial velocity V_R (due to the source) to give the resultant velocity vector V. The streamline at B is tangent to the velocity vector V. Extension of this vector addition to other points in the field gives a series of streamlines as shown in Fig. 12-2(b). A solid body can be substituted for any of the streamlines because no fluid crosses a streamline. In Fig. 12-2(b) the boundary C has been selected as a body. This superposition of a rectilinear flow on a source flow gives a resultant flow, the flow around the *half-streamline* body C. Various shapes of bodies can be obtained by using different values of the velocity V_0 and the source strength Q.

The flow around the oval in Fig. 12-3 is obtained by combining a rectilinear flow V_0 (along the line AB) with a source at A and a sink at B. The discharge rate Q from the source exactly equals the input rate to the sink. Different oval shapes can be obtained by employing different values of Q, V_0, and distances AB. If the point A is brought closer to B, the oval approaches a circle as its limit, like that shown in Fig. 12-4. This limiting combination of a source and a sink is a *doublet*.

Various graphical techniques can be developed for expediting the vector addition, for plotting streamlines, and for obtaining a picture of the flow. For computation purposes involving a large class of fluid motions, the veloc-

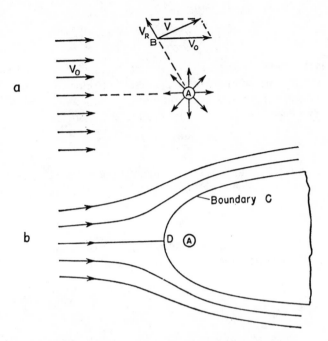

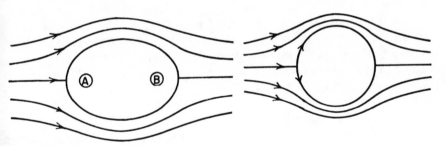

Fig. 12–2. Flow around a half-streamline body.

Fig. 12–3. Two-dimensional frictionless flow around an oval.

Fig. 12–4. Two-dimensional frictionless flow around a cylinder.

ity distribution can be obtained analytically by means of *potential functions.* The following articles will discuss a method for determining the vector field analytically. The general method of approach employed in this fluid flow problem is also used in other branches of applied physics as in rigid body mechanics, elasticity, electricity, and magnetism.

Potential functions have been devised for expediting the analytical investigation of velocity fields. One type of function is a velocity potential. Let Φ (Greek letter phi) represent the velocity potential; Φ is some function of x and y, and is defined such that

$$u = \frac{\partial \Phi}{\partial x} \qquad v = \frac{\partial \Phi}{\partial y} \tag{12-2}$$

where u is the velocity component of the resultant velocity in the x-direction and v is the velocity component of the resultant velocity in the y-direction. We will consider functions such that differentiation with respect to two variables is independent of the order of differentiation; that is,

$$\frac{\partial}{\partial x}\left(\frac{\partial \Phi}{\partial y}\right) = \frac{\partial}{\partial y}\left(\frac{\partial \Phi}{\partial x}\right) \tag{12-3}$$

Thus Equation (12–3) imposes a restriction upon the type of flow that can be described by the potential function. This restriction is

$$\frac{\partial u}{\partial y} = \frac{\partial v}{\partial x}$$

The use of the potential function Φ is limited to irrotational motion; this type of motion will be discussed further in the next article.

12–3. Types of motion

A fluid element may undergo four types of movement: (1) a pure translation, (2) a linear deformation, (3) an angular or shearing deformation, and (4) a rotation. The solid lines in Fig. 12–5(a) represent an infinitesimal fluid element at a certain instant. The dotted lines represent the element for a pure translation; there is no change in the lengths of parallel sides. The dotted lines in Fig. 12–5(b) represent a condition after linear deformation has taken place. Figure 12–6(b) indicates a rotation of the element from the position initially shown in Fig. 12–6(a). The dotted diagonal lines in Fig. 12–6 have turned through an angle.

Figure 12–7(b) indicates the position of an element after an angular or shearing deformation has taken place; the element was initially in the position shown in Fig. 12–7(a). During a time interval dt the infinitesimal element line initially along the x-axis has experienced an angular change equal to $(\partial v/\partial x)\, dt$, whereas the element line initially along the y-axis has experienced

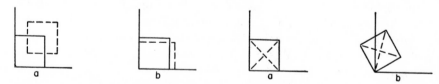

Fig. 12–5. Translation and linear deformation of a fluid element.

Fig. 12–6. Rotation of a fluid element.

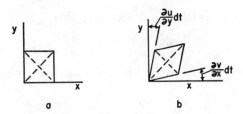

Fig. 12–7. Angular or shearing deformation of a fluid element.

an angular change equal to $(\partial u/\partial y)\, dt$. The *mean rotation* of an element is equal to the average or

$$\frac{1}{2} \left(\frac{\partial v}{\partial x} - \frac{\partial u}{\partial y} \right)$$

If $\partial v/\partial x = \partial u/\partial y$, then the rotation is zero. The sides of the element may undergo an angular deformation, but the diagonals of the infinitesimal element have not rotated, and the motion is classed as irrotational.

12–4. Velocity potential and stream function

The remainder of this chapter will be devoted to irrotational motion. Some examples and preliminary features will be given first.

It was pointed out that the velocity potential Φ is a function of x and y. The feet and second units will be employed in presenting a few examples. If the velocity potential for a certain flow is $\Phi_1 = 10x$, then $u = 10$ feet per second ($u = \partial \Phi/\partial x$). The velocity is 10 feet per second in the x-direction for all points in the field of flow. If the velocity potential for another flow is $\Phi_2 = 20y$, then $v = 20$ feet per second. The velocity at all points is 20 feet per second in the y-direction. These two flows can be superimposed or combined mathematically by simply adding the two potential functions. The *algebraic* addition of two potential functions is equivalent to a *vector* addition of the velocities. The potential function

$$\Phi = \Phi_1 + \Phi_2 = 10x + 20y$$

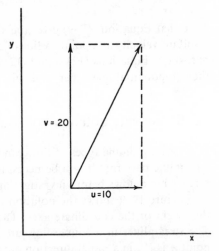

Fig. 12–8. Vector addition of velocities.

represents the combined flow pattern and is illustrated in Fig. 12–8 for one point. This simple example illustrates the convenience of potential functions.

The velocity potential Φ is defined such that

$$u = \frac{\partial \Phi}{\partial x} \qquad v = \frac{\partial \Phi}{\partial y} \tag{12-4}$$

The "stream function" Ψ (Greek psi) is a function of x and y, which is defined such that

$$u = \frac{\partial \Psi}{\partial y} \qquad v = -\frac{\partial \Psi}{\partial x} \tag{12-5}$$

For the function Φ we can put the total derivative in the form

$$d\Phi = \frac{\partial \Phi}{\partial x} dx + \frac{\partial \Phi}{\partial y} dy = u\,dx + v\,dy \tag{12-6}$$

We will define an "equipotential" line as a line along which the function Φ is constant; along this line, then, $d\Phi = 0$, and Equation (12-6) yields

$$\frac{dy}{dx} = -\frac{u}{v} \tag{12-7}$$

Equation (12-7) gives the slope of the equipotential line.

For the function Ψ we can put the total derivative in the form

$$d\Psi = \frac{\partial \Psi}{\partial x} dx + \frac{\partial \Psi}{\partial y} dy = -u\,dx + u\,dy \tag{12-8}$$

For a line of constant stream function Ψ, $d\Psi = 0$, and Equation (12-8) yields

$$\frac{dy}{dx} = \frac{v}{u} \tag{12-9}$$

Note that Equation (12-9) gives the slope of a curve which is tangent to the resultant velocity at a point, A line of constant Ψ is thus a streamline. A comparison of Equations (12-7) and (12-9) shows that at each point in the field the equipotential line is perpendicular or orthogonal to the streamline.

12-5. Velocity potential for source and sink

Certain simple types of flow can be combined to give various resultant streamline patterns. It will be necessary to derive the velocity potentials for a few simple flows before investigating combined flows.

Figure 12-9 shows the notation for the flow due to a source placed at the origin of the coordinate axes. The coordinates of point A are x and y. The streamlines in a source flow are radial lines. The resultant velocity at point A is V_R, in a radial direction outward. The x- and y-components of the resultant velocity are

$$u = V_R \cos \theta \qquad v = V_R \sin \theta \tag{12-10}$$

Let the radial distance from the origin O to point A be represented by R.

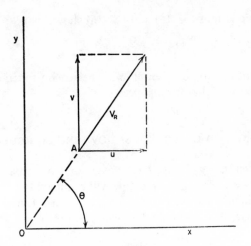

Fig. 12-9. Notation for source flow.

Assume a distance equal to unity perpendicular to the plane of flow, and let Q represent the volume of flow per unit time or the flux from the source. Then $Q = 2\pi R V_R$. Let Φ_S represent the velocity potential for the source. Then

$$u = \frac{Q}{2\pi R}\cos\theta = \frac{Q}{2\pi}\left(\frac{x}{x^2 + y^2}\right) = \frac{\partial\Phi_S}{\partial x} \qquad (12\text{-}11)$$

$$v = \frac{Q}{2\pi R}\sin\theta = \frac{Q}{2\pi}\left(\frac{y}{x^2 + y^2}\right) = \frac{\partial\Phi_S}{\partial y} \qquad (12\text{-}12)$$

Integration of Equations (12-11) and (12-12) shows that

$$\Phi_S = \frac{Q}{4\pi}\log(x^2 + y^2) \qquad (12\text{-}13)$$

which can be checked by differentiation. Since $R^2 = x^2 + y^2$,

$$\Phi_S = \frac{Q}{2\pi}\log R \qquad (12\text{-}14)$$

The potential function for a sink can be derived in a similar fashion. The velocity potential for a sink is simply the negative of that for a source.

12-6. Stream function for source and sink

Let Ψ_S represent the stream function for a source. Then, using the notation shown in Fig. 12-9, the velocity components are

$$u = \frac{Q}{2\pi}\left(\frac{x}{x^2 + y^2}\right) = \frac{\partial\Psi_S}{\partial y} \qquad (12\text{-}15)$$

$$v = \frac{Q}{2\pi}\left(\frac{y}{x^2 + y^2}\right) = -\frac{\partial\Psi_S}{\partial x} \qquad (12\text{-}16)$$

Integration of Equations (12–15) and (12–16) shows that

$$\Psi_s = \frac{Q}{2\pi} \tan^{-1} \frac{y}{x} \qquad (12\text{–}17)$$

which can be checked by differentiation. The stream function for a sink is simply the negative of that for a source.

12–7. Combination of rectilinear flow and source

As illustrated in Fig. 12–10, imagine a source placed at the origin O of the co-ordinate axes. To this source flow will be added vectorially the rectilinear flow U, which is parallel to the positive x-axis. The velocity potential for the rectilinear flow is simply Ux, whereas the stream function for the

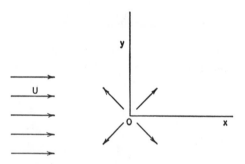

Fig. 12–10. Notation for combination of source and rectilinear flow.

rectilinear flow is simply Uy. The total velocity potential Φ for the combined flow and the total stream function Ψ for the combined flow are

$$\Phi = Ux + \frac{Q}{4\pi} \log (x^2 + y^2) \qquad (12\text{–}18)$$

$$\Psi = Uy + \frac{Q}{2\pi} \tan^{-1} \frac{y}{x} \qquad (12\text{–}19)$$

Since $u = \partial\Phi/\partial x$ and $v = \partial\Phi/\partial y$, the x- and y-components of the resultant velocity at any point in the field of flow can be found by employing Equation (12–18).

Using Equation (12–19) we can plot curves of constant Ψ values. These curves are streamlines. Figure 12–11 shows various constant Ψ curves for the combination of a source and rectilinear flow. A solid body can be substituted for any of the streamlines because no fluid crosses a streamline. For example, the line $\Psi = 0$ could be selected as a solid body. Then Equations (12–18) and (12–19) specify the flow around this half streamline body. Using Equation (12–18) we can calculate the resultant velocity at any point on the body.

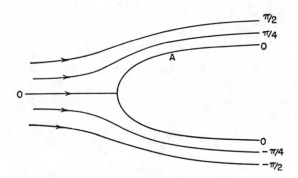

Fig. 12-11. Streamlines for a combination of source and rectilinear flow. Numerical values are values of Ψ.

As an illustration, say we calculate the resultant velocity V_A for any point A on the half-streamline body. If we know the velocity U a large distance upstream (at a large negative x-value), the corresponding upstream pressure p, and the density ρ, then the pressure p_A at a point A can be calculated by the dynamic equation

$$p_A + \tfrac{1}{2}\rho V_A^2 = p + \tfrac{1}{2}\rho U^2 \tag{12-20}$$

12-8. Doublet

Figure 12-12 shows the notation that will be used in deriving the potential function for the doublet. A source is placed at A, and a sink of equal strength (same Q) is placed at B. The distance ds is infinitesimal. The potential function Φ_D at any point P is the sum of the source and sink potential functions. Thus

$$\Phi_D = \frac{Q}{2\pi} \log R - \frac{Q}{2\pi} \log (R + dR)$$

$$\Phi_D = -\frac{Q}{2\pi} \log \left(1 + \frac{dR}{R}\right) \tag{12-21}$$

Expanding Equation (12-21) in terms of a power series gives

$$\Phi_D = -\frac{Q}{2\pi}\left[\frac{dR}{R} - \frac{1}{2}\left(\frac{dR}{R}\right)^2 \cdots\right] \tag{12-22}$$

If infinitesimals of higher order than

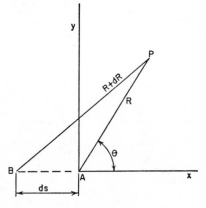

Fig. 12-12. Notation for a combination of source and sink.

the first are neglected, then

$$\Phi_D = -\frac{Q}{2\pi}\left[\frac{dR}{R}\right] = -\frac{Qds\cos\theta}{2\pi R} \qquad (12\text{-}23)$$

Imagine the distance ds to be continuously decreased while the term $Q/2\pi$ increases so that the product remains constant. Such a combination of flows at the limit is a *doublet*. Let $Qds/2\pi = F$. Then the velocity potential function for the doublet is

$$\Phi_D = -\frac{F\cos\theta}{R} = -\frac{Fx}{x^2 + y^2} \qquad (12\text{-}24)$$

Let Ψ_D represent the stream function for the doublet. Equation (12-24) and the relations for velocity components give

$$u = \frac{\partial\Phi_D}{\partial x} = -\frac{F(y^2 - x^2)}{(x^2 + y^2)^2} = \frac{\partial\Psi_D}{\partial y} \qquad (12\text{-}25)$$

$$v = \frac{\partial\Phi_D}{\partial y} = \frac{2Fxy}{(x^2 + y^2)^2} = -\frac{\partial\Psi_D}{\partial x} \qquad (12\text{-}26)$$

Integration of Equation (12-25) and (12-26) gives the stream function for the doublet

$$\Psi_D = \frac{Fy}{x^2 + y^2} \qquad (12\text{-}27)$$

which can be checked by differentiation.

12–9. Combination of rectilinear flow and doublet

It was pointed out in Sec. 12–7 that the combination of a source and a rectilinear flow gives the flow around a half-streamline body.

Assume that a doublet of the type discussed in the foregoing article is placed at the origin O of the co-ordinate axes as shown in Fig. 12–13. A uniform rectilinear flow with a velocity $-V_0$ along the x-axis is superimposed on this doublet flow. The velocity potential for the rectilinear flow is simply $-V_0 x$. The total velocity potential for the combined flow and the total stream function for the combined flow are

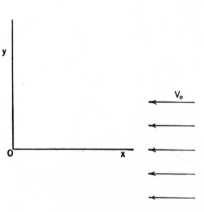

$$\Phi = -V_0 x - \frac{Fx}{x^2 + y^2} \qquad (12\text{-}28)$$

Fig. 12–13. Doublet at point O with. uniform rectilinear flow.

$$\Psi = -V_0 y + \frac{Fy}{x^2 + y^2} \qquad (12\text{-}29)$$

Using Equation (12–29) we could plot curves of constant Ψ and get a pattern of streamlines, as indicated in Fig. 12–14. Note, from Equation (12–29), that for the streamline $\Psi = 0$ we get the relation

$$0 = -V_0 y + \frac{Fy}{x^2 + y^2}$$

$$x^2 + y^2 = \frac{F}{V_0}$$

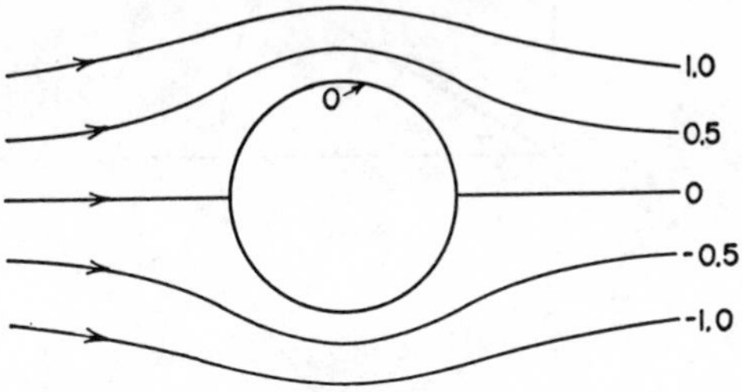

Fig. 12–14. Curves of constant Ψ for combination of rectilinear flow and doublet. Numerical values are values of Ψ.

Let the constant F/V_0 be represented by a^2. Then, for the streamline $\Psi = 0$, the foregoing equation is

$$x^2 + y^2 = a^2 \qquad (12\text{–}30)$$

Equation (12–30) is that for a circle with the radius a. Thus the combination of a rectilinear flow and a doublet gives the flow around a circular cylinder.

Noting that $F = a^2 V_0$, the total velocity potential for the combined flow and the total stream function for the combined flow become

$$\Phi = -V_0 x - \frac{x V_0 a^2}{x^2 + y^2} \qquad (12\text{–}31)$$

$$\Psi = -V_0 y + \frac{V_0 a^2 y}{x^2 + y^2} \qquad (12\text{–}32)$$

12–10. Pressure distribution around a circular cylinder

In order to determine the pressure at points along the surface of the cylinder, it is necessary to find the resultant velocity V_s at any point at the surface of the cylinder. Let θ be the angle between the radial line to any point A and the x-axis; the notation is shown in Fig. 12–15. At the cylinder surface, $x = a \cos \theta$ and $y = a \sin \theta$. The magnitude of the resultant velocity V_s is given by the relation

$$V_s = \sqrt{u_s^2 + v_s^2} \qquad (12\text{–}33)$$

where u_s and v_s are the x- and y-components, respectively, of the velocity

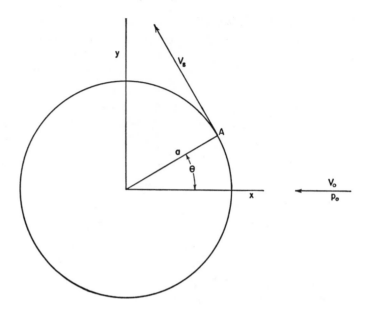

Fig. 12–15. Notation for cylinder in potential flow.

at the cylinder surface; these components can be calculated by using Equation (12–32). The final result is simply

$$V_s = 2V_0 \sin \theta \qquad (12\text{--}34)$$

Note that the velocity V_s is a maximum when θ is 90 degrees.

The velocity some distance ahead of the cylinder is V_0. Let the static pressure in the undisturbed stream ahead of the cylinder be represented by p_0. Let p be the static pressure at any point A on the surface of the cylinder. Applying the dynamic equation between a point in the undisturbed stream and the point A gives

$$p + \rho \frac{V_s^2}{2} = p_0 + \rho \frac{V_0^2}{2} \qquad (12\text{--}35)$$

Since $V_s = 2V_0 \sin \theta$, Equation (12–35) becomes

$$p = p_0 + \rho \frac{V_0^2}{2}(1 - 4 \sin^2 \theta) \qquad (12\text{--}36)$$

Thus the pressure at any point on the surface of the cylinder can be calculated by Equation (12–36).

12–11. Line integral and circulation

In a *vector field* there is a vector at each point in the field or region. Imagine the x–y plane as a vector field. Further, picture some path or curve in this field extending from point A to point B. The *line integral L* is defined as the integral of the vector component tangent to the path times the differential path length, such as

$$L = \int_A^B F \cos{(F, s)} \, ds \qquad (12\text{-}37)$$

where F is the vector at a point on the curve. (F, s) is the angle between F and the path tangent, and ds is the differential length along the path. The line integral is not to be confused with the area under the curve.

In many fluid-flow problems a particular type of line integral is important, namely a line integral in a velocity field. Figure 12-16 shows a closed contour arbitarily drawn in a field of flow. Let V be the resultant velocity at any point, and θ the angle between V and the contour element ds. The

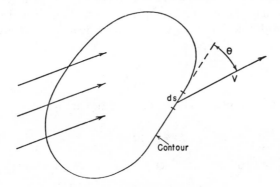

Fig. 12-16. Contour line in a field of flow.

line integral of the velocity is the integral of the product contour element ds and the component of the velocity in the direction of ds. The *circulation* Γ is defined as the line integral of the velocity around a closed contour line in the field; it is represented by the integral

$$\Gamma = \int_c V \cos \theta \, ds \qquad (12\text{-}38)$$

The vector field may be a force field, as a gravitational force field. The term "work," as a combination of force times distance, is an example of a line integral. An analogy might be drawn between work along a closed path and circulation; the force in the work integral corresponds to the velocity in the circulation integral, whereas the path element in one integral corresponds to the path element in the other.

In some cases it is convenient to form the line integral by studying an infinitesimal element of sides dx and dy. The diagonal is ds. Instead of following the path ds, let us form the velocity and distance combination along the sides separately, as

$$dL = V \cos \theta \, ds = u\,dx + v\,dy \qquad (12\text{-}39)$$

The velocity potential Φ is defined by the relations

$$u = \frac{\partial \Phi}{\partial x} \qquad v = \frac{\partial \Phi}{\partial y}$$

Thus the line integral of the velocity can be written as

$$dL = \frac{\partial \Phi}{\partial x} dx + \frac{\partial \Phi}{\partial y} dy = d\Phi$$

$$\int_A^B udx + vdy = \Phi_B - \Phi_A \tag{12-40}$$

The line integral along the path from A to B can be interpreted as a change in velocity potential. Note also that the difference in velocity potential is solely a function of point or position and not a function of the path between points.

Let us form the line integral completely around the sides of the infinitesimal element with sides dx and dy. Along one side dx the velocity is u. Along the parallel side the velocity is

$$u + \frac{\partial u}{\partial y} dy$$

Likewise, for the parallel dy sides, the velocities are v and

$$v + \frac{\partial v}{\partial x} dx$$

Thus the line integral around a series of infinitesimal elements, or the circulation, is

$$\int_c d\Phi = \int_c udx + \left(v + \frac{\partial v}{\partial x}\right) dy - \left(u + \frac{\partial u}{\partial y} dy\right) dx - vdy$$

$$= \int\int_c \left(\frac{\partial v}{\partial x} - \frac{\partial u}{\partial y}\right) dxdy \tag{12-41}$$

The "curl" of the velocity vector is defined as the combination of terms

$$\frac{\partial v}{\partial x} - \frac{\partial u}{\partial y} \tag{12-42}$$

The curl is twice the mean rotation. If the curl or mean rotation is zero, then the circulation is zero;

$$\frac{\partial v}{\partial x} = \frac{\partial u}{\partial y}$$

and the motion is classed as irrotational.

Picture a free or irrotational flow around a circular cylinder of radius a as illustrated in Fig. 12–17. The term "free or irrotational" vortex is used to distinguish from the case of a fluid rotating as a solid body. At the radius a the fluid velocity is V_p. At the variable radius r the fluid velocity is V. From the free or irrotational vortex relation, $V_p a = Vr =$ the constant K. The

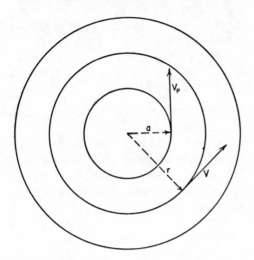

Fig. 12–17. Circulation around a circular cylinder.

surface of the cylinder will be selected first as the contour for the calculation of the circulation. Then

$$\Gamma = 2\pi a V_p = 2\pi V r = 2\pi K$$

Any other arbitrarily selected contour would give the same result. Thus the circulation around the center of a free or irrotational vortex is constant and independent of the contour selected.

12–12. Flow around a cylinder with a free vortex

The word *lift* means a force at right angles to the line of undisturbed flow. The flow around the cylinder discussed in Secs. 12–9 and 12–10 is a combination of a doublet and a uniform rectilinear flow. As illustrated in Fig. 12–14, such a flow is symmetrical with respect to the line of undisturbed motion (the x-axis). For one thing, there is no lift on a cylinder in such a flow. A dynamic lift can be developed if the streamline pattern is nonsymmetrical with respect to the line of undisturbed motion. A nonsymmetrical flow can be obtained by adding a free vortex to the doublet and rectilinear flow.

Let V_p represent the velocity at the surface of the cylinder due to a counterclockwise free vortex alone. Then the circulation Γ around the cylinder is given by the relation

$$\Gamma = 2\pi a V_p$$

Thus the magnitude of the resultant or total velocity V_T at the surface of the cylinder is

$$V_T = V_s + V_p = 2V_0 \sin \theta + \frac{\Gamma}{2\pi a} \qquad (12\text{–}43)$$

Applying the dynamic equation between a point in the undisturbed stream and a point on the cylinder gives

$$p_0 + \tfrac{1}{2}\rho V_0^2 = p + \tfrac{1}{2}\rho V_T^2$$

$$p = p_0 + \frac{1}{2}\rho V_0^2 - \frac{\rho}{2}\left(2V_0 \sin\theta + \frac{\Gamma}{2\pi a}\right)^2 \qquad (12\text{-}44)$$

The static pressure p at any point on the surface of the cylinder can be calculated by Equation (12-44).

The next step is to calculate the net lift force at right angles to the undisturbed motion. This lift force can be determined by integrating the pressure forces over the cylinder in the y-direction. The static pressure p acts normal to the surface of the cylinder. Let $d\theta$ be the differential change in the angle θ; the notation is illustrated in Fig. 12-18. Assume a distance equal to unity, and perpendicular to the plane of flow. The pressure force on the element of area $a\,d\theta$ is $pa\,d\theta$. The component of this force at right angles to V_0 is $pa\,d\theta \sin\theta$. The lift on the elementary area is $-pa\,d\theta \sin\theta$; the minus sign indicates the direction of the force. The total lift L on the cylinder is

$$L = -\int_0^{2\pi} pa \sin\theta\, d\theta \qquad (12\text{-}45)$$

The relation for p as given by Equation (12-44) can be substituted in Equation (12-145) and the resulting expression integrated. The final result is

$$L = \rho\Gamma V_0 \qquad (12\text{-}46)$$

The lift force per unit length of cylinder equals the product of density, circula-

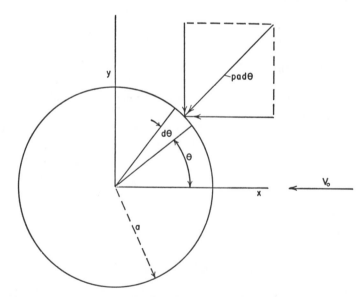

Fig. 12-18. Notation for lift force on a cylinder.

tion, and the undisturbed velocity. Equation (12–46) is called the Kutta-Joukowski relation.

12–13. Vortex and source

Three simple types of flow are used to form combinations: the rectilinear flow; the source and sink flow; and the vortex flow. The velocity potential and stream functions have been presented for the rectilinear flow and the radial flow; there remains the case for the vortex.

As illustrated in Fig. 12–19, let V represent the resultant velocity at some radius R in the free-vortex flow. The x- and y-components of V are u and v respectively. By using the constant value of the circulation Γ for the vortex, the velocity V can be arranged in the form

$$V = \frac{\Gamma}{2\pi R}$$

Let Φ represent the velocity potential and Ψ the stream function. Then the following relations hold:

$$u = -V \sin \theta = -\frac{\Gamma}{2\pi} \frac{y}{(x^2 + y^2)} = \frac{\partial \Phi}{\partial x} = \frac{\partial \Psi}{\partial y} \qquad (12\text{–}47)$$

$$v = V \cos \theta = \frac{\Gamma}{2\pi} \left(\frac{x}{x^2 + y^2} \right) = \frac{\partial \Phi}{\partial y} = -\frac{\partial \Psi}{\partial x} \qquad (12\text{–}48)$$

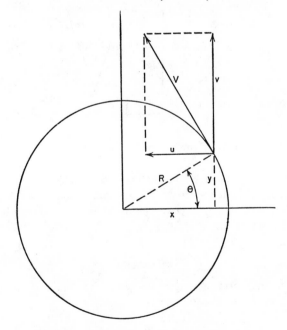

Fig. 12–19. Notation for vortex flow.

Integration of Equations (12–47) and (12–48) yields

$$\Phi = \frac{\Gamma}{2\pi} \tan^{-1} \frac{y}{x} \qquad (12\text{–}49)$$

$$\Psi = -\frac{\Gamma}{4\pi} \log (x^2 + y^2) \qquad (12\text{–}50)$$

which can be checked by differentiation.

The following relations illustrate a similarity in form

source $\Phi_s = \dfrac{Q}{4\pi} \log (x^2 + y^2)$ source $\Psi_s = \dfrac{Q}{2\pi} \tan^{-1} \dfrac{y}{x}$

vortex $\Psi = -\dfrac{\Gamma}{4\pi} \log (x^2 + y^2)$ vortex $\Phi = \dfrac{\Gamma}{2\pi} \tan^{-1} \dfrac{y}{x}$

The velocity potential for the source is similar to the stream function for the vortex; the stream function for the source is similar to the velocity potential for the vortex.

In some studies of fluid machinery such as the flow through impellers of centrifugal machines, attention is given to the combination of a source and vortex. Imagine a source at the origin of the coordinate axes, as illustrated in Fig. 12–20. Added to the source flow is a free vortex. At the point A, for

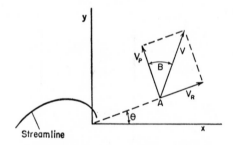

Fig. 12–20. Combination of vortex and source.

example, the resultant velocity V consists of two components: V_R, that due to the source, and V_P, that due to the vortex. Assume unit distance perpendicular to the plane of flow. Then, if Q represents the source strength or volume rate of flow, V_R is given by the relation.

$$V_R = \frac{Q}{2\pi R} \qquad (12\text{–}51)$$

Let Γ represent the vortex strength or circulation. Then V_P is given by the relation

$$V_P = \frac{\Gamma}{2\pi R} \qquad (12\text{–}52)$$

The angle B is given by the relation

$$B = \tan^{-1} \frac{V_R}{V_P} = \tan^{-1} \frac{Q}{\Gamma} \tag{12-53}$$

For a constant value of Γ and a constant value of Q, the angle B is constant for all values of R.

The stream function Ψ for the combination of a vortex and source is

$$\Psi = \frac{Q}{2\pi} \tan^{-1} \frac{y}{x} - \frac{\Gamma}{4\pi} \log (x^2 + y^2) \tag{12-54}$$

$$\Psi = \frac{Q}{2\pi} \theta - \frac{\Gamma}{2\pi} \log R \tag{12-55}$$

The equation for a streamline is obtained by taking constant values for Ψ. Then the equation of a streamline has the form

$$\frac{Q}{2\pi} \theta - \frac{\Gamma}{2\pi} \log R = \text{constant}$$

$$R = Ce^{(Q/\Gamma)\theta} \tag{12-56}$$

where C is a constant. Equation (12-56) is the equation of a logarithmic spiral; a streamline is indicated in Fig. 12-20.

12-14. Properties of velocity and stream functions

The foregoing articles bring out certain features, primarily from a physical point of view. A number of further analytical features may be helpful in understanding the type of flow under consideration.

The general relations for frictionless flow are given in Chapter Eleven. For steady two-dimensional flow, the general continuity equation and the general dynamic relations reduce to the special forms

$$\frac{\partial u}{\partial x} + \frac{\partial v}{\partial y} = 0 \tag{12-57}$$

$$P\rho - \frac{\partial p}{\partial y} = \rho \frac{dv}{dt} \tag{12-58}$$

$$Q\rho - \frac{\partial p}{\partial x} = \rho \frac{du}{dt} \tag{12-59}$$

The velocity potential Φ is defined by the relations

$$u = \frac{\partial \Phi}{\partial x} \qquad v = \frac{\partial \Phi}{\partial y}$$

Then the continuity Equation (12-56) can be arranged as

$$\frac{\partial^2 \Phi}{\partial x^2} + \frac{\partial^2 \Phi}{\partial y^2} = 0 \tag{12-60}$$

An equation in the form of Equation (12-60) is the Laplace equation. The various velocity potentials discussed in this chapter are solutions of the Laplace equation.

Let the resultant velocity at any point in the flow be V; this velocity has the components u and v. Neglecting body forces, the dynamic Equations (12–58) and (12–59) become

$$-\frac{\partial p}{\partial x} = \rho \frac{du}{dt} \qquad (12\text{–}61)$$

$$-\frac{\partial p}{\partial y} = \rho \frac{dv}{dt} \qquad (12\text{–}62)$$

Multiplying both sides of Equation (12–61) by dx, multiplying both sides of Equation (12–62) by dy, adding the resultant equations, and integrating gives the familiar result

$$p + \tfrac{1}{2}\rho V^2 = \text{constant}$$

Note that the problem of irrotational motion of a frictionless incompressible fluid is more or less a problem of pure geometry. The pressure does not enter the problem directly, because in the case of an incompressible fluid the pressure is not related to the physical state of the fluid.

PROBLEMS

12-1. The velocity potential for a certain flow is $(x^2 - y^2)$. Give the general equation of the streamlines in terms of x, y, and some arbitrary constant A.

12-2. At what points on the surface of a cylinder in potential flow with no circulation is the static pressure p equal to p_0?

12-3. For the potential flow around a cylinder with no circulation, plot a diagram of the dimensionless ratio $(p - p_0)/(\tfrac{1}{2}\rho V_0^2)$ against θ.

12-4. A source at the origin and a rectilinear flow in the positive x-direction give the flow around a half-streamline body. The stagnation point on the body is one foot from the origin of the coordinate system. Consider a straight path from point A at $x = -2$ feet, $y = 0$ to point B at $x = 0$, $y = 3$ feet. What is the line integral of the velocity along the path A to B? Assume standard air approaches the body with a velocity of 110 feet per second.

12-5. Water approaches a circular cylinder of 24 inches in diameter, with a velocity of 14 feet per second. The center of the cylinder is at the origin of the coordinate axes. A free vortex with a circulation of 44 in units of feet and seconds is superposed on the flow around the cylinder. At what points on the surface of the cylinder is the gage pressure zero? What is the lowest gage pressure on the surface of the cylinder?

12-6. A circular cylinder, 12 inches in diameter, has its center at the origin of the co-ordinate axes. Standard air approaches the cylinder with a velocity of 120 feet per second parallel to the x-axis. Point A is along the x-axis 12 inches away from the surface of the cylinder. What is the gage pressure at A?

12-7. Standard air flows around a circular cylinder with its center at the origin of the co-ordinate axes. The undisturbed velocity is 50 feet per second; the diameter of the cylinder is 2.0 feet. What is the magnitude of the circulation in order to have the stagnation point on the cylinder at $x = 0$? What is the gage pressure at this point?

12-8. Standard air flows between two concentric circular streamlines. The radius of one is 2 feet; the radius of the other is 3 feet. The pressure difference between the streamlines is 0.033 pounds per square inch. What is the value of the circulation?

12-9. Starting with Equation (12-45), derive Equation (12-46).

Selected
References

Prandtl, L. and O. G. Tietjens, *Fundamentals of Hydro- and Aeromechanics*. New York: McGraw-Hill, 1934.

Prandtl, L. and O. G. Tietjens, *Applied Hydro- and Aeromechanics*. New York: McGraw-Hill, 1934.

Lamb, H., *Hydrodynamics*. New York: Cambridge University Press, 1932.

Aerodynamic Theory, ed. W. F. Durand, Vols. I, II, III, IV, V, VI. Pasadena, California: Durand Reprinting Committee, California Institute of Technology, 1943.

Modern Developments in Fluid Dynamics, ed. S. Goldstein, Vols. I, II. New York: Oxford University Press, 1943.

Fundamentals of Gas Dynamics, ed. H. W. Emmons, Vol. III. Princeton, New Jersey: Princeton University Press, 1958.

Oswatitsch, K., *Gas Dynamics*. New York: Academic Press, 1956.

Milne-Thomson, L. M., *Theoretical Hydrodynamics*. New York: Macmillan, 1950.

von Kármán, T., *Aerodynamics*. Ithaca, New York: Cornell University Press, 1954.

Hoerner, S. F., *Fluid Dynamic Drag*. Midland Park, New Jersey: 1958.

Wilson, Warren E., *Positive-Displacement Pumps and Fluid Motors*. New York: Pitman, 1950.

Wislicenus, George F., *Fluid Mechanics of Turbomachinery*. New York: McGraw-Hill, 1947.

Collected Works of Theodore von Kármán, Vols. I, II, III, IV. London: Butterworth's Scientific Publications, 1956.

Fluid Meters, American Society of Mechanical Engineers. New York 1959.

386

Handbook of Noise Control, ed. C. M. Harris. New York: McGraw-Hill, 1957.

Liepmann, H. W. and A. Roshko, *Elements of Gas Dynamics*. New York: Wiley, 1957.

Stepanoff, A. J., *Turboblowers*. New York: Wiley, 1955.

Langhaar, H. L., *Dimensional Analysis and Theory of Models*. New York: Wiley, 1951.

Bakhmeteff, B., *Hydraulics of Open Channels*. New York: McGraw-Hill, 1932.

Jakob, Max, *Heat Transfer*. New York: Wiley, 1949.

Hermann, R., *Supersonic Inlet Diffusers and Introduction to Internal Aerodynamics*. Minneapolis, Minnesota: Minneapolis-Honeywell Regulator Co., 1956.

Schlicting, H., *Boundary Layer Theory*. New York: McGraw-Hill, 1955.

Wilson, E. B., *Vector Analysis*, founded upon the lectures of J. W. Gibbs. New York: Scribner, 1907.

Purday, H. F. P., *An Introduction to the Mechanics of Viscous Flow*. New York: Dover Press, 1949.

Equations, Tables, and Charts for Compressible Flow, NACA Technical Report 1135. 1953.

Humphreys, W. J., *Physics of the Air*. New York: McGraw-Hill, 1940.

Creager, W. P., J. D. Justin, and J. Hinds, *Engineering for Dams*. New York: Wiley, 1945.

Bingham, E. C., *Fluidity and Plasticity*. New York: McGraw-Hill, 1922.

Abbott, I. H. and H. E. von Doenhoff, *Theory of Wing Sections*. New York: McGraw-Hill, 1949.

Kuchemann, D. and J. Weber, *Aerodynamics of Propulsion*. New York: McGraw-Hill, 1953.

Answers

to Problems

2-18. 0.216 inch of water

2-19. 437 inches

2-20. 4710 pounds; 3930 pounds; 2950 pounds; 3.33 feet below free surface

2-21. $2/3h$

2-22. 1635 pounds; 8500 pounds; 3.0 feet below free surface

2-23. 5 feet

2-24. 4915 pounds

2-25. 40.7 pounds

2-26. 34,500 pounds, 0.346 foot below center of disk in plane of disk

2-27. 5820 pounds

2-28. 716 pounds

2-29. $R_x = 10,600$ pounds at 17.5 feet below free surface; $R_y = 12,380$ pounds at a distance of 4.7 feet from point A

2-30. 31,400 pounds

2-31. $R_x = 15,720$ pounds at 21.57 feet below free surface; $R_y = 14,352$ pounds at a distance of 4.56 feet from point A

2-32. 3.03 pounds

2-33. 842 pounds

2-34. 7770 tons

2-35. 78,624 foot-pounds

2-36. 7840 foot-pounds

2-37. 27,720 pounds at a depth of 6.74 feet

2-38. 3950 foot-pounds

2-39. 3120 pounds at 0.33 foot from bottom

2-40. 11,150 pounds

2-41. 48,600 pounds

2-42. 720 pounds at 4 feet below top

2-43. 2910 foot-pounds

2-44. 0.449 cubic foot; 3.215

2-45. 128.6 long tons

2-46. 11,700 pounds

2-47. 3.21 feet; 369 tons

2-48. 4310 pounds at 2.62 feet below free surface

2-49. 0.083

2-50. 67.0 tons; 1,850,000 cubic feet

2-51. 6.41 tons

2-55. No

2-56. 53.4 pounds

2-57. 33.8 pounds

2-60. 1027 feet

2-61. (*a*) 8.34 pounds per square inch
(*b*) 9.53 pounds per square inch
(*c*) 9.26 pounds per square inch

2-62. 7.86 pounds per square inch

2-63. Up

2-64. 10.10 pounds per square inch absolute

2-65. 303 degree Kelvin (absolute on centigrade scale)

2-66. 6170 feet

2-67. 16.1 feet

2-68. 47 degrees Fahrenheit at 3820 feet

2-69. 31.90 inches of mercury

3-1. 12-inch section $V = 3.75$ feet per second; 10-inch section $V = 5.4$ feet per second

3-2. 8-inch section $V = 2.81$ feet per second; 14-inch section $V = 1.17$ feet per second

3-3. $V = 2/3V_m$

3-4. 4.4 feet per second; 43.1 gallons per minute

3-5. 20 feet per second

3-6. 0.0785 cubic foot per second; 0.073 foot per second

3-7. 7.85 cubic feet per second

3-8. 6.55 feet per second

3-9. 76 feet per second

3-10. 18.6 pounds

3-11. 25.2 pounds

3-12. 24.2 pounds

3-13. 650 pounds

3-14. 11,250 pounds

3-15. 662 pounds

3-16. 0.093 cubic foot per second; 0.0159 foot per second

3-17. 56.3 degrees

3-18. 200 pounds

3-19. 24.97 pounds per cubic foot

3-20. 20.8 pounds per square inch absolute; 90.5 degrees Rankine

3-21. 1042 pounds

3-22. 154 pounds

3-24. 12.4 feet

3-25. 8.34 feet per second

3-26. 10.7 feet per second per second

3-27. 4.12 feet; 1.88 feet; 1675 pounds

3-28. 2.12 inches

3-29. 13.4 cubic feet

3-30. 27.3 pounds per square foot

3-31. 410

3-32. 27.8 radians per second; 1.64 pounds per square inch gage

3-33. 115 pounds per square inch

3-35. 35 inch-pounds

3-36. 2.02 pounds per square inch

3-37. 0.1136 pound per square inch

3-38. 385 pounds

3-39. 1325 pounds

3-41. 14.98 pounds per square inch absolute

3-42. 14.5 feet per second

3-43. 251 cubic feet per second

3-44. 7.55 cubic feet per second

3-45. 0.0052 foot

3-46. 29.91 inches of mercury

3-47. 30.4 horsepower

3-48. 50.12 Btu added per pound

3-49. 18.6 Btu per pound

3-51. 2.06 Btu per pound

3-52. 47.6 pounds per minute

3-53. 2670 feet per second

3-54. 61,600 foot-pounds per pound

3-55. 110.6 degrees Fahrenheit

3-57. 17,280 foot-pounds

3-58. 0.216 pound per square inch

3-59. From 10-inch to 5-inch

3-60. 1.94 pounds per second

3-61. 818 pounds

3-62. 1.16 degrees Fahrenheit

3-63. 7950 horsepower

3-64. 3.73 pounds per square inch below atmospheric

3-65. 39.5 feet

3-67. 14 feet

3-68. 4.84 feet

3-69. 113.5 feet per second

3-70. $-155,500$ foot-pounds per pound

3-71. 105.5 horsepower

3-72. 235.5 horsepower

3-73. 203 degrees Fahrenheit

3-74. 72.3 Btu per pound

3-75. -10.5 Btu per pound

3-76. -49.6 horsepower

3-77. -25.7 Btu per pound

3-78. 24.3 horsepower

3-79. 755 feet per second

3-80. 120.6 Btu

3-81. 107,300 foot-pounds

3-82. 0.314 horse power

4-1. $c = \text{constant} \sqrt{p/\rho}$

4-2. $R/\rho V^2 l^2$; $\rho V l/\mu$

4-3. $R/\rho V^2 b^2$; h/b; $\mu/\rho V b$

4-4. $\dfrac{Q}{D^3 N}$; $\dfrac{\rho N D^2}{\mu}$; $\dfrac{\mu N D}{\sigma}$; $\dfrac{\gamma D}{\mu N}$

4-5. l/D; V_1/V_2

4-6. $h^2 g \rho/t$; $R^2 g \rho/t$

4-7. 40 feet per second

4-8. 7.85 feet per second

4-9. 70.7 miles per hour

4-10. 95 knots; 1210 knots

4-11. 20 atmospheres

4-12. 2.32 revolutions per minute

4-13. 60 knots

4-14. 218 revolutions per minute

4-15. 2 knots

4-16. 432 miles per hour

4-17. 0.209 foot per second

4-18. 50,000

4-19. 1660 miles per hour

4-20. $V^2/D^2 N^2$

4-21. V/ND; $p/\rho(ND)^2$

4-22. $\mu V/pl$

4-23. $\rho l g/p$

4-24. l/D; $\mu/\rho V D$

4-26. $H\mu/Q\rho$; $H^3\sigma/Q^2\rho$

4-27. $\rho V^2 l/\sigma$; $\mu V/\sigma$

4-28. $\mu V d/\rho$; ρ/ρ_1; $\rho V^2 d/\sigma$

4-29. $\dfrac{C}{D}$; $\dfrac{D}{d}$; $\dfrac{p}{\rho V^2}$; $\dfrac{l}{\rho V}$

5-1. 126.5 pounds per square inch

5-2. 3.38 pounds per square inch

5-3. 1.29 pounds per square inch

5-4. 1.14 horsepower

5-5. 4.66 pounds per square inch

5-6. 1.39 pounds per square inch

5-7. 0.53 pound per square inch

5-8. 52.5 pounds per square inch

5-9. 58.6 horsepower

5-10. 30.4 horsepower

5-11. 417 pounds per square inch

5-12. 27.1 pounds per square inch absolute

5-13. Circular

5-14. 0.354

5-19. 106 pounds per square inch

5-20. 17.3 pounds per square inch

5-21. 29.3 pounds per square inch

5-22. 49.1 horsepower

5-23. 0.029

5-24. 280 pounds per square inch

5-25. 10.4 pounds per square inch

5-26. 3.48 pounds per square inch

5-27. 7740

5-28. 0.6 horsepower

5-29. $E = \dfrac{2}{\dfrac{A_2}{A_1} + 1}$

5-30. 25 per cent

5-31. 6 inches

5-32. 41.9 cubic feet per second

5-33. 0.71

5-34. 0.015 foot per second

5-35. 37.6×10^{-4} slugs per foot-second

5-36. Up; up; down

5-37. 0.64 pound

5-38. 4.81 pounds

5-39. 4.92 pounds; 4010 pounds

5-40. 13.08 horsepower; 1.63 horse-power

5-41. 76,300 pounds

5-42. 89 horsepower

5-43. 145.3 horsepower

5-44. 1150 pounds

5-45. 2215 pounds

5-46. 0.0084 horsepower

5-49. 78.7 degrees

5-50. 101,000,000 pound-feet

5-51. 0.098 pound

5-52. 0.16 horsepower

5-53. 138 pound-feet

5-54. 0.0094

5-55. 8.34 pounds

5-56. 202 pounds

5-57. 0.64; 0.15

6-1. 1.12 seconds

6-2. 1500 pounds per square inch

6-3. 0.91

6-4. 1058 feet per second; 1004 feet per second

6-5. 3440 feet per second

6-6. 2.5 for rocket; 0.85 for plane

6-7. 17.97 pounds per square inch absolute: 17.78 pounds per square inch absolute

6-8. 7.35 pounds per square inch gage; 123.5 degrees Fahrenheit

6-9. At A, $M = 0.40$; at B, $M = 1.07$

6-10. 2386 feet per second

6-11. 1.55 pounds per second

6-12. 1.41

6-15. 0.94

6-16. 880 feet per second

6-17. 64.0 pounds per square inch absolute

6-18. One

6-19. 4550 feet per second

6-20. 4280 feet per second

6-21. 0.38; 0.80

6-22. 143 degrees Fahrenheit; 10.2 pounds per square inch gage

6-23. 0.5; 0.96

6-24. 52.6 pounds per square inch absolute

6-25. 0.488

6-27. 301 degrees Fahrenheit

6-28. 42.400 foot-pounds per pound

6-31. 7.13 pounds per second

6-32. $c^2 \rho l / \sigma$

6-33. $p / \rho c^2$; $\mu V / \rho l c^2$

6-34. 23.0 pounds per second

6-35. 735 feet per second

6-36. 435 feet per second

6-37. 349 degrees Rankine

6-38. 2.22 pounds per square inch absolute

6-39. 0.452; 604 feet

6-40. 169 pounds per square inch absolute, 131 pounds per square inch absolute

6-41. 1390 Btu per second; 2440 pounds

6-42. 0.384 pounds per square foot; 1070 pounds

6-43. 476 degrees Fahrenheit; 368 degrees Fahrenheit

6-44. 0.606

7-1. 28.1 feet per second

7-2. 3.6 cubic feet per second

7-4. 1.17 cubic feet per second

7-5. 0.85 cubic foot per second

7-6. $C_v = 0.90$; $C_c = 0.65$

7-7. 0.060 cubic foot per second

7-8. 1.36 cubic feet per second

7-9. 4.76 inches

7-10. 8.86×10^{-5} slug per foot-second

7-11. 4.23×10^{-4} slug per foot-second

7-12. 0.0028 pound per square inch gage

7-13. 12.2 cubic feet per second

7-14. 4.36 pounds

7-15. 3210 pounds

7-16. 111 feet per second

7-17. 4 inches of water

7-18. 2.97 inches

7-19. 10.1 cubic feet per second

7-21. 13.9 pounds per second

7-22. 46.5 pounds per second

8-1. 37.1 horsepower

8-2. 1.34 cents

8-3. 1218 gallons per minute; 87.0 feet

8-4. 518 gallons per minute; 36.3 feet

8-5. 1695

8-7. Single-suction centrifugal

8-8. Axial flow

8-9. 0.42 horsepower

8-10. 82.3 per cent

8-11. $\dfrac{Q}{(gH)^{1/2}D^2}$; $\dfrac{ND}{(gH)^{1/2}}$; $\dfrac{\mu}{\rho D(gH)^{1/2}}$

8-13. 6510 revolutions per minute

8-18. 99 feet per second

8-19. Forward; 58.9 degrees

8-20. 133 feet per second

8-21. 285 horsepower

8-22. 27.6 feet

8-23. 0.0324 pound per square inch

8-24. 63.5 foot-pounds per pound

8-25. 18.2 feet per second

8-26. 15.2 feet per second

8-27. 17.6 feet

8-28. 31.1

8-29. 0.885

8-30. Impulse

8-31. $\gamma^{1/2}g^{-3/4}$

8-32. Axial flow

8-33. 431 foot-pounds per pound

8-35. 164 horsepower; 219 horsepower

8-36. 93.5 per cent; 797 horsepower

8-37. 0; 33.3 horsepower

8-40. 3120 horsepower

8-41. 80 per cent; 1239 feet per minute

8-42. 1075 feet per second

8-43. 377 pounds

8-44. 1387 horsepower

8-45. 7.85 feet per second; 18 degrees 54 minutes

8-46. 1350 horsepower; $39\frac{1}{2}$ degrees; 84.4 per cent

8-47. 0.061 pound

8-48. 1590 pounds

9-1. 2880 feet per second

9-2. (a) 232 psi, (b) 380 psi.

9-3. 196 feet, 178 feet, 205 feet.

10-1. 33.6 cubic feet per second

10-2. 0.000084

10-3. 980 cubic feet per second

10-4. $b = 2D$

10-5. 3.88 feet

10-6. Tranquil

10-7. 876 cubic feet per second

10-11. No

10-12. 54,000 foot-pounds per second

10-13. 21.4 per cent

10-14. 60 cubic feet per second

10-15. 1.25 feet

10-17. 6.14 feet per second

10-18. 95.1 cubic feet per second

10-19. 2.72 feet per second

10-20. 34.5 degrees

11-1. 1, 1, 0, 0

11-2. i, j, k

11-4. $\mathbf{i}y^2z + \mathbf{j}2xyz + \mathbf{k}xy^2$

11-5. 4

11-6. $2k\sqrt{x^2 + y^2}$

11-7. V/u

12-1. $xy = A$

12-4. 264.6 feet squared per second

12-5. $\theta = 14.5$ degrees; -6.94 pounds per square inch

12-6. 0.025 pound per square inch

12-7. 628 feet squared per second; 0.021 pound per square inch.

12-8. 1068 feet squared per second

Index

Index

Jan-Tai Kuo